GENETIC ENGINEERING OF ANIMALS

An Agricultural Perspective

BASIC LIFE SCIENCES

Alexander Hollaender, General Editor

Council for Research Planning in Biological Sciences, Inc., Washington, D.C.

A Continuation Order Plan is available for this series A continuation order will bring delivery of each new volume immediately upon publication Volumes are billed only upon actual shipment For further information please contact the publisher

GENETIC ENGINEERING OF ANIMALS

An Agricultural Perspective

Edited by
J. Warren Evans

College of Agriculture and Environmental Sciences
University of California, Davis
Davis, California

and
Alexander Hollaender

Council for Research Planning in Biological Sciences, Inc.
Washington, D.C.

Technical Editor
Claire M. Wilson

Council for Research Planning in Biological Sciences, Inc.
Washington, D.C.

PLENUM PRESS • NEW YORK AND LONDON

Library of Congress Cataloging in Publication Data

Symposium on Genetic Engineering of Animals (1985: University of California, Davis)
  Genetic engineering of animals.

  (Basic life sciences; v. 37)
  "Proceedings of a Symposium on Genetic Engineering of Animals, held September 9–
12, at the University of California, Davis" — T.p. verso.
  Includes bibliographies and index.
  1. Domestic animals — Genetic engineering — Congresses. 2. Animal genetic engineer-
ing — Congresses. I. Evans, J. Warren (James Warren), date. II. Hollaender,
Alexander, date. III. Title. IV. Series.
SF756.5.S97 1985   636.08′24   86-559
ISBN-13: 978-1-4684-5112-2   e-ISBN-13: 978-1-4684-5110-8
DOI: 10.1007/ 978-1-4684-5110-8

Proceedings of a symposium on Genetic Engineering of Animals,
held September 9–12, 1985, at the University of California, Davis,
Davis, California

© 1986 Plenum Press, New York
Softcover reprint of the hardcover 1st edition 1986
A Division of Plenum Publishing Corporation
233 Spring Street, New York, N.Y. 10013

ACKNOWLEDGEMENTS

We are grateful for the generous support of the Cooperative State Research Service of the U.S. Department of Agriculture; Genentech, Inc.; Gould, Inc., Imaging and Graphics Division; the National Cancer Institute; the National Science Foundation; and the Office of the University of California, Davis, all of whom made possible the conduct of this Symposium.

We are especially indebted to Charles Hess for his encouragement and support throughout the planning of this first symposium on Genetic Engineering of Animals.

Our thanks also to our co-members of the National Organizing Committee composed of W. French Anderson, Douglas J. Bolt, Robert H. Foote, William Hansel, Harold Hawk, Tsune Kosuge, Robert Scibienski, Howard S. Teague, and Robert Wall for their scientific suggestions and development of the program.

Thanks also to the Local Organizing Committee consisting of Gary Anderson, Eric Bradford, Robert Cardiff, Sally DeNardo, Roy Doi, and Kathryn Radke who helped plan the local arrangements for these proceedings.

We are indebted to the services of Carroll Miller and Donna Hyatt, assisted by Carolyn Norlyn, who managed the mechanics of the meeting.

This Volume was technically edited and assembled by Gregory Kuny in cooperation with Claire M. Wilson.

J. Warren Evans and

Alexander Hollaender

CONTENTS

INTRODUCTION

J. Warren Evans

Department of Animal Science
Texas A&M University
College Station, Texas 77843

In the near future, improvement of domestic animals for the production of food and fiber is poised to undergo a revolution by the utilization of recent breakthroughs and advances in molecular genetics, embryo manipulations, and gene transfer systems. Utilization of these techniques will have a wide impact on animal agriculture by improvement of production efficiency via manipulation and control of many physiological systems. The end result will be to decrease production costs, increase food production and quality, and lower food costs. Health and well being of domestic and other animals will be improved as a result of new methods of disease diagnosis, vaccine production, and disease prevention practices. Genetic engineering also offers the possibility of utilizing animals for the development of pharmaceutical products to benefit society. Research progress will be enhanced via manipulation of the gene pool.

The objectives of this Conference were to discuss the current status of animal bioengineering and to realistically assess the potential applications of current and future genetic technologies for the production of food and fiber to meet the needs of our hungry world, and to provide animal scientists who may wish to utilize bioengineering in current or future research programs with current background information regarding concepts, applications, and methodologies.

To accomplish these objectives, an initial conceptual section is presented in the Volume to give an overview of genetic engineering in animals and to discuss the kinds of genes of interest in domestic animals as well as the basic techniques involved in genetic engineering.

Potential and actual sources of genetic material are generally discussed in the second section, followed by the presentation by leading scientists of procedures and concepts used to actually obtain genes.

The third section of this Volume addresses actual applications of genetic engineering in animals for modifying growth rates, improving disease resistance, diagnosing diseases, and producing vaccines. Limitations to molecular genetic approaches to improving domestic animals are also discussed as are problems related to gene expression and the physiology of transgenic animals.

The final component of this book is a discussion of the moral, legal, and future implications of genetic engineering of animals. Therefore, throughout the Volume, there is an integration of traditional and new genetic approaches for the improvement of domestic animals.

The credit for organizing the program goes to the National Organizing Committee composed of: W. French Anderson, National Heart, Lung and Blood Institute, Bethesda, Maryland; Douglas J. Bolt, U.S. Department of Agriculture, Beltsville, Maryland; Robert H. Foote, Cornell University, Ithaca, New York; William Hansel, Cornell University, Ithaca, New York; Harold Hawk, U.S. Department of Agriculture, Beltsville, Maryland; Alexander Hollaender, Council for Research Planning in Biological Sciences, Washington, D.C.; Tsune Kosuge, Department of Plant Pathology, University of California, Davis; Robert Scibienski, Department of Medical Microbiology, University of California, Davis; Howard S. Teague, U.S. Department of Agriculture, Washington, D.C.; Robert Wall, U.S. Department of Agriculture, Beltsville, Maryland.

The Local Organizing Committee had responsibility for all local arrangements and solving logistical problems. Those serving on the Committee were: Gary Anderson, Department of Animal Science; Eric Bradford, Department of Animal Science; Robert Cardiff, Department of Medical Pathology; Sally DeNardo, Department of Nuclear Medicine; Roy Doi, Department of Biochemistry & Biophysics; Kathryn Radke, Department of Avian Sciences.

Two people who assumed major responsibility for the mechanics of presenting the program and actually making arrangements were Carroll Miller, coordinator of special events section in the Dean's Office, College of Agricultural and Environmental Sciences, and her assistant Donna Hyatt. They were assisted by Carolyn Norlyn from the Campus Events and Information Office. Without their help, it would not have been possible to present the Conference at the University of California, Davis. Claire Wilson and Gregory Kuny from the Council for Research Planning in Biological Sciences have also had a very important role in planning the Conference and editing this proceedings Volume.

WELCOME: CHALLENGES ENCOUNTERED BY THE COLLEGE OF

AGRICULTURAL AND ENVIRONMENTAL SCIENCES*

Charles E. Hess

College of Agricultural and Environmental Sciences
University of California, Davis
Davis, California 95616

It is a pleasure to welcome you to the campus of the University of California, Davis. This Conference is one of a series in which we are exploring the current state of the art in genetic engineering or biotechnology, particularly as it relates to agriculture. In 1982, we considered genetic engineering in plants; last year, we focused our attention on the molecular basis of plant disease; and now we turn to the advances in genetic engineering in animals.

Before we begin, I'd like to share with you a minihistory of the U.C. Davis campus and then talk about some of the challenges faced by land grant colleges in the near future and over the longer term. U.C. Davis is now about 77 years old. It was established as the farm school for the University of California, Berkeley. It was felt that there was a need to provide a forum for practical "hows" to complement the theoretical "whys" that were being taught at Berkeley. That need continues to exist today, particularly since many of our students come from urban and suburban areas and have had no direct agricultural experience. Today's students are calling for more hands-on experiences in laboratories and through internships so that they can combine theory and practice, and add to their storehouses of marketable skills upon graduation.

The farm school idea was very successful and other components were added to the agricultural studies program here. In 1946, the School of Veterinary Medicine was established, followed by the College of Letters and Science and the College of Engineering. By 1968, Davis also had a School of Medicine and a School of Law. Our most recent addition, established about 3 years ago, is the School of Administration.

The student population at U.C. Davis is about 19,000. In the College of Agricultural and Environmental Sciences, we have 3,900 undergraduate and 1,200 graduate students. From the statistics I've seen, we are the largest undergraduate college of our kind, although, I believe, Texas A&M has now moved ahead of us in the number of graduate students. However, we graduated the largest number of students with B.S., M.S,, and Ph.D. degrees in the last several years.

---

*Address given at the opening of the symposium proceedings.

With these bright enrollment figures, I don't want to give the impression that we do not face some very serious problems. Enrollment in our college has decreased in much the same pattern as in other land grant colleges across the United States--a 19 to 20% decrease since enrollment was at its peak back in 1977. This can be traced in part, I think, to the fact that student interest in environmental issues and the "return to nature" which flourished in the late 1960s and '70s has fallen off substantially--particularly during the last recession when job opportunities in environmental careers were among the first to be cut back.

Students have become much more concerned about their future financial stability and have been entering majors such as engineering, computer science, and business, which appear to promise more stability and financial opportunity. The present economic problems in agriculture become a major deterrent to students in both urban and rural communities. And I know many parents who are leaders in California agriculture who urge their sons and daughters to enter professions other than agriculture.

It has become our special challenge to better communicate the fact that our colleges offer studies that go far beyond the traditional agricultural programs. I'm sure there are many high school graduates who would be amazed by what we will be doing at these proceedings. The opportunities opened by the new biotechnology provide outstanding challenges for bright young minds, and part of our job is to be sure that high school and community college students are aware of these opportunities.

Let us now turn closer to the subject of the Conference. Our goal is to review the advances in genetic engineering in animal agriculture. We want to determine where we have data gaps and identify the research areas with the greatest potential for development. We want to set the stage for collaboration among animal scientists and molecular biologists. We also want to project what the potential applications are in agriculture so we can begin to assess the future economic and social impacts of our scientific advances. An example of this latter point is the development of the bovine growth hormone at Cornell University. The potential use of the hormone could increase productivity by 20 to 25%. Given our current milk surplus, though, the question could arise, "Why should this new technology be introduced?" But, if you can decrease the number of cows needed to meet the current demand for milk, production efficiency would be increased.

This has been the history of agricultural research. For example, in 1945, there were 27 million dairy cows in the United States producing 120 billion pounds of milk. In 1983, however, there were 11 million dairy cows producing 135 billion pounds of milk. In this way, the consumer has received an important commodity at a fair price. But now, with the potential of introducing a technology that, in a very short time, could dramatically change productivity, what will be the impact on the dairy industry--particularly on the smaller dairy farmer who may not have the management sophistication or capital to successfully use the new technology? Does the new technology place him at an unfair competitive disadvantage? A Cornell University agricultural economics study indicated that a number of smaller dairy farms would no longer be able to compete.

So, now we have a scientific discovery evolving into a social issue. Is it more important to increase the efficiency of milk production, potentially lowering the cost to the consumer? Or is it more important to preserve the livelihood of the smaller dairy farmer? This is a very difficult question which will no doubt have to be settled eventually in state legislatures or, perhaps, in Congress. Ultimately, strategies will have to be developed for equitable compromise--ways to minimize the negative impact of technologies which also bring us positive effects.

Issues such as these have, in part, led to the greater interest society has taken in our research activities and to the desire on the part of the public to play a greater role in deciding what science should or should not be doing. And we should not really discourage that interest. It is good for the public to be aware of and interested in what scientists are doing, how they are doing it, and why they are doing it. The danger, of course, lies in a well-intentioned but uninformed or misinformed public interest in science. At one end of the spectrum, there are those whose efforts and concerns would lead to over-regulation of the research we do, which can be a serious threat to our freedom of inquiry.

What can we do as part of the scientific community to forestall restrictions to free inquiry? First, we must be sure our own house is in order. Beyond the traditional objective that technology must be profitable for the user, we must continue to rigorously apply additional criteria in agricultural research and development. Then, we must be sure that the public is aware of the criteria we are using.

It is understandable that as science progresses--especially with biotechnology--certain segments of the public become fearful. Fear of the unknown by the layman is as natural a reaction as is curiosity about the unknown to the scientist. When the subject is biotechnology, we all know of the public suspicion that we may be creating deadly new bacteria which will plague the planet or be used to conduct war, or of creating monstrous new creatures through genetic engineering. It becomes part of our task, then, to educate a fearful public.

Let us let the public know that we are concerned not only with useful technology, but with energy efficiency and that we address whether or not long-term physical impact on the environment is acceptable. Let us let the public know that we have studied and minimized health and safety risks, and that we are aware of any social consequences of our developments. Let us commit ourselves to sharing in the task of finding solutions to any expected social costs. Let us let the public know that the research we conduct is both necessary and conscientious, and that if a risk is involved, such as the release of a pathogen, we have taken all possible precautions. When animals are used in research, we must ensure that they are properly cared for and that any pain involved is kept to a minimum.

Let us additionally remind the public that the scientific community established its own stringent regulations at the very beginning of recombinant DNA research, giving reassurance to the public that we place--and always have placed--highest value on their safety and concerns. In return for that commitment, we were given public trust and there was self-policing rather than over-regulation from outside groups. Finally, when doing research supported in large part by public funds, we must provide assurance that the work we do is for the benefit of the greater society, and not just for specific interest groups at the expense of the public.

Sometimes it can be frustrating to have to turn from the business at hand in science, and turn to patiently educate the layman. But, it has become part of our responsibility in these times of very sophisticated science to do so. Doing so will give us public trust and, in the final analysis, it is far better that we regulate ourselves than have external agencies do it for us. To be able to do that, we must commit ourselves to the very highest standards and we must let the public know that no one watches us with greater scrutiny than we ourselves.

I am pleased to see that this conference will be addressing the regulation of application and the moral impact of genetic engineering because certainly we have a responsibility to consider these issues as well as the technical issues.

I wish to acknowledge the organizational work of Dr. J. Warren Evans, who chaired the Scientific Organizing Committee for us, even though he has since left us for Texas A&M. We are happy for Warren and wish him every success, but certainly Texas A&M's gain is our loss.

I wish also to recognize Dr. Alexander Hollaender of the Council for Research Planning in Biological Sciences who, I must say, is the key instigator of the conferences we have held here at Davis as well as the ones we have sponsored at other locations.

At this time, I also want to recognize a very important group, whose contributions were instrumental in making this conference a success, and in fact, necessary in making it possible: our sponsors. We are grateful for the generous support of the Cooperative State Research Service of the U.S. Department of Agriculture; Genentech, Inc.; Gould, Inc., Imaging and Graphics Division; the National Cancer Institute; the National Science Foundation; and the Office of the Vice-Chancellor, Academic Affairs, here at U.C. Davis.

Finally, I want to thank Carroll Miller and Donna Hyatt, our special events coordinators for the college, who worked with the Campus Events and Information Office in handling the logistics of the conference.

We are very pleased to be your hosts for this immense information exchange, and we hope you have a pleasant stay, in spite of our unseasonable rainfall. I'm certain you will gain from this symposium a lot of valuable information.

GENETIC ENGINEERING OF ANIMALS

W. French Anderson

Laboratory of Molecular Hematology
National Heart, Lung, and Blood Institute
Bethesda, Maryland 20892

INTRODUCTION

Genetic engineering of animals is becoming a reality as the later chapters in this Volume will illustrate. Genetic engineering is the field of manipulating the DNA of a cell or of an animal in order to alter the genetic information contained within the organism's genome. The standard techniques of recombinant DNA are used. A number of excellent textbooks are now available that explain in clear fashion how to work with DNA to clone genes, to alter known genes, and so on. One convenient aspect of DNA is that the same techniques can be used whether the source of material is from humans, bacteria, yeast, plants, cattle, or chickens. Essentially any DNA can be recombined with any other DNA.

My major charge is to examine the techniques for gene transfer into cells and to look in detail at those techniques useful for inserting genes into whole animals (1). There are 4 well-established groups of techniques--viral (e.g., simian virus 40), chemical (e.g., calcium phosphate), physical (e.g., microinjection), and fusion (e.g., liposomes or protoplasts)--and one new procedure (i.e., retroviruses).

VIRAL TECHNIQUES

DNA viruses, such as simian virus 40 (SV40) with DNA as the nucleic acid in their core, have been employed for several years as gene transfer vectors. An adenovirus vector has recently been developed that will efficiently infect animal and human cells (including hematopoietic cells) with the result that one or a few copies of the recombinant virus are integrated into the host cell's genome. One subcategory of DNA viruses should be mentioned: bovine papilloma virus. This viral DNA replicates extrachromosomally so that bovine papilloma virus-based vectors may prove to be useful for maintaining genes in cells in a nonintegrated manner. The advantage of DNA viruses is that they are effective and well-studied vectors for gene transfer in plants, animals, and bacteria. The disadvantages are that many of the most widely studied viruses can replicate within cells and some can kill the host cell. In addition, they are much less efficient at transformation than retroviruses (see below). Thus, because they are a potential hazard and they have low efficiency, their wide-spread use in animals appears unlikely at the present.

Another procedure actively used for insertion of genes into tissue culture cells is calcium phosphate-mediated DNA uptake. The original procedure of Graham and van der Eb (6) was modified by Wigler et al. (12) in order to insert into the genome of mammalian cells growing in culture a fragment of DNA carrying one or more genes. A large number of different genes have been transferred into tissue culture cells using this technique.

Transfection is carried out by pipetting a suspension of DNA, complexed into small precipitates with calcium phosphate, onto a monolayer of cells growing in a tissue culture dish. A number of protocols are used to increase the efficiency of transfection in different cell types: for example, diethylaminoethyl dextran can be employed instead of calcium phosphate or the cells can be shocked with glycerol after 2 hours of incubation. The efficiency of the process varies with the cell line. Under optimal conditions and very receptive cells (i.e., mouse L cells), one cell in 10^2 to 10^3 can be obtained that has integrated and expressed the exogenous DNA. Because the usual efficiency is 10^{-5} to 10^{-7}, a procedure is required to detect the occasional transfected cell. In other words, a gene must be present that can protect the cell from a lethal selective agent that is added to the incubation medium or that complements a genetic defect [i.e., hypoxanthine-guanine phosphoribosyl transferase (HPRT) or thymidine kinase]. The transfected cell will survive while all others are killed. Attempts to obtain transfected cells without selective pressure have generally been unsuccessful.

Transfection appears to work poorly in suspension cells, as, for example, bone marrow cells. Efficiencies can only be estimated, but the value is probably one cell in 10^6 or 10^7. Using the powerful selection system offered by the mutant dihydrofolate reductase gene (isolated from 3T6-R400 cells) that provides exceptional resistance to methotrexate, Carr et al. (3) reported that the calcium phosphate transfer technique can be successfully employed to obtain mouse bone marrow cells that contain a functional exogenous DHFR gene. The permanently transfected cells can partially repopulate a lethally irradiated mouse. These results support the studies of Cline et al. (2,4) who reported successful transfer of a functional dihydrofolate reductase gene into the bone marrow of mice. However, the presence of the dihydrofolate reductase gene has not been confirmed with DNA hybridization studies and, until such experiments are reported, the efficiency of the calcium phosphate procedure is uncertain.

The advantages of chemical techniques are that they are easy and inexpensive to perform, require no special equipment, and they do not require an infectious agent, so they are very safe. The disadvantages are that uptake and expression of exogenous DNA is low and they insert multiple copies of the transferred gene linked head-to-tail. Thus, the low efficiency makes their use in animals unlikely unless the procedure is coupled with a strong selective advantage.

PHYSICAL TECHNIQUES

Microinjection and electroporation are the 2 principal classes of physical techniques. Electroporation, a relatively new procedure, is the transport of DNA directly across a cell membrane by means of an electric current (9). It has been used to transfer a variety of genes into a number of different cells including the immunoglobulin κ gene into B cells (10). Its potential for gene transfer into animals is uncertain.

Microinjection has been used for a number of years and has the advantage of high efficiency (up to one cell in 5 injected can be permanently transfected). However, the distinct disadvantage is that only one cell at a time can be injected. Transfection of a large number of hematopoietic stem cells, for example, is not feasible.

An area where microinjection has had spectacular success is in transferring genes into fertilized mouse eggs (see Fig. 1). Gordon et al. (5) first demonstrated that if plasmid DNA is microinjected into one of the 2 pronuclei of a recently fertilized mouse egg, and the ovum is then placed into the oviduct of a pseudopregnant female, the egg could develop into a normal mouse carrying the plasmid DNA in every cell of its body. Furthermore, the injected DNA can be transmitted to offspring in a normal Mendelian manner. Mice carrying an exogenous gene in their genome are called "transgenic."

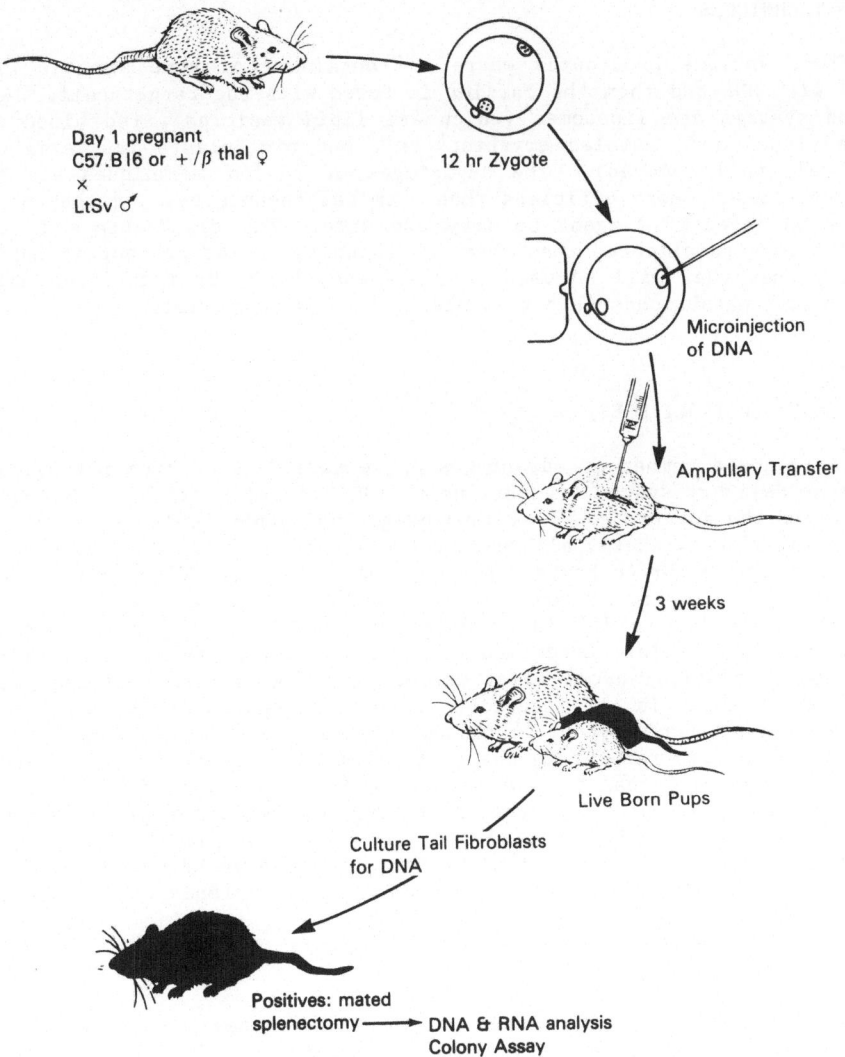

Fig. 1. Schematic flowchart for microinjection of murine zygotes.

Hammer et al. (7) used this technique to partially correct a mouse with a defect in its growth hormone production. By attaching a rat growth hormone gene to an active regulatory sequence (specifically, the promoter that normally directs the synthesis of metallothionein messenger RNA in mice), they obtained a recombinant DNA construct that actively produces growth hormone in the genetically defective mouse. Although the level of growth hormone production is inappropriately controlled--that is, influenced by signals that normally regulate metallothionein synthesis--these experiments do show that microinjection can be used as a delivery system that can put a gene into every cell of an animal's body. The major advantage of microinjection of fertilized eggs is that transgenic animals can be obtained. The disadvantages are that the procedure still results in many eggs being damaged during the injection process, and that regulation of the inserted gene is still not controlled well.

FUSION TECHNIQUES

These include procedures where a "container" of some sort is first loaded with DNA and then the carrier is fused with the target cell. Well-studied systems are liposomes (which are lipid vesicles), red blood cell ghosts (lysed and resealed erythrocytes), and protoplasts (bacteria with their cell wall removed). The advantages of fusion techniques are that they are somewhat more efficient than chemical techniques, and they do not require an infectious agent so they are safe. The disadvantage is that they are more complicated than chemical techniques. At present it appears unlikely that they will be used in preference to other techniques unless further work establishes that they are in some way superior.

RNA VIRUSES (RETROVIRUSES)

There are a number of advantages of vectors derived from retroviruses as a gene delivery system. First, up to 100% of cells can be infected and can express the integrated viral (and exogenous) genes; this is in contrast to chemical methods where, although most cells take in the DNA, as shown by positive assays after 48 hours, only one cell in 10^3 to 10^7 stably expresses the exogenous gene. Second, as many cells as desired can be infected simultaneously; 10^6 to 10^7 is a convenient number for a simple protocol. Third, under appropriate conditions the DNA can integrate as a single copy at a single, albeit random, site, whereas the chemical and physical techniques often result in the insertion of multiple copies of the transferred gene, all linked head-to-tail in tandem repeats. Fourth, although integration is random with respect to the host genome, it is precise with respect to the viral genome--that is, the structure of the integrated DNA is known. Fifth, the infection and long-term harboring of the retroviral vector usually does not harm cells. Finally, a wide and controllable host range is available. A number of retroviral vector systems have been developed, the most common one being the use of vectors based on Moloney murine leukemia virus.

Life Cycle and Structure

The details of the life cycle of retroviruses have been reviewed recently (11). In brief, the retrovirus, composed of an RNA-protein core and a glycoprotein envelope, enters a cell where the RNA acts as a template for the reverse transcription of the genetic information into a double strand of DNA. This DNA can precisely integrate as a single copy, called a provirus, at a random location in the genome of the host.

10

Although much has been learned about the regulatory features of retroviruses, uncertainties remain. Those features of the proviral structure that are thought to be necessary for transcription and transmission of the viral genome are (see Fig. 2): a long terminal repeat (LTR) sequence on each end, containing regulatory signals for initiating and terminating transcription; sequences required for reverse transcription and others for proviral integration; short sequences (called here, for short, r^- and r^+) immediately adjacent to each LTR and necessary for reverse transcription; the packaging sequence, called ψ in Moloney murine leukemia virus, necessary for the viral RNA to be packaged into an infectious viral particle; and the donor (D) and acceptor (A) splice sites.

Retroviral RNA is synthesized from the proviral DNA by the host cell's own RNA polymerase. A portion of this RNA is used in the cell's translational machinery to synthesize the viral proteins that go into the final viral particles along with the genomic RNA. These viral particles bud off from the cell and can then infect other cells.

From experimental studies as well as from the existence of a number of naturally occurring defective viruses, it is known that almost all of the regions coding for viral proteins (gag, pol, and env in Fig. 2) can be deleted and some or all of these sequences replaced with other DNA. Once the viral genes are deleted, the retroviral vector becomes defective. In order to obtain infectious viral particles, a cell harboring a defective provirus must be infected with a "helper" virus, which carries all the viral functions needed--that is, the genes for gag, pol, and env.

Use as Gene Delivery System

The proviral DNA for the desired retrovirus is isolated and inserted into a convenient plasmid. The viral genes can then be replaced with the exogenous genes of choice by standard recombinant DNA techniques. This construct is used to transfect tissue culture cells (i.e., NIH 3T3 cells) by a convenient gene transfer procedure (i.e., calcium phosphate). After infecting the cells with a helper virus (such as intact Moloney murine leukemia virus), infectious viral particles, possessing both the retroviral vector and the helper virus, bud off from the cells into the surrounding medium. This particle-containing supernatant is collected and used to infect bone marrow cells in culture or, more simply, freshly extracted bone marrow is incubated directly with the cells budding the viral particles. The marrow cells are removed and injected intravenously into a mouse whose bone marrow has been killed by X-rays (lethally irradiated). The animal is

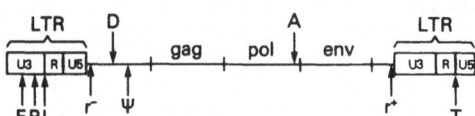

Fig. 2. Simplified structure of Moloney murine leukemia provirus DNA. Abbreviations: E, enhancer; P, promoter; I, initiation (Cap) site for viral RNA synthesis; r^-, replication initiation site for minus DNA strand (transfer RNA binding site); D, donor splice site; ψ, packaging sequence; A, the major acceptor splice site; r^+, replication initiation site for plus DNA strand (purine-rich site); T, terminal [poly(A) addition] site for viral RNA synthesis; LTR, long terminal repeat; U3, R, and U5 are portions of the LTR; gag, group-specific (that is, viral core) antigens: p15, p12, p30, and p10; pol, RNA-dependent DNA polymerase (reverse transcriptase); and env, envelope proteins: gp70, p15E, and R. (Not drawn to scale.)

11

then studied to determine if the transferred marrow cells express the desired gene from the vector.

An improvement of this procedure is to treat bone marrow cells with a retroviral particle that can deliver the vector but which will not itself produce a spreading infection. Mann et al. (8) have developed a technique for accomplishing this goal. The regulatory signal ψ (Fig. 2) contains a sequence, the exact size and structure of which are not yet known, that must be present in the viral RNA for it to be packaged into a viral particle. A helper virus was constructed with this sequence deleted (ψ^-) by making use of convenient restriction endonuclease sites (BalI and PstI) flanking the ψ sequence in Moloney murine leukemia virus. The ψ^- helper is able to produce all the viral proteins required to make a particle, but the particle does not package its own RNA. Since the retroviral vector has a ψ sequence, it is packaged. Consequently, the particle can just infect once; it is only a delivery system for the vector, not an infectious agent.

In order to use the ψ^- helper virus conveniently, a line of NIH 3T3 cells was established with the helper proviral DNA permanently integrated; ψ^- helper viral RNA is produced constitutively. The transfection of this cell line (called ψ-2) with the retroviral vector DNA results 48 hours later in a supernatant that contains viral particles with only the vector (13). I will discuss in a subsequent chapter (W.F. Anderson et al., this Volume) the use of this system to transfer functional genes into intact animals.

Properties still needed for an optimal delivery system can be visualized. An ideal delivery system not only would be stable but also would be tissue-specific. When transferring a gene into the hematopoietic system, the isolated bone marrow can be treated. But no other tissue, except skin cells, can be removed, treated, and replaced at present. Since many viruses are known to infect only specific tissues (that is, to bind to receptors that are present only on certain cell types), a retroviral particle containing a coat glycoprotein that recognizes only hematopoietic stem cells would permit the retroviral vector to be given intravenously with little danger that cells other than those in the marrow would be infected. Such specificity could permit the transfer of genes into an animal's liver or pancreas, for example. One problem, however, is that cell replication appears to be necessary for integration. It would not be possible to infect nondividing brain cells, for example, as far as we now know.

The optimal system not only would deliver the vector specifically into the cell type of choice but would also direct the vector to a predetermined chromosomal site. Specific insertion into a selected site of a chromosome by means of homologous recombination can be readily achieved in lower organisms but appears to be a formidable task in mammals, whether retroviral vectors or plasmid-based vectors are used. Present evidence suggests that homologous, site-specific integration occurs at a very low level, when it occurs at all, in mammals.

CONCLUSION

All 5 types of gene transfer procedures discussed above are of use for various types of tissue culture experiments. However, only 2 procedures, microinjection and retroviruses, have proven to be effective in inserting functional genes into intact animals. Microinjection is valuable for producing germline gene transfer while retroviruses are at present only useful for somatic cell gene transfer into bone marrow cells. Progress is rapid, however, and improved techniques for genetic engineering of animals will undoubtedly arise over the next several years.

REFERENCES

1. Anderson, W.F. (1984) Prospects for human gene therapy. *Science* 226: 401–409.
2. Bar-Eli, M., H.D. Stang, K.E. Mercola, and M.J. Cline (1983) Expression of a methotrexate-resistant dihydrofolate reductase gene by transformed hematopoietic cells of mice. *Somatic Cell Genet.* 9:55–67.
3. Carr, F., W.D. Medina, S. Dube, and J.R. Bertino (1983) Genetic transformation of murine bone marrow cells to methotrexate resistance. *Blood* 62:180–185.
4. Cline, M.J., H. Stand, K. Mercola, L. Morse, R. Ruprecht, J. Browne, and W. Salser (1980) Gene transfer in intact animals. *Nature* (London) 284:422–425.
5. Gordon, J.W., G.A. Scangos, D.J. Plotkin, J.A. Barbosa, and F.H. Ruddle (1980) Genetic transformation of mouse embryos by microinjection of purified DNA. *Proc. Natl. Acad. Sci., USA* 77:7380–7384.
6. Graham, F.L., and A.J. van der Eb (1973) A new technique for the assay of infectivity of human adenovirus 5 DNA. *Virology* 52:456–467.
7. Hammer, R.E., R.D. Palmiter, and R.L. Brinster (1984) Partial correction of murine hereditary growth disorder by germ-line incorporation of a new gene. *Nature* (London) 311:65–67.
8. Mann, R., R.C. Mulligan, and D. Baltimore (1983) Construction of a retrovirus packaging mutant and its use to produce helper-free defective retrovirus. *Cell* 33:153–159.
9. Neumann, E., M. Schaefer-Ridder, Y. Wang, and P.H. Hofschneider (1982) Gene transfer into mouse lyoma cells by electroporation in high electric fields. *EMBO J.* 1:841–845.
10. Potter, H., L. Weir, and P. Leder (1984) Enhancer-dependent expression of human κ immunoglobulin genes introduced into mouse pre-B lymphocytes by electroporation. *Proc. Natl. Acad. Sci., USA* 81:7161–7165.
11. Varmus, H.E. (1982) Form and function of retroviral proviruses. *Science* 216:812–820.
12. Wigler, M., S. Silverstein, L.-S. Lee, A. Pellicer, Y.-C. Cheng, and R. Axel (1977) Transfer of purified herpes virus thymidine kinase gene to cultured mouse cells. *Cell* 11:223–232.
13. Williams, D.A., I.R. Lemischka, D.G. Nathan, and R.C. Mulligan (1984) Introduction of new genetic material into pluripotent haematopoietic stem cells of the mouse. *Nature* (London) 310:476–480.

MANIPULATION OF GENES IN VITRO AND IN VIVO

Dean H. Hamer

Laboratory of Biochemistry
National Cancer Institute
Bethesda, Maryland 20892

INTRODUCTION

The genetic engineering of animals is by no means a new concept. Since the beginnings of history, man has attempted to improve his livestock and domestic animals through selective breeding. Such classical breeding schemes, though successful in many aspects, are limited by at least 2 major constraints. First, genetic information can be exchanged only between members of the same or very closely related species. Second, these schemes depend on naturally occurring mutations or variants that may arise at very low frequencies.

Recombinant DNA technology holds the promise of overcoming both of these limitations. Because recombination is performed in vitro rather than in vivo, genetic information can be exchanged between completely divergent species. Also, the ability to manipulate genes in vitro makes it possible to generate mutants in days or weeks rather than in millions of years of evolutionary history. In this brief chapter, I describe 2 types of gene manipulations that may in the future be of special use in agricultural genetic engineering.

GENE REPLACEMENT

This method involves exchanges between chromosomal and cloned DNA sequences. The endogenous chromosomal sequences can either be completely removed (a deletion) or they can be replaced by closely related sequences (e.g., a point mutant) or completely unrelated sequences. A major purpose of such replacement experiments is to determine gene function. Suppose that we think that a given gene plays a critical role in some physiological function, e.g., growth. Then obviously animals lacking this gene should show abnormal growth characteristics. If instead they grow normally, then the original hypothesis is disproven.

An example of a typical gene replacement experiment in the lower eukaryote Saccharomyces cerevisiae is shown in Fig. 1. In this experiment, we were interested in determining the function of the gene CUP1 which encodes a small, cysteine-rich, copper-binding protein called copperthionein. First, the CUP1 gene, together with 5' and 3' flanking sequences, was cloned into a bacterial plasmid. Second, all of the CUP1 coding sequences

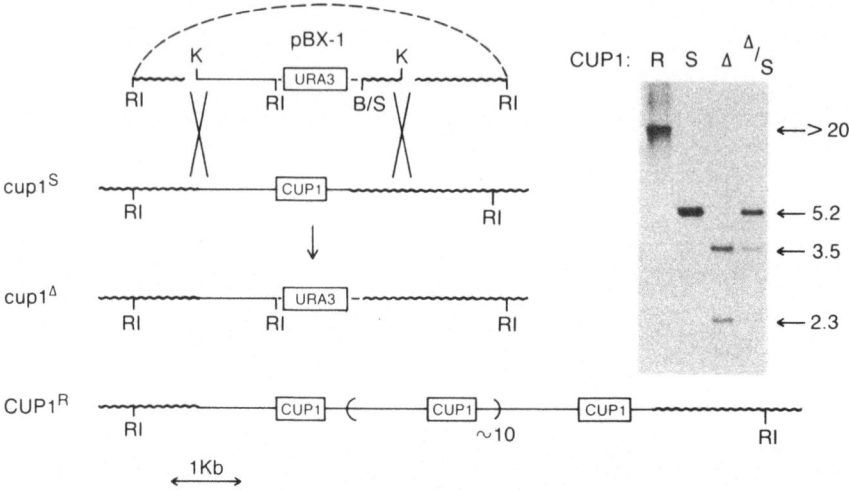

Fig. 1.   Replacement of the yeast CUP1 gene. (Left) Gene structures. Dashed lines represent plasmid DNA, wavy and straight lines represent flanking DNA, and boxes represent coding sequences. (Top line) A plasmid in which CUP1 coding sequences have been replaced by URA3 coding sequences. (Second line) Chromosomal DNA of a cup1^S strain carrying a single copy of CUP1. (Third line) Chromosomal DNA of a cup1Δ strain generated by gene replacement. (Bottom) Chromosomal DNA of a naturally occurring CUP1^R variant containing amplified CUP1 genes. (Right) Gel transfer hybridization analysis of chromosomal DNA samples. For details see Ref. 1.

(but not the flanking sequences) were removed from the plasmid and replaced by the heterologous URA3 gene. Third, the plasmid was digested with a restriction enzyme that cleaves in the flanking sequences and thereby generates recombinogenic termini. Finally, the plasmid DNA fragment was transformed into a ura3$^-$ recipient strain carrying a single copy of CUP1. We anticipated that a double recombination event would result in eviction of the resident CUP1 sequences and their replacement by URA3 sequences. Gel transfer hybridization analysis of the DNA from URA3$^+$ transformants confirmed this prediction. As expected, the resulting strains were incapable of copperthionein synthesis as judged by polyacrylamide gel electrophoresis of [^{35}S]-cysteine-labeled proteins (1).

Because there has been considerable debate about the possible detoxifying or homeostatic functions of metallothionein, it was of great interest to study the physiology of these strains. Three major findings were obtained. First, copperthionein protects cells against poisoning by excess copper in the medium. Second, copperthionein feedback regulates the transcription of its own structural gene; in the absence of copperthionein, there is a high, constitutive level of transcription from an episomal CUP1 promoter. We currently believe that this is simply due to the ability of thionein to chelate intracellular copper that would otherwise be available to activate transcription factors. Third, copperthionein is not required for normal growth, mating, sporulation, or germination. Surprisingly, strains lacking copperthionein also accumulated normal levels of total cellular copper and of the copper-dependent form of superoxide dismutase. From these results, we conclude that the major function of copperthionein is to act as a scavenger that maintains low levels of free intracellular copper. At high copper concentrations this protects the cells against toxicity while at low concentrations it prevents futile transcription of the <u>CUP1</u> gene. Previous speculations as to the role of thionein in copper transport or enzyme activation are ruled out by these results (1).

At present, gene replacement experiments have been conducted successfully only in yeast and bacteria, but the recent demonstration of low-level, homologous recombination in cultured mammalian cells may soon make it possible to conduct similar experiments in animals. If so, a number of important applications can be envisioned. For example, it might be possible to manipulate complex, feedback-regulated physiological systems, such as fat production, by eliminating one hormone gene while amplifying another. It might also be possible to improve the nutritional value of milk by replacing the endogenous gene with one containing a more desirable amino acid composition.

FUSION GENES

In this method, transcriptional control sequences from one gene are joined to structural sequences from a second gene that encodes a desirable gene product. The fusion gene is then introduced into a host cell that is capable of carrying out any post-translational modifications required to render the gene product biologically active. Ideally, the host cell should also be capable of prolonged, high-density growth.

Vectors containing the bovine papilloma virus replicon and metallo-thionein promoter sequences have proven especially useful for the overproduction of mammalian gene products. In a typical experiment, coding sequences for human growth hormone were joined to mouse metallothionein-I gene upstream sequences. The fusion was made so as to retain all of the metallothionein transcriptional control sequences and the transcription initiation site, and all of the growth hormone translated sequences, translation initiation codon, and 3' translated sequences. When the fusion gene was introduced into mouse C127 cells on a bovine papilloma virus vector, essentially all of the transformed foci produced detectable levels of growth hormone. Moreover, the hormone was appropriately modified in that the amino-terminal leader sequence was precisely removed and the processed polypeptide was secreted into the culture medium. The highest producers secreted as much as 100-200 mg/l per day of mature hormone and were capable of continued growth and production for up to 40 days in culture (2).

The practical applications of this technology are obvious since essentially any protein, including specifically altered polypeptides derived from mutant genes, can be overproduced at will. Special interest will focus on products such as animal growth hormones and fertility hormones.

PROSPECTS

Molecular biologists have developed a large arsenal of techniques to isolate, analyze, recombine, mutate, and transfer genes. Nevertheless, we are still a long way from realizing all the possible benefits from this technology in practical areas such as agriculture. One important step will be to move from simple model systems, such as yeast and cultured mammalian cells, to livestock animals. A second will be to start studying the complicated questions of the effects of gene-gene and gene-environment interactions on phenotypes. Clearly, these endeavors will greatly benefit from interactions between molecular biologists and the more traditional practitioners of animal engineering.

17

ACKNOWLEDGEMENTS

I thank George Pavlakis, Dennis Thiele, and J. Lemontt for their fruitful collaboration and Carolyn Ray for editorial assistance.

REFERENCES

1. Hamer, D.H., D.J. Thiele, and J.F. Lemontt (1985) Function and auto-regulation of yeast copperthionein. <u>Science</u> 228:685-690.
2. Pavlakis, G.N., and D.H. Hamer (1983) Regulation of a metallothionein-growth hormone hybrid gene in bovine papilloma virus. <u>Proc. Natl. Acad. Sci., USA</u> 80:397-401.

EXPRESSION OF THE BOVINE GROWTH HORMONE GENE

IN CULTURED RODENT CELLS

John J. Kopchick, Francoise Pasleau,
and Frederick C. Leung

Department of Animal Drug Discovery
Merck, Sharp & Dohme Research Laboratories
P.O. Box 2000
Rahway, New Jersey   07065

INTRODUCTION

Expression of recombinant DNA has been achieved in both prokaryotic and eukaryotic cells. This chapter deals with expression of bovine growth hormone recombinant DNA in cultured mammalian cells.

Bovine growth hormone (bGH), a protein of $\sim$22,000 daltons and 191 amino acids, is synthesized by the anterior pituitary gland (1,8,9). Growth hormones in various species are responsible for growth (28,35,40) as well as for a variety of metabolic processes, including nitrogen, lipid, mineral, and carbohydrate metabolism (28). In vitro assays in adipose tissue show that growth hormone is involved in glucose uptake and conversion to carbon dioxide, lipolysis, glycerol release, and amino acid incorporation into protein (51). In addition, bGH will stimulate lactation in the cow (37,38).

Use of recombinant DNA technology has resulted in the cloning of a variety of growth hormone genes and complementary DNAs, including those derived from human, rat, bovine, and porcine tissue (10,30,31,32,34,44,45). Plasmid DNAs complementary to human, bovine, and porcine mRNA have been expressed in Escherichia coli (10,45) whereas rat and human growth hormone genes have been expressed in mouse L cells (5,42).

For the possible use of growth hormone in animal husbandry, we were interested in generating stable rodent cell lines that express and secrete relatively large quantities of biologically active bGH.

In this study a genomic clone of bGH was used. The following steps were employed to generate mouse or rat cell lines that express bGH: (a) in vitro construction of recombinant DNA molecules containing eukaryotic transcriptional regulatory elements ligated to the bGH gene; (b) screening for the ability of plasmid constructs to direct the synthesis and secretion of bGH in a transient eukaryotic assay system; (c) generation of stable mouse or rat cell lines containing active plasmid constructs that continuously secrete bGH; and (d) determination of the biological activity of bGH secreted from these cells.

Using an avian retroviral long terminal repeat (LTR) to direct expression of bGH, we established stable mouse and rat cell lines that secrete large amounts of bGH. Bovine GH produced by these rodent cells was found to be biologically active.

Also, we were interested in comparing 2 eukaryotic viral promoters in their ability to direct bGH expression in cultured rat or mouse cells. The Rous sarcoma viral (RSV) LTR contains control sequences that direct the synthesis of progeny viral RNA and viral mRNAs (52). The human cytomegalovirus (CMV) genome contains a major immediate early (IE) gene, located in the long, unique region of the viral genome (49,50), which is expressed in large amounts after infection (54). Recombinant DNA molecules were constructed that contain these two viral promoters ligated to the bGH gene.

The CMVIE promoter was found to be 3- to 4-fold more efficient than the RSV-LTR in directing expression of the bGH gene by cultured rat or mouse cells.

MATERIALS AND METHODS

Cell Culture

Mouse L cells [thymidine kinase-negative (TK$^-$) and adenine phosphoribosyl transferase negative (APRT$^-$)] were maintained in Dulbecco's Modified Eagle's Medium (DMEM) containing 10% calf serum (HyClone) and 50 µg/ml 2,6-diaminopurine (Sigma). TK$^+$ APRT$^-$ mouse cell transformants were selected and maintained in DMEM, 10% calf serum, 15 µg/ml hypoxanthine, 1 µg/ml aminopterin, and 5 µg/ml thymidine (39,57). Rat GH$_3$ cells were maintained in DMEM supplemented with 10% Nu-Serum (Collaborative Research).

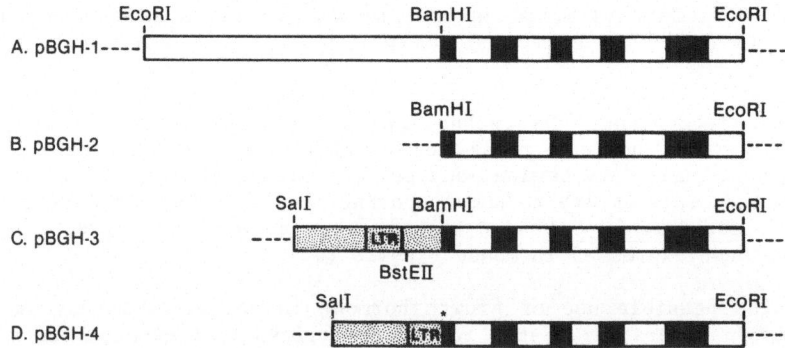

Fig. 1.   The RSV-bGH recombinant DNA clones. A. pbGH-1. An ∿4.0-kb DNA fragment from a λ bacteriophage DNA library was subcloned into pBR322. Bovine GH exons are represented by black boxes. The region upstream of exon 1 contains approximately 2.0 kb of 5'-flanking sequences. B. pbGH-2. The bovine GH gene with the 5'-flanking sequences removed. C. pbGH-3. A SalI–BamHI fragment of Rous sarcoma virus-derived DNA containing the viral LTR (shown in stippling) ligated to the bGH gene. D. pbGH-4. Identical to pbGH-3 with the region between the BstEII and BamHI sites removed. Blunt-end ligation resulted in the destruction of the BamHI and BstEII sites. The asterisk denotes the region in which the ligation occurred.

Construction of Bovine Growth Hormone DNA Expression Vectors

Bovine pituitary DNA fragments, partially cleaved with EcoRI, were used in the construction of a Charon 4A genomic library. A 4.0-kilobase DNA molecule isolated from one of the λ clones that hybridized to human GH DNA was subcloned into pBR322. This plasmid, termed pbGH-1 (Fig. 1A), possessed restriction enzyme cleavage sites and nucleotide sequences similar to the bGH gene reported previously (59).

Eukaryotic, transcriptional regulatory signals occur upstream or within the bGH gene (59). To remove the 5'-flanking segment of genomic DNA containing the putative transcriptional regulatory sequences, pbGH-1 was cleaved with BamHI and a 2.0-kb DNA fragment was isolated by agarose gel electrophoresis. Following ligation and transfection into E. coli (strain RRI), a pBR322 derivative was selected that contained the bGH gene with the 5'-flanking region removed. This recombinant DNA plasmid was termed pbGH-2 (Fig. 1B).

The RSV-LTR or CMVIE promoter was attached to the bGH gene as previously described (17,36) (Fig. 2). The nucleotide sequence at the junction of the CMVIE or RSV-LTR bGH fragments was determined. Also, a summary of the structures of the CMVIE promoter and RSV-LTR is shown in Fig. 3A and B, respectively. We presume that transcription initiates from the normal Cap site found within the LTR and CMVIE promoters and continues into the nearby bGH sequences. The locations of the thymidine-adenine-thymidine-adenine (TATA) box, the cytosine-adenine-adenine-thymidine (CAAT) box, and the Cap site are shown in Fig. 3A and B. A consensus promoter sequence [thymidine-adenine-thymidine-thymidine-thymidine-adenine-adenine (TATTTAA)] has been located in the U3 region of the LTR, 30 nucleotides upstream from the viral RNA start point found in the R region of the LTR (Fig. 3B) (14,43). Typical TATA and CAAT boxes (Fig. 3A) are found 28 base pairs and 61 base pairs, respectively, upstream of the Cap site, in the CMVIE promoter (53). The 5'-mRNA nontranslated leader sequences, extending from the viral RNA Cap site to the bGH adenine-thymidine-guanine (ATG) translational start codon, are 164 bp and 128 bp long, for pbGH-4 and pCMVIE-bGH, respectively.

Transient Eukaryotic Expression System

A transient, eukaryotic expression assay system was employed. A general protocol for the assay system was as described previously (16,17,36,

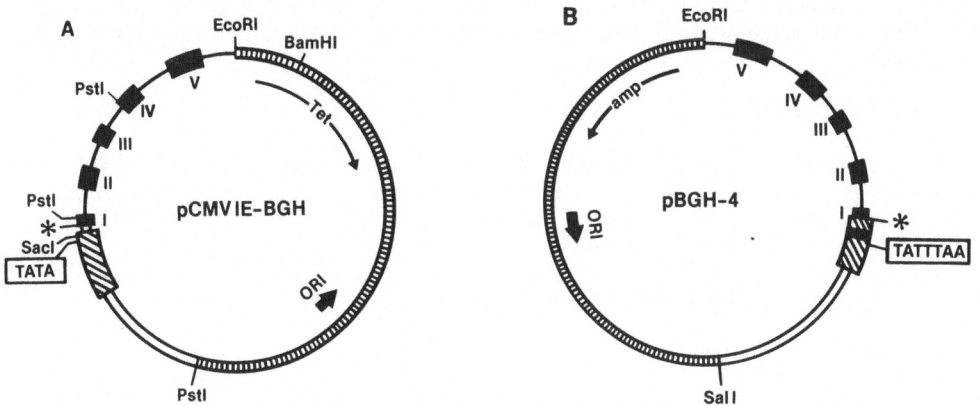

Fig. 2.   Plasmids pCMVIE-bGH and pbGH-4. The tetracycline-resistant plasmid, pCMVIE-bGH (A), and the ampicillin-resistant plasmid, pbGH-4 (B), are depicted. The construction of these plasmids is described in "Materials and Methods" section in text.

50). Briefly, approximately 5.0 x 10^5 cells were plated onto 30-mm tissue culture plates. Cell lines included mouse L cells or rat GH_3 cells, a rat pituitary cell line that synthesizes and secretes rat growth hormone (rGH). The rat GH_3 expression assay was employed as follows. Following overnight incubation, the cells were rinsed with culture fluid minus serum. A 1.0-ml solution containing DNA (50-1,500 ng) and DEAE-Dextran (200 µg) was added to the cells. Following incubation at 37°C for 30-45 min, the DEAE-Dextran-DNA solution was removed and the cells were rinsed with complete medium. Cells were incubated for various lengths of time in the presence of 2.0 ml of culture fluid which was removed at 24-hr intervals. Bovine GH was assayed in the culture fluid. For mouse L cells, a DEAE-Dextran method coupled to a dimethyl sulfoxide shock protocol was used (26).

Transformation of TK⁻ APRT⁻ Cells

TK⁻ APRT⁻ cells were co-transformed with 100 µg of plasmid DNA containing the herpes simplex thymidine kinase gene (pTK), 10 µg pbGH-4 DNA, and 10 µg TK⁻ APRT⁻ L-cell carrier DNA as described elsewhere (39,57). Following selection for the TK⁺ phenotype, the culture fluids were screened by radioimmunoassay for bGH.

Amplification of DNA sequences in TK⁻ APRT⁻ L cells was accomplished by a co-transformation procedure using plasmid DNA encoding an intact APRT gene along with a truncated TK gene lacking its promoter sequences (pDλAT-3), as described by Roberts and Axel (41). Briefly, 5 x 10^5 LTK⁻ APRT⁻ cells per 100-mm tissue culture dish were plated in DMEM plus 10% calf serum. Cells were co-transformed 24 hr later with 20 ng of pDλAT-3 DNA and 22 µg of pbGH-4 DNA using the $Ca_2(PO_4)$ method (58). Cells were first selected in DMEM, 10% calf serum, 4 µg/ml azaserine, and 15 µg/ml adenine for expression of the APRT⁺ phenotype. These APRT⁺ colonies were expanded and then selected in DMEM, 10% calf serum, 15 µg/ml hypoxanthine, 1 µg/ml aminopterin, and 5.15 µg/ml thymidine (HAT) for expression of the TK⁺ phenotype. Bovine GH secretion into the culture fluid was determined by radioimmunoassay. A diagrammatic representation of this co-transformation procedure is shown in Fig. 4.

Radioimmunoassay

Culture fluids were assayed for bGH by a standard double-antibody radioimmunoassay procedure similar to that previously described for chicken GH (23). Bovine GH antibody was purchased from Chemicon (Los Angeles, California). Results obtained are expressed as ng/ml using a bGH standard (NIH-B18). In this assay, rGH crossreacts with bGH at less than 5% (data not shown).

NaDodSO₄-Polyacrylamide Gels

Bovine GH produced by mouse cells was immunoprecipitated and analyzed by NaDodSO₄-polyacrylamide gel electrophoresis as described previously (18). Immunoprecipitations were analyzed on 12.5% acrylamide-NaDodSO₄ gels (20). Fluorography was performed as described previously (4,22).

Culture Fluid Collection and Growth Hormone Bioassay

Collection of culture fluid from stably transformed mouse L cells for biological activity determinations was described previously (17). Bovine GH synthesized by these mouse cells will be referred to as rbGH. The reconstituted culture medium was assayed for GH biological activity in a hypophysectomized rat tibia bioassay (11). Highly purified bGH (1.57 i.u./mg) derived from bovine pituitary glands was used as a standard.

22

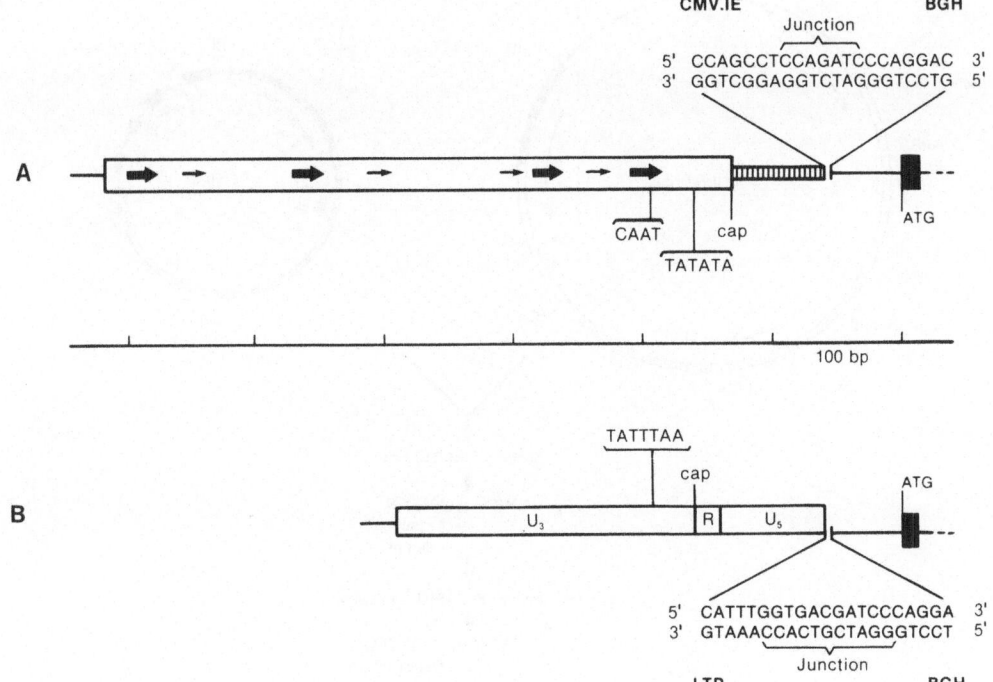

Fig. 3.   Nucleotide sequence of the CMVIE-bGH and RSV-bGH junctions. A.
pCMVIE-bGH. The structure of the CMVIE promoter and its junction
with the first exon of the bGH gene are summarized. The relative
positions of the viral mRNA Cap site (cap), the TATA and CAAT
regions, and the 19-bp (thick arrows) and 18-bp (thin arrows)
direct repeat sequences are presented as previously reported
(53). The exact nucleotide sequence at the junction between the
viral (CMVIE) and bovine (bGH) DNA fragments was determined.
Briefly, pCMVIE-bGH was cleaved with Pstl. A 1.2-kb DNA fragment
containing the CMV sequence and a portion of the first exon of
the bGH gene (Fig. 1) was isolated from a 1% low-melting agarose
gel (seaplaque from FMC, Marine Colloid Division, Rockland,
Maine). The Pstl fragment was 3'-end-labeled with dideoxyadeno-
sine 5'-[α^{32}P]-triphosphate (ddATP, 250 μCi, 300 Ci/mmol; Amer-
sham) and terminal transferase. Fragments with a single 3'-
labeled end were produced by digesting the Pstl fragment with
SacI. The SacI-Pstl fragments (∿133 bp) were separated on an 8%
polyacrylamide gel, purified, and sequenced (29). B. pbGH-4.
The structure of the RSV-LTR is summarized: U3, portion of the
LTR derived from the 3' terminus of viral RNA; U5, portion of the
LTR derived from the 5' terminus of the retroviral RNA; R, 21-bp
terminal redundancy found in viral genomic RNA. The viral mRNA
Cap site (cap) and Hogness box (TATTTAA) are shown (43). The
strategy used for sequencing the junction between the RSV promoter (LTR) and the bovine genomic DNA (bGH) has been described
elsewhere (17). The position of the translation start codon
(ATG) in the first exon of bGH is indicated (59).

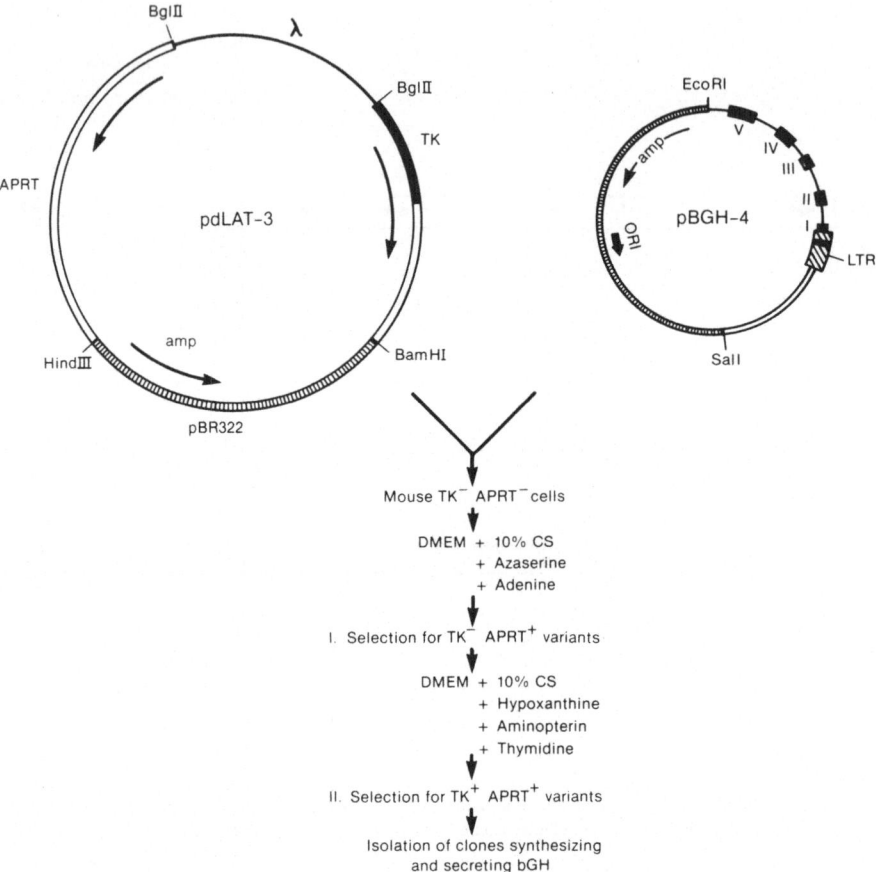

Fig. 4. The protocol for selection of mouse cells that secrete bGH. The co-transformation procedure involves exposing cells to a mixture of pDλAT-3 and pbGH-4 DNA. The representation of pDλAT-3 DNA, shown above (left), was presented originally by Roberts and Axel (41). L = λ (pDLAT-3 = pDλAT-3). Mouse cells are selected for either the TK$^+$ phenotype (I) or TK$^+$, APRT$^+$ phenotype (II). Bovine GH (bGH) levels were determined by radioimmunoassay.

RESULTS

Transient Expression of Bovine Growth Hormone DNA
Recombinants in Cultured Rat Cells

To determine the biological activity of recombinant DNA plasmids containing the bGH gene, a rapid and sensitive expression assay system was developed in rat GH$_3$ cells. Gene expression can be detected at the transcriptional or translational levels. Using a radioimmunoassay, we detect gene expression at the translational level. Previous studies have shown that introduction of a DNA-DEAE-Dextran complex into cultured fibroblasts results in the expression of the gene approximately 2-3 days following transfection (16). Depending on the cell type, expression will continue for a few days and then cease. Apparently the DNA does not integrate stably into the rat cell's genome.

Following introduction of DNA into cultured rat cells, culture fluids were collected and assayed for bGH expression. As can be seen, pbGH-1,

Tab. 1.   Bovine growth hormone expression by transiently transfected rat cells.

DNA	Bovine growth hormone (ng/ml)
None	8.5
pbGH-1 (500 ng)	7.4
pbGH-2 (500 ng)	6.2
pbGH-3 (500 ng)	5.7
pbGH-4 (500 ng)	403.0

pbGH-2, and pbGH-3 did not direct synthesis and secretion of bGH at 5 days following transfection, whereas pbGH-4 DNA did (Tab. 1). We have optimized the conditions necessary for the transient expression of bGH by the rat pituitary cells (Tab. 2). Exposure of between 250-500 ng of pbGH-4 to GH_3 cells for 30 min to 1 hr results in optimal levels of bGH expression.

Production of Rodent Cell Lines Secreting Bovine Growth Hormone

Previous work has shown that the co-transfer of individual plasmids encoding the herpes simplex virus TK DNA and the human growth hormone (hGH) gene produced a large number of TK^+ colonies that contained one or more copies of hGH DNA (42). Many of the TK^+, hGH-positive colonies also produced hGH mRNA and protein (42). This approach was employed to introduce bGH DNA stably into mouse L cells. TK^+ colonies were generated and subcultured, and the amount of bGH secreted into the culture fluid was determined for each colony (Tab. 3A). Cultures L-bGH-4-3 and L-bGH-4-5 secrete ∿3.0 µg bGH per 5 x 10^6 cells per day.

In order to generate a mouse cell line that secretes a larger quantity of bGH, an alternative approach was attempted. The protocol involves co-transformation of mouse L cells (TK^-, $APRT^-$) by plasmid DNA that encodes the APRT gene and a truncated TK gene, pDλAT-3 (44), along with plasmid pbGH-4 DNA (Fig. 4). Results from these types of co-transformation experiments have revealed that plasmid DNAs are amplified within the mouse cell following transformation. Amplification of this plasmid DNA results in a corresponding amplification of gene product (41). Using this procedure, we have obtained 14 mouse L-cell lines (TK^+, $APRT^+$). The amounts of bGH expressed by these cells are shown in Tab. 3B. In general, mouse L cell pDλ clones (1-14), which contain amplified copies of pbGH-4 DNA (data not shown), also secrete relatively large quantities of bGH. For example, L-pDλ-bGH-4-13 secretes 75 µg of bGH per 24 hr per 5.0 x 10^6 cells. This value is approximately 25 times higher than that of mouse cell line L-bGH-4-3.

In a similar manner, rat GH_3 cell lines were generated that secrete relatively high levels of bGH (up to 175 µg per 24 hr per 5 x 10^6 cells). In this case, a neomycin-resistant plasmid DNA was used in the co-transformation protocol (data not shown).

Detection of Bovine Growth Hormone Produced by L-pDλ-bGH-4-13 Cells

Cell line L-pDλ-bGH-4-13 was pulse-labeled for 4 hr with [^{35}S]-methionine, after which the culture fluids and cellular lysates were immunoprecipitated and analyzed on a 12.5% $NaDodSO_4$-polyacrylamide gel. The results are shown in Fig. 5. Lane G represents 25 µl of [^{35}S]-methionine-labeled

Tab. 2. Optimization of transfection conditions that lead to bovine growth hormone production by rat GH_3 cells.

Exp. #	Plasmid	DNA (ng)	Duration of transfection (min)	Bovine growth hormone (ng/ml) following transfection (hr)					
				24	48	72	96	120	144
1	--	--	120	15.1	14.1	10.0	20.6	17.1	18.9
2	pbGH-4	50	30	17.9	20.1	25.0	40.0	46.2	68.1
3	"	100	30	19.1	16.0	36.5	87.1	85.6	118.6
4	"	250	30	26.0	29.0	115.7	205.1	748.2	764.8
5	"	500	30	29.2	41.5	89.5	259.6	308.3	475.1
6	"	50	60	23.4	12.9	30.0	33.2	34.0	56.0
7	"	100	60	17.5	21.1	26.4	62.6	82.5	212.1
8	"	250	60	18.2	22.5	96.5	222.1	481.2	506.1
9	"	500	60	33.1	34.3	92.2	200.4	397.1	545.2
10	"	50	120	18.8	16.3	26.1	29.2	30.1	44.4
11	"	100	120	21.1	17.9	39.2	44.3	52.6	60.5
12	"	250	120	13.0	23.3	44.7	60.2	105.7	120.0
13	"	500	120	30.4	39.9	43.9	65.0	67.0	147.3

Tab. 3. Bovine growth hormone expression by stably transformed mouse L cells.

Stable mouse L-cell lines	Bovine growth hormone (ng/5 x 10^6 cells/24 hr)
A. TK^+, $APRT^-$	
L^-A^- (control)	0.94
L-bGH-4-1	30.24
L-bGH-4-2	86.24
L-bGH-4-3	3,000.00
L-bGH-4-4	0.68
L-bGH-4-5	3,000.00
L-bGH-4-6	0.52
L-bGH-4-7	49.71
L-bGH-4-8	0.78
L-bGH-4-9	341.72
L-bGH-4-10	267.87
B. TK^+, $APRT^+$	
L-pDλ-bGH-4-1	300
L-pDλ-bGH-4-2	16,000
L-pDλ-bGH-4-3	18,000
L-pDλ-bGH-4-4	15,000
L-pDλ-bGH-4-5	12,000
L-pDλ-bGH-4-6	4,000
L-pDλ-bGH-4-7	6,000
L-pDλ-bGH-4-8	20,000
L-pDλ-bGH-4-9	4,000
L-pDλ-bGH-4-10	20,000
L-pDλ-bGH-4-11	12,000
L-pDλ-bGH-4-12	14,000
L-pDλ-bGH-4-13	75,000
L-pDλ-bGH-4-14	47,000

culture fluid (derived from L-pDλ-bGH-4-13 cells) loaded directly onto the gel without immunoprecipitation. A protein of ∿21,000 daltons (see arrow) can easily be distinguished. Immunoprecipitation of 1.0 ml of this culture fluid using rabbit anti-bGH serum and analysis of the proteins on NaDodSO$_4$-polyacrylamide gel is depicted in lane D. A protein of 21,000 daltons is found in the immunoprecipitate. In a 4-hr pulse-labeling period, 12.5 μg of bGH is secreted by L-pDλ-bGH-4-13 cells. Lane E represents immunoprecipitation as described above in the presence of 5.0 μg of bGH isolated from bovine pituitary glands. As can be seen, [^{35}S]-methionine-labeled bGH found in culture fluids derived from L-pDλ-bGH-4-13 cells is not completely competed by addition of 5.0 μg of cold bGH to the immunoprecipitation reaction (lane E). Similar competition experiments, in which 15.0 μg of cold bGH is included in the immunoprecipitation reaction, show complete displacement of the [^{35}S]-methionine-labeled bGH in the reaction (data not shown). Immunoprecipitation using normal rabbit serum is depicted in lane F. No radioactive material of apparent molecular mass of 21,000 daltons is seen.

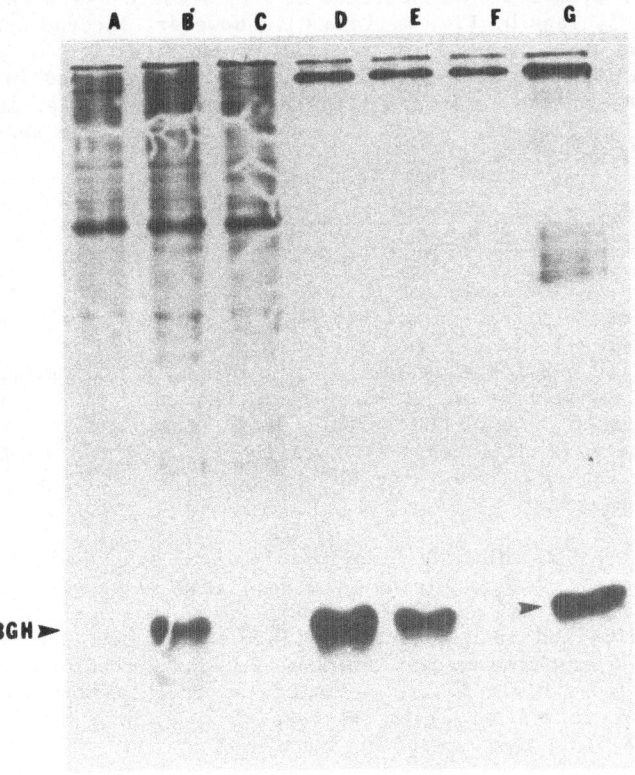

Fig. 5. Fluorography of a 12.5% NaDodSO$_4$-polyacrylamide slab gel of bGH immunoprecipitated with anti-bGH serum. L-pDλ-bGH-4-13 cells (10^7 cells) were pulse-labeled for 4 hr with [^{35}S]-methionine (60 μCi/4.0 ml). Culture fluids were removed and 0.025 ml applied directly to the gel (lane G). The remainder of the culture fluids was divided into 3 equal aliquots (1.3 ml) and immunoprecipitated with anti-bGH serum (lane D), anti-bGH serum plus 5.0 μg of cold bGH (lane E), or normal rabbit serum (lane F). Cellular lysates were immunoprecipitated with normal rabbit serum (lane A), anti-bGH serum (lane B), or anti-bGH serum plus 5.0 μg of cold bGH (lane C).

Effect of β-Hydroxyleucine on Bovine Growth Hormone Processing

Bovine GH is synthesized via a precursor protein that contains a 26-amino-acid, amino-terminal signal peptide (32) that is removed from the growth hormone molecule during the secretory process (55). Immunoprecipitation of bGH from an intracellular lysate derived from mouse cells (pDλ-pbGH-4-13), followed by NaDodSO$_4$ gel electrophoresis, resulted in detection of a radioactive band of the same apparent size as that secreted by the mouse cells (Fig. 5, lane B). A bGH molecule containing a 26-amino-acid signal peptide was not observed. This is not unusual since the signal peptide of a variety of secreted proteins is removed by a nascent chain proteolytic cleavage reaction (55). Inhibition of this cleavage has been shown to occur by incorporation of leucine analogs, such as β-hydroxy leucine (β-OH leucine), into the respective precursor proteins (13,19). We have incorporated [35S]-methionine in the presence or absence of β-OH leucine into bGH, immunoprecipitated the cellular lysates and culture fluids, and analyzed the immunoprecipitates on NaDodSO$_4$-polyacrylamide gels. In the absence of β-OH leucine, secreted bGH is detected as a band of ∿21,000 daltons (Fig. 5, lane D; Fig. 6, lane C). However, incorporation of [35S]-methionine and β-OH leucine into L-cell protein results in inhibition of bGH secretion (Fig. 6, lane B). Intracellular bGH labeled in the presence of β-OH leucine exists in a form with an apparent molecular mass of 25,000 daltons (Fig. 6, lane D), ∿4,000 daltons greater than the secreted form of bGH (Fig. 6, lane C).

NH$_2$ Terminal Amino Acid Sequence of Recombinant and Native Bovine Growth Hormone

The above results indicate that a protein of 21,000 daltons, which is similar in size to pituitary-derived bGH, is found in culture fluids from mouse L-pDλ-bGH-4-13 cells, and is immunoprecipitated by rabbit anti-bGH serum. Purified bGH competes with this protein in an immunoprecipitation reaction. Also, a protein of this size is found in cellular extracts immunoprecipitated by anti-bGH serum. Immunoprecipitation of this intracellular protein is diminished by addition of cold bGH to the reaction. During the secretory process, the NH$_2$ terminal "signal" peptide is cleaved from the bGH molecule.

In order to determine the fidelity of signal peptide cleavage by mouse L cells expressing rbGH, a micro amino acid (aa) sequence analysis comparing native pituitary bGH and rbGH was performed. Results (Tab. 4) indicate that both native and rbGH possess similar heterogenous NH$_2$ termini. Approximately 50% of the molecules possess alanine as their first amino acid, with the remainder divided between molecules possessing either phenylalanine (aa 2) or proline (aa 3) in the case of rbGH, or phenylalanine (aa 2) or alanine (aa 4) in the case of native bGH (Tab. 4).

Biological Activity of Bovine Growth Hormone Derived from L-bGH-4-3 Mouse Cells

Culture fluids collected from L-pDλ-pbGH-4-13 fibroblasts or control L cells (TK⁻, APRT⁻) were concentrated and suspended in distilled water, and adjusted to a bGH concentration of 40 μg/ml or 10 μg/ml (determined by radioimmunoassay). These solutions were injected intraperitoneally into hypophysectomized rats. Bovine GH derived from L-pDλ-bGH-4-13 mouse cells was as potent as bGH purified from bovine pituitary glands in the biological assay (Fig. 7). Concentrated culture fluid derived from control L (TK⁻, APRT⁻) cells showed no biological activity in the growth hormone bioassay.

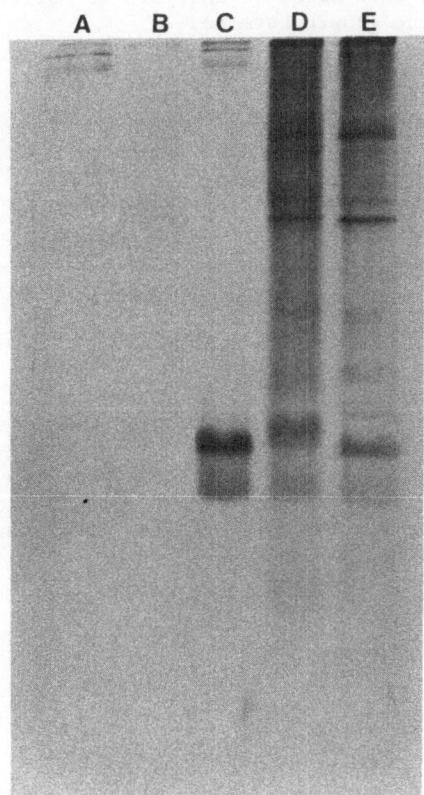

Fig. 6. Effect of β-OH leucine on bGH processing. L-pDλ-pbGH-4-13 cells were labeled with [^{35}S]-methionine in the presence (lanes B and D) or absence (lanes A, C, and E) of β-OH leucine. Cellular lysates (lanes D and E) or culture fluids (lanes A, B, and C) were immunoprecipitated using anti-bGH serum as described in the legend to Fig. 3. The immunoprecipitated proteins were fractionated on a 12.5% polyacrylamide-NaDodSO$_4$ gel. Lane A represents immunoprecipitation using normal rabbit serum.

Comparison of the RSV-LTR and CMVIE Promoters in Directing Synthesis of Bovine Growth Hormone by Cultured Rodent Cells

We were interested in comparing the RSV-LTR and the CMVIE promoters in their ability to direct bGH gene expression in a similar host cell environment. Recombinant plasmid DNA molecules were constructed that contain viral promoters ligated to the coding sequence of the bGH gene (see "Materials and Methods" above). pbGH-4 and pCMVIE-bGH plasmid DNAs were introduced transiently into cultured rat GH$_3$ or mouse L cells.

The CMVIE promoter appeared more efficient than the RSV-LTR in its ability to direct expression of the bGH gene in GH$_3$ cells (Fig. 8) or in mouse L cells (Fig. 9) at all times post-transfection. Also, pCMVIE-bGH plasmid DNA led to the synthesis of 3 to 5 times more bGH than did pbGH-4 DNA, independent of the dose of transfected DNA or time post-transfection.

Since the hybrid DNA molecules RSV-bGH and CMVIE-bGH were cloned into pBR322 at different restriction sites and in different orientations, a possibility exists that plasmid sequences may influence transcription of the

Tab. 4.   Micro amino acid sequence analysis of native pituitary and recombinant bovine growth hormone.

Peptide	Amino acid #										Native pituitary bGH	Recombinant bGH
	1	2	3	4	5	6	7	8	9	10		
bGH (1-191)	N-Ala-Phe-Pro-Ala-Met-Ser-Leu-Ser-Gly-Leu--										45%	51%
bGH (2-191)	N-Phe-Pro-Ala-Met-Ser-Leu-Ser-Gly-Leu--										43%	30%
bGH (3-191)	N-Pro-Ala-Met-Ser-Leu-Ser-Gly-Leu--										--	16%
bGH (4-191)	N-Ala-Met-Ser-Leu-Ser-Gly-Leu--										14%	3%

bGH gene. DNAs encoding the hybrid genes were excised from the pBR322 vector and purified by gel electrophoresis. Transfection of the linear purified DNA fragments into GH_3 cells or mouse L cells yields results identical to those presented above, indicating that pBR322 nucleotide sequences do not influence transcription of the bGH gene in either of the hybrid molecules (data not shown).

DISCUSSION

The experiments presented above describe methods by which the bGH gene can be altered, ultimately resulting in expression of biologically active bGH by rodent cells.

Alterations in the bGH gene include removal of the 5'-flanking region of the authentic gene. This 5'-flanking region of a variety of genes has been shown to contain important transcriptional regulatory sequences (2,3, 6,7,12,27,33,56,60). The CMVIE promoter, when added to this 5'-flanking region on bGH, was found to be 3 to 5 times more efficient than the RSV-LTR in directing expression of bGH by GH_3 or mouse L cells. The molecular characteristics that enable the $CMVIE^3$ promoter/enhancer to be more efficient than the RSV-LTR or the natural bGH 5'-regulatory flanking sequences in directing expression of bGH in GH_3 or mouse L cells are unknown. The precise locations of the TATA box, the CAAT box, and the viral mRNA Cap site in the CMVIE promoter-regulatory region have been defined (53). Sequence analysis of the human CMVIE promoter revealed nucleotide repeating units of 16, 18, and 19 bp (53). The 19-bp repeats appear highly conserved between different strains of human and simian CMV. Thomsen et al. (53) suggested a correlation between these repeat sequences, the formation of cruciform structures, and the CMVIE promoter strength. In the case of retroviruses, viral RNA synthesis has been shown to initiate at the Cap site in the R region using promoter sequences located in U3 (52). A putative Hogness box (TATA) appeared 30 nucleotides upstream from the Cap site in the RSV genomic DNA sequence (14,43). A functional CAAT box was described in the U3 region of the Abelson murine leukemia virus (48), but no homologous transcriptional regulation signal appeared in the RSV-LTR sequence (43). The RSV-LTR has been shown to contain an efficient promoter as well as an enhancer (21,24,46). Short sequences in the RSV-LTR were found homologous to the tandem repeat enhancer element of simian virus 40 (SV40) (14). Also, enhancers have been identified within the U3 region of various retroviral LTRs, and different studies suggest a correlation between the LTR enhancer activity and the activation of cellular proto-oncogenes by retroviruses (15,48). Further experiments that explicitly compare the role of these particular sequences in the LTR and CMVIE promoter/enhancer are ongoing in our laboratory. Also, other eukaryotic transcriptional regula-

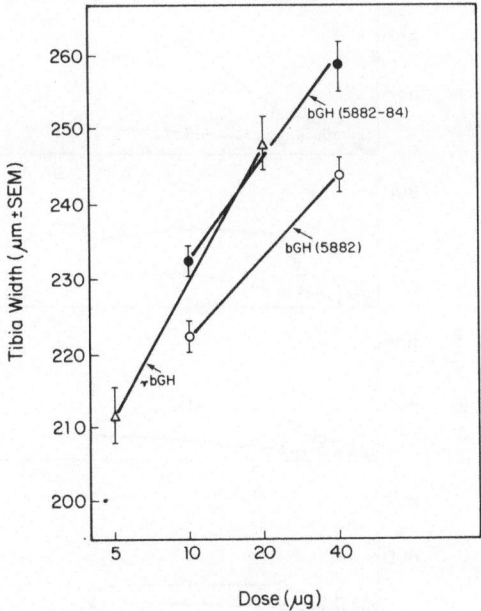

Fig. 7.   Biological activity of bGH derived from L-bGH-4-3 cells. Culture
fluids were collected as described in "Materials and Methods"
section in text and suspended to a bGH concentration of 10 μg/ml
or 40 μg/ml, or to a rbGH concentration of 5 μg/ml or 20 μg/ml.
Values presented represent the mean value of tibia growth of 4 to
6 hypophysectomized rats injected either with concentrated culture fluid (Δ) or 2 preparations, 5882 (●) and 5882-84 (o), of
purified bGH derived from bovine pituitary glands.

tory sequences have been attached to the bGH gene. Of such sequences, the
SV40 early promoter has been found to direct the highest levels of bGH expression in mouse L cells (data not shown). Thus, a transient expression
system as described above may be used in assaying a variety of transcription regulatory sequences in their ability to direct expression of bGH in
cultured rodent cells.

A goal of this research was to produce rodent cell lines that secrete
bGH. To reach this objective we have employed a co-transformation protocol
to generate mouse cell lines that contain pbGH-4 DNA and secrete bGH. The
co-transformation process used to generate TK^+, bGH^+ mouse L-cell lines appears very efficient in that 80% of the TK^+ clones also expressed detectable bGH. Cell lines have been selected that secrete bGH at rates ranging
from 30-3,000 ng per 24 hr per 5 x 10^6 cells.

Amplification of the co-transformed DNA sequences (41) results in stably transformed, mouse L-cell lines that secrete large amounts of bGH. We
have generated 14 clones that secrete between 4 and 75 μg of bGH per 24 hr
per 5 x 10^6 cells. The efficiency of co-transformation of the pDλAT-3 DNA
and pbGH-4 DNA is quite high. Of 14 L-cell (TK^+, $APRT^+$) clones generated,
all express elevated bGH. L-cell clone L-pDλ-bGH-4-13 secreted ∿75 μg of
bGH per 24 hr per 5 x 10^6 cells. This cell line contained multiple copies
of the bGH gene (J. Kopchick and M. Silberklang, manuscript in prep.). Also, stable rat GH_3 cell lines have been generated (data not shown) that
secrete higher levels of bGH as compared to mouse L cells. This may be due
to increased copy number of bGH-4 DNA in these cells or to the fact that
GH_3 cells normally secrete rat GH and, therefore, may be able to secrete

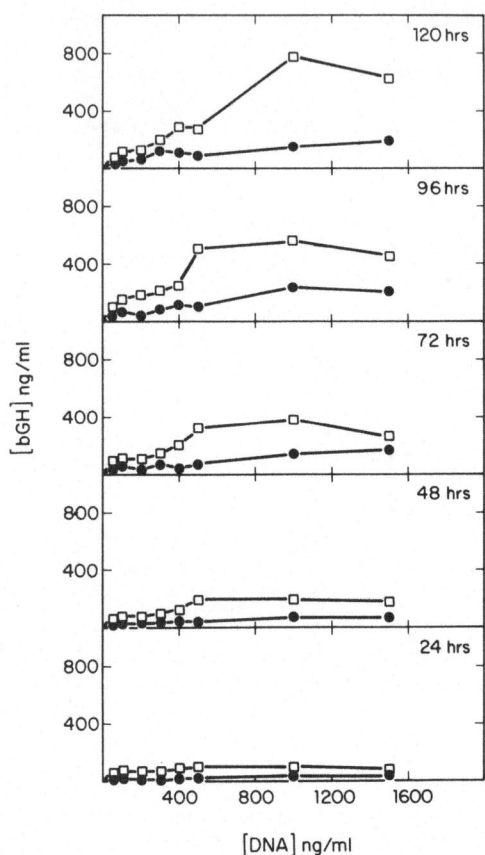

Fig. 8.   Bovine GH expression by transiently transformed GH$_3$ cells. Equal
amounts (ranging from 50 to 1,500 µg/ml) of purified pbGH-4 or
pCMVIE-bGH DNA were transfected into rat GH$_3$ cells in the presence of DEAE-dextran (200 µg/ml) as described in "Materials and
Methods" section in text. Secretion of bGH into the culture
fluid was assayed by a standard double antibody radioimmunoassay
(23). The production of bGH by the cells transformed with pbGH-4
(●————●) or with pCMVIE-bGH (□————□) was measured 24, 48, 72,
96, and 120 hr after transfection, and is expressed in ng/ml of
culture medium (DMEM). A total of 2 ml of culture fluid was used
in each cellular incubation.

bovine GH more efficiently than mouse L cells.   Experiments to elucidate
this difference between cell lines in bGH secretion are ongoing.

Bovine GH is made from a precursor protein that contains 26 extra
amino acids labeled at the amino terminus of the molecule (32). This 26-
amino-acid signal peptide is involved in translocation of the nascent protein across the membrane of the endoplasmic reticulum (55). Results of
leucine analog incorporation experiments revealed that L-pDλ-pbGH-4-13
cells do synthesize bGH as a precursor protein. Importantly, incorporation
of β-OH leucine into bGH inhibits secretion of the protein by mouse cells.
This indicates that bGH synthesized within the mouse cell contains extra
amino acids (a signal sequence), and that cleavage occurs prior to secretion of bGH into the culture fluid.

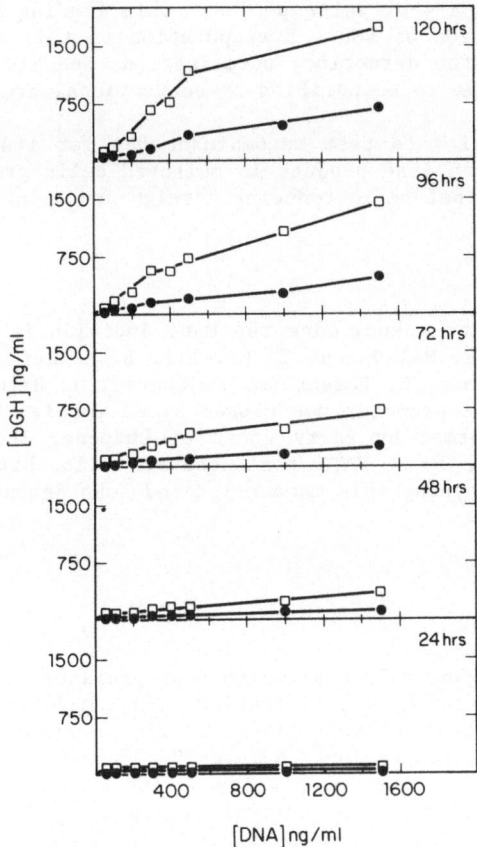

Fig. 9.   Bovine GH (bGH) expression in transiently transfected mouse L
          cells. Equal amounts of purified pbGH-4 (●———●) or pCMVIE-bGH
          (□———□) DNA were transfected into mouse L cells using the
          DEAE-Dextran-dimethyl sulfoxide shock procedure as described in
          "Materials and Methods" section in text. Bovine GH expression
          was determined as described in the legend to Fig. 8.

     Amino acid sequencing of native bGH as compared to rbGH revealed a
similar degree of amino-acid heterogeneity at the NH_2 terminus. Native bGH
was reported by previous investigators (24,25) to have a heterogenous NH_2
terminal. Since rbGH possesses the same type of heterogeneity as authentic
bGH (see "Results" above), a possibility exists that the information necessary for cleavage is an inherent part of the bGH molecule and not cell
type-specific. Further experiments to test the observation are proceeding.

     It was of importance to determine whether bGH secreted by these mouse
cells was biologically active. Concentrated culture fluids derived from
L-pDλ-bGH-4-13 cells were found to be biologically active in the rat tibia
GH bioassay. Culture fluid containing bGH or authentic bGH derived from
bovine pituitary glands was found to be of equal potency in the GH bioassay. Concentrated culture fluids from control cells (LTK⁻, APRT⁻) were
found to be inactive in this assay. Also, we have recently shown that recombinant and authentic bGH possess equal potency in a GH receptor-binding
assay.

     The mouse L-cell line L-pDλ-bGH-4-13 secreted large quantities of bGH,
representing between 50 and 60% of the total protein secreted by the cells

(Fig. 5, lane G) in a 4-hr pulse period. This finding has greatly facilitated the purification of bGH. Extrapolation of this technology to other
systems may aid in the detection, purification, and biological analysis of
novel peptides unable to be purified by conventional procedures.

Also, the ability to test recombinant DNA for its ability to direct
expression of a given gene product in cultured cells provides a convenient
and necessary assay before introducing foreign genes into animals.

ACKNOWLEDGEMENTS

We would like to acknowledge the many individuals involved with this
project, including R. Malavarca, T. Livelli, B. Kelder, H. Chen, M. Bayne,
J. Taylor, S. Steelman, C. Rosenblum, E. Convey, J. Birnbaum, H. Hafs, and
J. Smith. The CMVIE promoter was cloned by M. Tocci. Amino acid sequence
analyses were performed by Barry Jones of Unigene, Inc. F. Pasleau was
supported, in part, by a NATO postdoctoral fellowship. Also, we thank
Marilyn Serson for typing this manuscript and John Reminger for photograph-
ic assistance.

REFERENCES

1. Andrews, P. (1966) Molecular weights of prolactin and pituitary growth
    hormones estimated by gel filtration. Nature 209:155-157.
2. Benoist, C., and P. Chambon (1981) In vivo sequence requirement of the
    SV40 early promoter region. Nature 290:304-310.
3. Benoist, C., K. O'Hare, R. Breatnach, and P. Chambon (1980) Ovalbumin
    gene-sequence of putative control regions. Nucl. Acids Res. 8:127-
    142.
4. Bonner, W.M., and R.A. Laskey (1974) A film detection method for
    tritium-labelled proteins and nucleic acids in polyacrylamide gels.
    Eur. J. Biochem. 46:83-88.
5. Cathala, G., N.L. Eberhardt, N.C. Lan, D.G. Gardner, A. Gutierrez-
    Hartmann, S.H. Mellon, M. Karin, and J.D. Baxter (1983) Structure and
    expression of growth hormone-related genes. In Perspective on Genes
    and the Molecular Biology of Cancer, D.L. Robberson and G.F. Saunders,
    eds. Raven Press, New York.
6. Cullen, B.R., P.T. Lomedico, and G. Ju (1984) Transcriptional inter-
    ference in avian retroviruses--Implications for the promoter insertion
    model of leukaemogenesis. Nature 307:241-245.
7. Darnell, J.E., Jr. (1982) Variety in the level of gene control in eu-
    caryotic cells. Nature 297:365-371.
8. Dellacha, J.M., J.A. Santome, and L. Faiferman (1966) Molecular weight
    of bovine growth hormone. Experimentia 22:16-17.
9. Ellis, G.J., E. Marler, H.C. Chen, and A.E. Wilhelmi (1966) Molecular
    weight of bovine, porcine and human growth hormone by sedimentation
    equilibrium. Fed. Proc. 25:348.
10. Goeddel, D.V., H.L. Heyneker, T. Hozumi, R. Arentzen, K. Itakura, D.G.
    Yansura, J.J. Ross, G. Miozari, R. Crea, and P.H. Seeburg (1979)
    Direct expression in Escherichia coli of a DNA sequence coding for hu-
    man growth hormone. Nature 281:544-548.
11. Greenspan, F.J., C.H. Li, M.E. Simpson, and H.M. Evans (1949) Bioassay
    of hypophyseol growth hormones: The tibia test. Endocrinology 45:
    455-463.
12. Grosveld, G.C., E. de Boer, C.K. Shewmaker, and R.A. Flavell (1982)
    DNA sequences necessary for transcription of the rabbit β-globin gene
    in vivo. Nature 295:120-126.
13. Hortin, G., and 1. Boime (1980) Inhibition of preprotein processing in

ascites tumor lysates by incorporation of a leucine analog. Proc. Natl. Acad. Sci., USA 77:1356-1360.

14. Ju, G., and A.M. Skala (1980) Nucleotide sequence analysis of the long terminal repeat (LTR) of avian retroviruses: Structural similarities with transposable elements. Cell 22:379-386.

15. Khoury, G., and P. Gruss (1983) Enhancer elements. Cell 33:313-314.

16. Kopchick, J.J., and D.W. Stacey (1984) Differences in intracellular DNA ligation after microinjection and transfection. Mol. Cell. Biol. 4:240-246.

17. Kopchick, J.J., R.H. Malavarca, T.J. Livelli, and F.C. Leung (1985) Use of avian retroviral bovine growth hormone DNA recombinants to direct expression of biologically active growth hormone by cultured fibroblasts. DNA 4:23-31.

18. Kopchick, J.J., W.L. Karshin, and R.B. Arlinghaus (1979) Tryptic peptide analysis of 'gag' and 'gag-pol' gene products of Rauscher murine leukemia virus. J. Virol. 30:610-623.

19. Kozak, M. (1983) Translation of insulin-related polypeptides from messenger RNAs with tandemly reiterated copies of the ribosome binding site. Cell 34:971-978.

20. Laemmli, U.K. (1970) Cleavage of structural proteins during the assembly of the head of bacteriophage T4. Nature 227:680-685.

21. Laimins, L.A., P. Tsichlis, and G. Khoury (1984) Multiple enhancer domains in the 3' terminus of the Prague strain of Rous sarcoma virus. Nucl. Acids Res. 12:6427-6442.

22. Laskey, R.A., and A.D. Mills (1975) Quantitative film detection of ^{3}H and ^{14}C in polyacrylamide gels by fluorography. Eur. J. Biochem. 56: 335-341.

23. Leung, F.C., J.G. Taylor, S.L. Steelman, C.D. Bennett, J.A. Rodkey, R.A. Long, R. Serio, R.M. Weppelman, and G. Olson (1984) Purification and properties of chicken growth hormone and the development of a homologous radioimmunoassay. Gen. Comp. Endocrinol. 56:389-400.

24. Li, C.H., and L. Ash (1953) The nitrogen terminal end groups of hypophyseal growth hormone. J. Biol. Chem. 203:419-424.

25. Lingappa, V.R., A. Devillers-Threry, and G. Blobel (1977) Nascent prehormones are intermediates in the biosynthesis of authentic bovine pituitary growth hormone and prolactin. Proc. Natl. Acad. Sci., USA 74:2432-2436.

26. Lopata, M.A., D.W. Cleveland, and B. Soelner-Webb (1984) High levels of transient expression of a chloramphenicol acetyl transferase gene by a DEAE-dextran mediated DNA transfection coupled with a dimethyl sulfoxide or glycerol shock treatment. Nucl. Acids Res. 12:5707-5717.

27. Luciw, P.A., J.M. Bishop, H.E. Varmus, and M.R. Capecchi (1983) Location and function of retroviral and SV40 sequences that enhance biochemical transformation after microinjection of DNA. Cell 33:705-716.

28. Martin, J.B. (1978) Neural regulation of growth hormone secretion. New Engl. J. Med. 288:1384-1388.

29. Maxam, A.M., and W. Gilbert (1977) A new method for sequencing DNA. Proc. Natl. Acad. Sci., USA 74:560-564.

30. Miller, W.L., and N.L. Eberhardt (1983) Structure and evolution of the growth hormone gene family. Endocrine Rev. 4:97-129.

31. Miller, W.L., D. Coit, J.D. Baxter, and J.A. Martial (1981) Cloning of bovine prolactin cDNA and evolutionary implications of its sequence. DNA 1:37-50.

32. Miller, W.L., J.A. Martial, and J.D. Baxter (1980) Molecular cloning of DNA complementary to bovine growth hormone mRNA. J. Biol. Chem. 255:7521-7524.

33. Minty, A., and P. Newmark (1980) Gene regulation: New, old and remote controls. Nature 288:210-211.

34. Niall, H.D., M.L. Hogan, R. Sayer, I.Y. Rosenblum, and F.C. Greenwood (1971) Sequences of pituitary and placental lactogenic and growth hormones: Evolution from a primordial peptide by gene replication. Proc. Natl. Acad. Sci., USA 68:866-869.

35. Palmiter, R.D., R.L. Brinster, R.E. Hammer, M.E. Taumbauer, M.G. Rosenfeld, N.C. Birnberg, and R.M. Evans (1982) Dramatic growth of mice that develop from eggs microinjected with metallothionein-growth hormone fusion genes. Nature 300:611-615.

36. Pasleau, F., M.J. Tocci, F. Leung, and J.J. Kopchick (1985) Growth hormone gene expression in eukaryotic cells directed by the Rous sarcoma viral long terminal repeat or cytomegalovirus immediate early promoter. Gene (in press).

37. Peel, C.J., D.E. Bauman, R.C. Gorewit, and C.J. Snifton (1981) Effect of exogenous growth hormone on lactational performance in high yielding dairy cows. J. Nutr. 111:1662-1671.

38. Peel, C.J., T.J. Frank, D.E. Bauman, and R.C. Gorewit (1983) Effect of exogenous growth hormone in early and late lactation on lactation performance of dairy cows. J. Dairy Sci. 66:776-782.

39. Pellicer, A., D. Robins, B. Wold, R. Sweet, J. Jackson, I. Lowy, J.M. Roberts, G.K. Sim, S. Silverstein, and R. Axel (1980) Altering genotype and phenotype by DNA-mediated gene transfer. Science 209:1414-1422.

40. Raben, M.S. (1958) Treatment of a pituitary dwarf with human growth hormone. J. Clin. Endocrinol. 18:901-904.

41. Roberts, J.M., and R. Axel (1982) Gene amplification and gene correction in somatic cells. Cell 29:109-119.

42. Robins, D.M., I. Paek, P.H. Seeburg, and R. Axel (1982) Regulated expression of human growth hormone genes in mouse cells. Cell 29:623-631.

43. Schwartz, D.E., R. Tizard, and W. Gilbert (1983) Nucleotide sequence of Rous sarcoma virus. Cell 32:853-869.

44. Seeburg, P.H., J. Shine, J.A. Martial, J.D. Baxter, and H.M. Goodman (1977) Nucleotide sequence and amplification in bacteria of the structural gene for rat growth hormone. Nature 270:486-494.

45. Seeburg, P.H., S. Sias, J. Adelman, H.A. De Boer, J. Hayflick, P. Jhurani, D.V. Goeddel, and H.L. Heyneker (1983) Efficient bacterial expression of bovine and porcine growth hormones. DNA 2:37-45.

46. Skalka, A.M., P.N. Tsichlis, R. Malavarca, B. Cullen, and G. Ju (1983) Viral sequences determining the oncogenicity of avian leukosis viruses. In Oncogenes and Retroviruses: Evaluation of Basic Findings and Clinical Potential, T.E. O'Connor and F. Rauscher, eds. Alan R. Liss, Inc., New York, pp. 105-118.

47. Sompayrac, L.M., and K.J. Danna (1981) Efficient infection of monkey cells with DNA of simian virus 40. Proc. Natl. Acad. Sci., USA 78:7575-7578.

48. Srinivasan, A., R.E. Premkumar, C.Y. Dunn, and S.A. Aaronson (1984) Molecular dissection of transcriptional control elements within the long terminal repeat of the retrovirus. Science 223:286-289.

49. Stenberg, R.M., D.R. Thomsen, and M.F. Stinksi (1984) Structural analysis of the major immediate early gene in human cytomegalovirus. J. Virol. 49:190-199.

50. Stinski, M.F., D.R. Thomsen, R.M. Stenberg, and L.C. Goldstein (1983) Organization and expression of the immediate early genes of human cytomegalovirus. J. Virol. 46:1-14.

51. Swislock, N.I., M. Sonnenber, and N. Yamasaki (1970) In vitro metabolic effects of bovine growth hormone fragment in adipose tissue. Endocrinology 87:900-904.

52. Temin, H.M. (1982) Function of the retrovirus long terminal repeat. Cell 28:3-5.

53. Thomsen, D.T., R.M. Stenberg, W.F. Goins, and M.F. Stinski (1984) Promoter-regulatory region of the major immediate early gene of cytomegalovirus. Proc. Natl. Acad. Sci., USA 81:659-663.

54. Wathen, M.W., and M.F. Stinski (1982) Temporal patterns of human cytomegalovirus transcript: Mapping the viral RNAs synthesized at immediate early, early and late times after infection. J. Virol. 4:462-477.

55. Walter, P., R. Gilmore, and G. Blobel (1984) Protein translocation across the endoplasmic reticulum. <u>Cell</u> 38:5-8.
56. Weiher, H., M. Konig, and P. Gruss (1983) Multiple point mutations affecting the simian virus 40 enhancer. <u>Science</u> 219:626-631.
57. Wigler, M., S. Silverstein, L.S. Lee, A. Pellicer, Y.C. Cheng, and R. Axel (1977) Transfer of purified herpes virus thymidine kinase gene to cultured mouse cells. <u>Cell</u> 11:223-232.
58. Wold, B., M. Wigler, E. Lacy, T. Maniatis, S. Silverstein, and R. Axel (1979) Introduction and expression of a rabbit β-globin gene in mouse fibroblasts. <u>Proc. Natl. Acad. Sci., USA</u> 76:5684-5688.
59. Woychick, R.P., S.A. Camper, R.H. Lyons, S. Horowitz, E.C. Goodwin, and F.M. Rottman (1982) Cloning and nucleotide sequencing of the bovine growth hormone gene. <u>Nucl. Acids Res.</u> 10:7197-7210.
60. Yaniv, M. (1982) Enhancing elements for activation of eucaryotic promoters. <u>Nature</u> 297:17-18.

MAPPING GENES IN DOMESTICATED ANIMALS

Frank H. Ruddle and Rudolf Fries

Department of Biology
Yale University
New Haven, Connecticut   06511

INTRODUCTION

During the last few decades, animal breeders have been very successful in breeding animals that are superior in the production of milk, eggs, meat, and wool. The strategies for optimized breeding of livestock have been based on the concepts of quantitative genetics which assume that there are several genes, possibly on different chromosomes, each of which has a certain effect on a particular quantitative trait. However, we do not know how many genes are actually involved in the expression of a given quantitative trait or how each gene contributes to the trait and where it is located on the chromosomes. Such knowledge will become crucial for improvements in the breeding of animals with a heightened resistance to diseases in unfavorable environments or that have better fertility but are still capable of producing good yields of milk, eggs, meat, or wool. Modern methods of gene mapping, so far mostly applied in mice and man, offer new ways towards a better understanding of the genetic determination of animal performance.

Human geneticists have been very successful in mapping the human chromosomes. Hundreds of genes have been assigned to particular chromosomes (12a,42). Knowledge gained by human gene mapping has been applied in prenatal diagnosis (50). As a result of gene mapping, the chromosomal regions responsible for Huntington's disease and Duchenne muscular dystrophy have been determined, and the isolation of the genes that cause these diseases will become possible in the near future (59). Human gene mapping has also produced important insights into the etiology of cancer by showing that genes that cause cancer are sometimes translocated near genes that enhance their activity (9).

In this chapter* we would like to give an overview of the methods and strategies involved in mapping genes and describe the present status of the gene maps of domestic species. Further, we will focus on a potential application of gene maps: the identification and isolation of genes for complex traits.

---

* This chapter is based on a previous review of this subject by Fries and Ruddle (15).

There are 3 different approaches towards the mapping of genes: those that are based on the study of the transmission of genes in families (family studies); those that employ gene transfer between mammalian somatic cells from different species (somatic cell hybridization, chromosome-mediated gene transfer); and those that involve the direct assignment of genes by in situ hybridization of DNA of a specific type to fixed chromosomes.

Family Studies

Up to the late 1960s family studies were the only means by which the location of genes on the same chromosome (linkage) and the assignation of genetic loci to particular chromosomes could be carried out. To determine whether 2 genetic loci are on the same chromosome, the meiotic recombination frequency is measured between the loci of interest in informative families, i.e., families in which one parent is heterozygous at the 2 loci. If the recombination frequency (r) is smaller than 0.50 and if the assumption of random segregation at the loci is correct, one can conclude that the 2 loci are on the same chromosome. The statistical test applied to determine whether an observed segregation fraction is significantly different from 0.50 is usually the logarithm of the odds (lod) score test (40,57). The lod score is calculated as shown in Fig. 1. The lod scores from different pedigrees can be added and are usually presented in standard lod score tables for the values of r = 0.05, 0.10, 0.20, 0.30, and 0.40. The recombination fraction giving the highest lod value is taken as the maximum likelihood estimate. Lod scores of +3.0 or more, meaning that the likelihood of linkage is at least 1,000 times greater than the likelihood of nonlinkage, are considered as sufficient evidence for linkage, while lod scores of -2.0 are sufficient for the rejection of linkage in human studies.

Family studies can also be used for the assignment of a gene to a specific chromosome. In such analyses a chromosomal variant present in heterozygous form is used as one locus in the linkage test. Linkage detected between a chromosomal variant on the long arm of chromosome 1 with a blood group locus made possible the first assignment of a gene to an autosome in man (12). Fries, Stranzinger, and Vögeli (17) took a similar approach in an attempt to determine the chromosomal position of several genes in swine. They studied the linkage relationships of natural, as well as radiation-induced marker chromosomes with genetic loci. The outcome of this study was exclusion of linkage up to a level of 30% recombination for several combinations of marker chromosomes and loci for blood groups, enzymes, and serum proteins, and the provisional assignment of the G-blood group locus to chromosome 15. The evidence for this assignment is presented in Tab. 1. There was a positive lod score above the significance limit of 3.0 for the

$$z = \log \frac{\text{Likelihood of a pedigree, assuming } r < 0.5}{\text{Likelihood of a pedigree, assuming } r = 0.5}$$

or

$$z = \log 2^{(a+b-1)} [r^a(1-r)^b + r^b(1-r)^a]$$

a = Number of recombinant progeny
b = Number of parental progeny
r = Recombination frequency

Fig. 1. Calculation of the lod score (z).

combination of the G-blood group locus and a natural marker chromosome found in hybrids between domestic pigs and European wild pigs [rob(15;17)]. The positive lod scores calculated for the combination of the G-blood group locus and a radiation-induced marker chromosome 15 [rcp(2p+;15Q-)] (16), as well as the negative lod scores from the combination of a chromosome 17 marker (17C+), suggested that genes encoding the G-blood antigens reside on chromosome 15. In a subsequent study (18), the provisional assignment of the G-blood group locus to chromosome 15 was confirmed based on positive lod scores calculated from segregation data of G-blood group alleles and a centromere variant of chromosome 15 (15C+).

The major drawback of gene mapping with family studies is that a locus can only be assigned to a particular chromosome when the following requirements are met:

(a)   An already assigned polymorphic marker locus or a polymorphic cytogenetic locus is not further from the locus to be mapped than a genetic distance characterized by 20% recombination. Linkage can only be detected with a reasonable number of animals when the recombination frequency is not much greater than 20%.

(b)   There must be at least 2 alleles at the locus to be mapped and they must both segregate with alleles at the marker locus in informative families. It is obvious that these requirements are met for only a few genes, mainly because of the lack of a sufficient number of polymorphic marker loci. The human genome is estimated to span 3,000 recombination units or centimorgans (cM) (49). One can assume that this number is approximately the same for most mammalian genomes (at least there is no contradictory evidence thus far). Therefore, at least 75 marker loci evenly spaced at 40-cM intervals are required to cover the genome and to place any new gene locus within 20 cM of a marker locus. However, before evenly spaced loci can be obtained, a much larger number of marker loci must first be placed on the map (59).

It has now become generally accepted that the best way to accommodate the required polymorphic markers is to make use of DNA restriction fragment length polymorphisms (RFLPs) which are readily detected at reasonably high

Tab. 1.   Lod scores for the G-blood group locus and several porcine marker chromosomes. [From Fries et al. (17) and Fries et al. (18).]

Marker combination	Recombination frequency (r)					N^a
	0.05	0.10	0.20	0.30	0.40	
G-rob(15;17)[b]	0.48	2.39	3.36	2.97	1.82	43
G-rcp(2p+;15q-)[c]	0.74	1.55	1.87	1.60	0.96	21
15C+[d]	-2.44	2.21	4.84	4.38	2.31	103
17C+[e]	-3.09	-1.79	-0.70	-0.25	-0.05	13

[a] Number of informative progeny.
[b] Centromeric ("Robertsonian") fusion of chromosomes 15 and 17.
[c] Reciprocal translocation of chromosomes 2 and 15.
[d] Enlargement of the centromeric region of chromosome 15.
[e] Enlargement of the centromeric region of chromosome 17.

frequencies (4,59). Physical methods to be described below can be used in conjunction with RFLP mapping to assign these fragments to particular chromosomal regions. Breeding studies serve to order the fragments over short distances.

The RFLP approach has the following 2 important advantages: (a) a uniform, simple methodology can be employed, and (b) sufficient highly polymorphic markers exist to readily saturate the gene map. Somatic cell genetic techniques and in situ techniques (see below) can be useful in establishing the physical position and distribution of the RFLP marker loci. Once a saturated RFLP map has been achieved, any other single locus-determined trait can be assigned a map position by this approach. In this way, traits important to the animal breeder can be mapped to a specific genetic locus.

Gene Mapping by Somatic Cell Genetics

Gene mapping by somatic cell genetics is based on parasexual events which facilitate the transfer of genetic material between cells (gene transfer) and the partial loss of the transferred genetic material from hybrid cells. The most common approach for the delivery of genes in a random fashion into recipient cells for the purpose of gene mapping is cell hybridization. Somatic cells of one species undergo spontaneous fusion with those from another species at a low frequency when they are cultured together. This frequency can be enhanced considerably by exposure of the cells to inactivated Sendai virus (60) and to chemical agents such as polyethylene glycol (PEG) (44). The use of mutant cell lines that lack hypoxanthine-guanine phosphoribosyl transferase (HPRT), an enzyme in the purine salvage pathway, or thymidine kinase (TK), an enzyme in the pyrimidine salvage pathway, as one fusion partner and cells that are TK^+ or $HPRT^+$ as the other, with subsequent culture in medium that contains HAT (hypoxanthine, aminopterin, thymidine), allows the efficient selection of hybrid cells from the $HPRT^-$ or TK^- parental cells (36). The somatic cell hybrid system is depicted schematically in Fig. 2. Some cell lines which are $HPRT^-$ or TK^- are resistant to Ouabain (3). The combination of $HPRT^-$ or TK^-, Ouabain-resistant cells with $HPRT^+$ or TK^+, Ouabain-sensitive cells and growth in medium that contains HAT allows selection against all parental cells that did not undergo fusion. If mouse and hamster cell lines are combined with human cells or cells from other species, such as swine or cattle, there is a progressive and preferential loss of the chromosomes of the non-mouse or non-hamster species, as first observed by Weiss and Green (58) in human-mouse hybrids. Clones of hybrid cells have only some of the chromosomes of the original parental cell.

It is possible to draw conclusions about synteny (location of genes on the same chromosome) or to assign gene loci to particular chromosomes after the investigation of the chromosomal complement and the determination of the presence or absence of gene products in several hybrid clones. When a gene probe is available, the presence of a gene can be directly determined by restriction fragment analyses (50). A collection of hybrid lines providing these kinds of information is called a hybrid panel. Kamarck et al. (29) reviewed the problems involved in optimizing a hybrid panel. The minimal number of hybrid lines with multiple chromosomes in unique configurations is specified by $C = 2^m$, where C is the haploid chromosome number and m is the minimum number of panel members needed to assign a gene to any of the chromosomes. In man, with 24 different chromosomes, and in swine, with 20 different chromosomes, the minimum number of panel members would only be 5 (Fig. 3). However, the panels that are currently used for mapping human genes usually consist of 20 to 30 members, because it is practically impossible to obtain hybrid cell lines with the specific chromosome complements needed for minimizing the number of hybrid lines of a mapping panel. More-

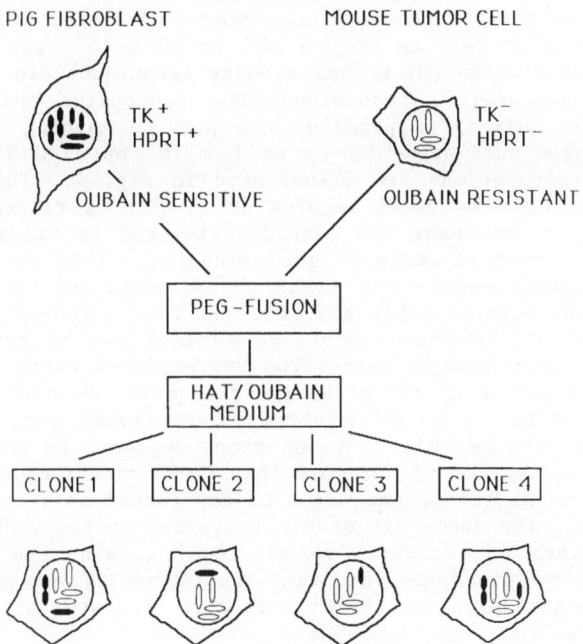

Fig. 2. Generation of somatic cell hybrids.

over, a panel with the minimum number of hybrid lines is subject to false positive and negative conclusions if a single phenotype determination is misinterpreted. Redundancy of cloned hybrid lines within a panel substantially reduces this possibility.

Once a gene is assigned to a chromosome, the next step is to localize the gene to a region within the chromosome. One way to achieve such a regional assignment is in situ hybridization, as we will see below. Another way to regionally map a gene is by using subchromosomal mapping panels. A subchromosomal mapping panel can be constructed with somatic cell hybrids that have deletions or translocations that involve the chromosome to which the gene has been mapped. It was for this reason that we chose porcine cell lines, known to carry well-characterized, heterozygous translocations, as one of the parental lines in fusions with hamster and mouse cell lines. We can expect that, after extensive segregation and subcloning of the hybrids, cell lines will be obtained that will have either a normal copy of certain chromosomes or only parts of them. A series of hybrid lines with different parts of a chromosome can then be used to delimit the position of the gene or, in other words, to determine the "shortest region of overlap" (SRO).

CHROMOSOME NO.

		1	2	3	4	5	6	7	8
HYBRID	A	+	+	+	+	−	−	−	−
PANEL	B	+	+	−	−	+	+	−	−
MEMBER	C	+	−	+	−	+	−	+	−

Fig. 3. Somatic cell hybrid mapping panel. Assignment to chromosome 7 in a hypothetical genome of only 8 chromosomes.

The mapping resolution that can be achieved with a suitable subchromosomal mapping panel is 5-10 cM or 5×10^6 to 10×10^6 base pairs at the most, when high-resolution chromosome banding techniques are applied (50). However, at the DNA level, recombinant DNA procedures that include so called "chromosome walking" approaches can provide mapping data from the level of the single nucleotide bp up to 1×10^5 bp [100 kilobase pairs (kb)]. These considerations are illustrated in Fig. 4. The gap between the level of resolution by panel mapping or in situ hybridization and the level achieved by recombinant DNA methodologies can be bridged by an approach based on chromosome-mediated gene transfer. This technique allows the transfer of intact segments of donor chromosomes, in the size range of 1,000 to 10,000 kb, into suitable murine cells (30). The procedure is illustrated in Fig. 5. Dominant selection markers can be introduced into the donor genome beforehand by retrovirus expression vectors (5,38). Segments of the donor genome carrying the inserted selectable markers can then be isolated in rodent cells after chromosome-mediated gene transfer. By this method one should be able to clone donor segments in the range of 10 to 10,000 kb. A collection of not more than 1,000 rodent clones would contain most of the donor genome and might be designated as a chromosome segment library (51). The donor DNA of clones, shown to carry donor genes of interest, could then be recovered by microbial cloning and selection by filter hybridization with donor-specific, repetitive DNA. A maximum of only about 300 cosmid inserts may be sufficient to saturate an entire segment which can span from 10 up to 10,000 kb.

In Situ Hybridization

In situ hybridization is a straightforward mapping technique involving the hybridization of radiolabeled DNA probes to fixed metaphase chromosomes and the subsequent visualization of the signal as silver grains after autoradiography. Hybridization in situ was originally developed by Gall and Pardue (19) and John, Birnstiel, and Jones (27). This technique first allowed only the localization of highly reiterated or amplified genes. Improvements in the hybridization procedure, including the use of Dextran sulphate in the reaction mixture, made possible the routine mapping of single copy genes in man by in situ hybridization with probes labeled to high specific activities by nick translation with [^{3}H]- or [^{125}I]-labeled nucleotides (25,46). An essential step in gene mapping by in situ hybridization is the unambiguous identification of the chromosomes. This can be achieved by prephotographing Giemsa(G)-banded chromosomes after trypsin treatment (52) or Quinacrine(Q)-banded chromosomes after quinacrine mustard staining (7). Q-bands can also be visualized by quinacrine mustard staining after hybridization (39). The resolution of mapping by in situ hybridization, like the maximum resolution achieved by subchromosomal panel mapping,

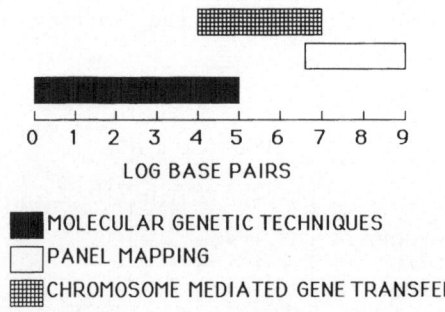

Fig. 4. Gene mapping resolution by different methods.

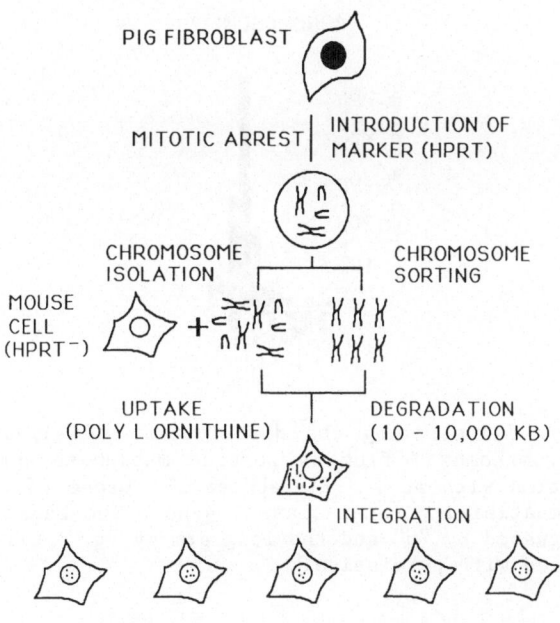

PIG FIBROBLAST

MITOTIC ARREST | INTRODUCTION OF MARKER (HPRT)

CHROMOSOME ISOLATION   CHROMOSOME SORTING

MOUSE CELL (HPRT⁻)

UPTAKE (POLY L ORNITHINE)   DEGRADATION (10 – 10,000 KB)

INTEGRATION

TRANSFORMED MOUSE CELLS (IN HAT-MEDIUM)

Fig. 5.   Chromosome-mediated gene transfer.

depends on the band resolution of the chromosomes and is, therefore, in the range of 5 to 10 cM (50).

Rabin et al. (45) assigned the major histocompatibility complex (SLA) of the domestic pig to chromosome 7 with an SLA class I-specific recombinant DNA probe. Autoradiographic silver grains were scored in a total of 84 metaphase spreads prepared from 2 different cell lines. Of the 300 silver grains associated with chromosomes, 97 were found to be concentrated within the region q12 → p12 of chromosome 7 (Fig. 6). Geffrotin et al. (20) previously used an [^{35}S]-labeled human major histocompatibility complex complementary DNA (cDNA) probe for hybridization in situ with pig metaphase chromosomes. They also found significant labeling on chromosome 7.

STATUS OF THE GENE MAPS

The gene maps of the economically most important domestic species (cattle, sheep, swine, goat, and chicken) are not far advanced when compared to the human or mouse gene maps. Updated information about linkage and synteny groups and gene assignments in some of these species can be found in Ref. 42. However, most of the gene maps of domestic animals consist of only a few linkage or synteny groups which in many cases are not assigned to specific chromosomes (Tab. 2). In this section we present the gene map of the pig as an example of a gene map of a domestic species (Tab. 2 and 3; Fig. 7).

A major drawback of gene mapping in domestic species is the lack of sufficient standardization of the karyotypes. At the First International Conference for the Standardization of Banded Karyotypes of Domestic Animals (48), agreement was reached only on how karyotypes should be arranged, and only a verbal description of the main G-bands was published. A standardized, numerical description of the bands, similar to the description of

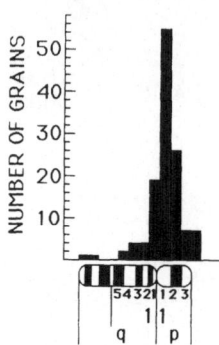

CHROMOSOME 7

Fig. 6.   Histogram illustrating the distribution of silver grains found
over chromosomes 7 from 84 porcine metaphase spreads after hybridization with an [^{125}I]-labeled DNA probe of a porcine major
histocompatibility (SLA) class I gene. The short chromosome arm
is designated by "p" and the long arm by "q." [From Rabin et al.
(45), used with permission.]

human banded chromosomes (43), is not yet available; however, standardization committees for the different species are at work and are expected to
publish standardized band idiograms in the near future. The cattle and
goat karyotypes pose special problems; all 58 autosomes in cattle and all
chromosomes except the Y-chromosome in the goat are telocentric. In each
of the two species the G-banding patterns of some chromosomes of similar
length look so similar that the chromosomes are not easy to identify.
Fluorescent banding techniques seem to provide better banding resolution
and therefore facilitate considerably the identification of the chromosomes
(10). Gene mapping of the chicken is restricted by the fact that this species has 39 pairs of chromosomes, of which only 10 pairs are large enough
to permit identification. The remaining 29 pairs cannot be identified by
any of the available banding techniques. In situ hybridization techniques
may ultimately be useful in solving this problem.

STRATEGIES FOR MAPPING GENES

A first step in obtaining more detailed gene maps of domestic species
should consist in studying the synteny of genes known to be closely linked
in man and mouse. Table 4 gives a summary of synteny groups that have

Tab. 2.   Status of the gene maps of some domestic species.

Species	Number of linkage or synteny groups	Number of chromosomal assignments
Cow (Bos taurus)	25	34
Sheep (Ovis aries)	12	22
Goat (Capra hircus)	0	0
Pig (Sus scrofa)	10	17
Chicken (Gallus gallus)	10	29

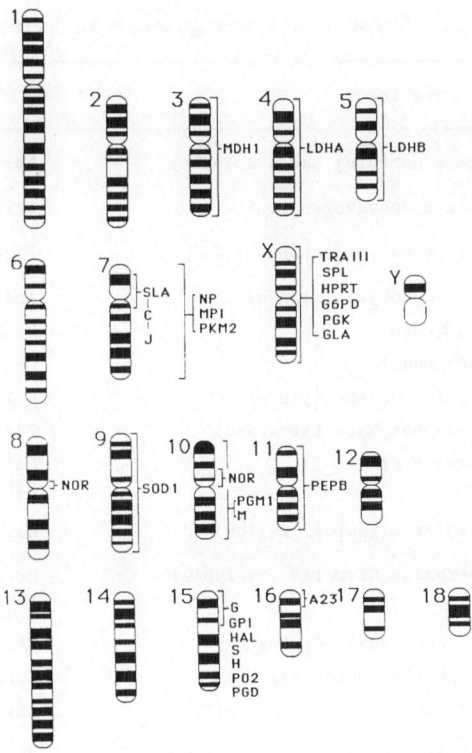

Fig. 7. The gene map of the pig. For explanation of gene symbols and references, see Tab. 3.

already been shown to be conserved in man, mouse, pig, sheep, and cow. The comparative mapping approach allows the identification of homologous chromosome regions in different species and, most importantly, it can provide preliminary information about the location of gene loci. Comparative mapping is particularly helpful in evolutionarily closely related species such as goat, cow, and sheep.

As described above, gene assignments to chromosomes and linkage groups by family studies are applicable in only a few cases. It is therefore important that somatic cell hybrid panels be established for all species. It can be expected that the bulk of mapping data will be collected by panel mapping. Confirmation and regional assignments will be accomplished by in situ hybridization in cases where gene probes are available. However, gene assignments by somatic methods (panel mapping, in situ hybridization) are part of the so-called "physical" gene map and do not provide information about genetic distances. The translation of the physical distance, in terms of base pairs, into the genetic distance, expressed as centimorgans, is not straightforward, because the frequency of crossing-over seems to vary throughout the genome. As the physical maps fill up as a result of assignments by somatic methods, family studies will become important again in the determination of genetic distances. The potential of in situ hybridization for gene map formulation in domesticated species should be emphasized. Thousands of unique probes have already been generated for the formulation of human and mouse gene maps. In most instances, these can be applied directly to gene mapping in other mammalian species by the applica-

Tab. 3. Chromosomal assignments in swine.

Chromosome	Gene locus	Reference(s)
3	MDH1 (malate dehydrogenase, soluble)	Förster and Hecht (13)
4	LDHA (lactate dehydrogenase A)	Förster and Hecht (13)
5	LDHB (lactate dehydrogenase B)	Förster and Hecht (13)
7	SLA (swine leukocyte antigen)	Andresen and Baker (1)
	C (C-blood group)	Hruban et al. (26)
	J (J-blood group)	Gellin et al. (22)
	NP (nucleoside phosphorylase)	Dolf (11)
	MPI (mannosephosphate isomerase)	Förster and Hecht (13)
	PKM2 (pyruvate kinase-3)	Geffrotin et al. (20) Rabin et al. (45)
8	NOR (nucleolar organizer region)	Mayr et al. (37)
9	SOD1 (superoxide dismutase, soluble)	Leong et al. (34) Förster and Hecht (13)
10	NOR (nucleolar organizer region)	Christensen (8)
	PGM1 (phosphoglucomutase-1)	Förster and Hecht (13)
	M (M-blood group)	Mayr et al. (37)
11	PEPB (peptidase B)	Förster and Hecht (13)
15	G (G-blood group)	Andresen and Jensen (2)
	GPI (glucosephosphate isomerase)	Rasmusen (47)
	HAL (halothane sensitivity)	Juneja et al. (28)
	S (A-O inhibition)	Tikhonov et al. (55)
	H (H-blood group)	Fries et al. (18)
	PO2 (postalbumin-2)	
	PGD (6-phosphogluconate dehydrogenase)	
16	A-23 (serum protein A-23)	Knyazev and Tikhonov (31)
X	TRAIII (paralytic tremor AIII)	Lax (33)
	SPL (splay leg condition)	Harding et al. (24)
	HPRT (hypoxanthine-guanine phospho-ribosyl transferase	Förster et al. (14) Gellin et al. (21)
	G6PD (glucose-6-phosphate dehydrogenase)	Leong et al. (35)
	PGK (phosphoglycerate kinase)	
	GLA (alpha-galactosidase A)	

tion of the in situ hybridization approach. This would appear to be a particularly efficacious way to map genes of known function in domesticated species.

Genes that have no known phenotype in cultured cells or for which there are no cloned DNA probes available cannot be mapped by somatic cell hybridization or by in situ hybridization. The map position of such genes, however, can be determined by studying linkage with genetic markers that have already been mapped by somatic methods. Genetic markers, based on polymorphisms detected at the level of the DNA by restriction enzymes, may be present in an abundant number in domestic animals. Botstein et al. (4) first described this new type of genetic marker, called RFLP. Over 200 loci of RFLPs were reported in man by 1983 (53). It has already been possible to map, indirectly, the defects responsible for muscular dystrophy (41) and Huntington's disease (23) based on associations found between loci of RFLPs and the inherited diseases. Searching for RFLPs will most likely become an integral part of gene mapping in domestic species. Randomly cloned DNA fragments, free of repetitive sequences, as well as cloned, unique DNA sequences that encode known genes, can be used as probes in Southern blot analyses of DNA samples from members of large families to determine loci of RFLPs. Sources for random DNA sequences are both cDNA and genomic libraries. Newly detected RFLP loci can be assigned chromosomally by physical methods (i.e., somatic cell hybridization, in situ hybridization) and can then be used as marker loci for the study of the map positions of genes that have no known phenotype in cell cultures.

Finally, it should be emphasized that any strategy for mapping genes in domestic species should adopt the general rule of human gene mapping: gene assignments must be confirmed by several groups independently before complete confidence can be placed on their accuracy.

APPLICATION OF GENE MAPPING

As the complexity of domesticated species gene maps advances, one can expect many direct applications to practical animal breeding of the knowledge gained by gene mapping. As the gene map fills up, one might come across further linkage groups similar to those found between the locus for Halothane sensitivity (HAL), as an indicator for the porcine stress syndrome, and blood group and enzyme loci (56), and Marek's disease and the major histocompatibility locus in chicken (6). These linkage groups have turned out to be very useful tools for selecting against porcine stress syndrome and Marek's disease, respectively.

However, we do believe that gene mapping will play an important role in the definition and isolation of genes that determine or modify complex traits such as lactation, fertility, growth, and disease resistance. It can be assumed that not more than 10 genes affect these characteristics to a useful extent (32). In spite of this limitation, most of the investigations designed to define associations between biochemical markers or blood group loci and productive traits have been unsuccessful. However, once a sufficient number of polymorphic loci have been placed throughout the genome--and the RFLPs can be expected to facilitate this undertaking considerably--it should become possible to locate many of these unknown genes. A few large families, typed for as many RFLPs and other marker loci as possible, and consisting of crosses and backcrosses of breeds with large phenotypic differences in productive traits or resistance against diseases, will provide the relevant information.

Once linkage between an RFLP locus, which has been localized chromosomally, and a gene that contributes to a complex trait has been established, the next step consists in searching within a chromosome segment library for a clone that carries the RFLP locus. A genomic library constructed from this clone could then serve as a source for unique DNA fragments to define more RFLP loci, all within not more than 10,000 kb from the gene of interest. From further linkage studies it should then be possible to localize the gene between 4 RFLP loci, 2 on either side. Next, one would again screen a chromosome segment library for a segment that contains the proximal, but lacks the distal, flanking markers. A cosmid library of less than 100 cosmid inserts would then be sufficient to cover the entire

Tab. 4. Some conserved synteny groups.

Human Chr	Human Locus	Mouse Locus	Mouse Mus musculus	Rabbit Oryctolagus cuniculus	Cattle Bos taurus	Sheep Ovies ovies	Pig Sus scrofa
1p	PGD	Pgd-1	4		U1	U1	15
1p	GDH	Gpd-1	4		U21		
1p	ENO1	Eno-1	4	U2	U1	U1	
1p	FUCA1	Fuca	4				
1p	AK2	Ak-2	4				
1p	PGM1	Pgm-2	4	U3	U6	U1	10
1p	PND	Pnd	4				
1p	AMY1	Amy-1	3				
1p	AMY2	Amy-2	3				
1p	NGFB	Ngf	3				
1q	ACTA	Acts	3				
1p	TSHB	Tshb	3				
1p	NRAS	Nras	3				
1q	PEPC	Pep-3	1		U13		
1q	SPTA	Spta(sph)	1				
1q	RNUI	Rnu-1	1				
1q	GUKI			15	U19		
1	REN	Ren-1,2	1				
1	APOA2	Alp-2	1				
1q	XPAC	Xpa	4				
1	ALPL	Akp-2	4				
2p	ACP1	Acp-1	12				
2p	POMC	Poc-1	12				
2p	MDH1			U5			3
2p	IGK	Igk	6	15			
2p	LEU2	Ly-2,3	6				
2q	IDH1	Idh-1	1		U17		
2	CRYG	Len-1	1				
3p	ACY1	Acy-1	9	9	U12		
3p	GLB1	Bgl-e	9				
3	TF	Trf	9		Un		
3q	CP				Un		
3q	SST	Smst	16				
3	GPX1			9			
3	RAF1	Raf-1	6				
4	PGM2	Pgm-1	5		U15	U5	
4	IL2						
4	PEPS	Pep-7	5				
4q	ALB	Alb-1	5				
4q	AFP	Afp	5				
4	EGF	Egf	3				
4q	ADH1	Adh-1	3				
4q	ADH3	Adh-3	3				
4	RAF2	Raf-2	6				
5	DHFR						
5	CHR						
5	RPS14						
5	LARS						
5	DTS						
5	FMS						
5	ARSB	As-1	13				
5q	GRL	Grl-1	18				
5	DHLAG	I1	18				
6p	PRL						
6p	HLA-A	H-2D	17			U6	7
6p	HLA-1	H-21	17				
6p	C4	C4	17			U6	7
6p	C2	C2	17				
6p	HLA-D	H-2K	17				
6p	PGK1	Pgk-2	17				

Tab. 4 (continued).

Human Chr.	Human Locus	Mouse Locus	Mouse	Rabbit	Cattle	Sheep	Pig
6p	GLOI	Glo-1	17		U21	U7	
6p	BF	Bf	17				
6	TCPI	Tcp-1	17				
6	CA2IHA	21Oh-1	17				
6	CA2IHB	21Oh-2	17				
6q	SOD2	Sod-2	17		U2	U8	
6q	PGM3	Pgm-3	9		U2	B	
6q	ME1	Mod-1	9		U2	B	
6q	CGA	Tsha	4				
6q	MYB	Myb	10				
6p	ME2	Mod-2	7				
7p	BLVR	Blvr	2				
7	GUSB	Gus	5				
7	MDH2	Mor-1	5	15	U8		
7	ASL	Asl	5				
7	PSP	Psph	5				
7	ERBB	Erbb	11				
7q	TRYI	Try-1	6				
7q	CPA	Cpa	6				
7q	TCRB	Tcrb	6				
7q	COL1A2	Col1a-2	16				
7	TCRG	Tcrg	13				
7	ASNS						
7q	H3FI	H3	13				
8p	GSR	Gr-1	8	19	U14		
8q	MOS	Mos	4				
8q	MYC	Myc	15				
8q	NIARD	Niard	15				
8q	TG						
8	CAI,2	Car-1,2	3				
9p	ACOI	Aco-1	4		C18		
9p	GALT	Galt	4				
9p	AK3						
9p	IFNA	Ifa	4				1q
9p	IFB	Ifb	4				
9	ALAD	Lv	4				
9q	ORM	Agp-1	4				
9q	AKI	Ak-1	2				
9q	ABL	Abl	2				
9q	FPGS	Fpgs	2				
9q	ASS	Ass	2				
10	NEU	Neu-1	17				
10	PP	Pyp	10				
10q	HKI	Hk-1	10				
10	LIPA	Lip-1	19				
10q	GOTI	Got-1	19				
10q	PGAMA	Pgam-1	19				
10q	ADK	Adk	14				
10	OAT	Oat	7				
11p	INS	Ins-2	7				
11p	HBB	Hbb	7				
11p	LDHA	Ldh-1	7	1	U7		4
11p	HRASI	Hras-1	7				
11p	CALCI	Ct	7				
11p	FSHB	Fshb	7				
11p	HBE	Hbey3	7				
11q	FNL2	Fn	7				
11p	ACP2	Acp-2	2	1			
11p	CAT	Cas-1	2		U20		
11q	APOAI	Apl-1	9				
11q	UPS	Ups	9				
11q	ESA4	Es-17	9				
11q	GST3	Gsta	9				
11	THYI	Thy-1	9				
11q	T3D	T3d	9				
11	NCAM	sg	9				
12p	GAPD	Gapd	6				
12p	TPI·I	Tpi-1	6	4	U3	U2	

(continued)

Tab. 4 (continued).

Human Chr.	Human Locus	Mouse Locus	Mouse	Rabbit	Cattle	Sheep	Pig
12	ALDH2	Ahd-1	4				
12q	ACH	cn	4				
12p	KRAS2	Kras-2	6				
12p	LDHB	Ldh-2	6	4	U3	U2,M3	5
12q	CS	Cs	10				
12q	PEPB	Pep-2	10	U1,17	U3	U2,M3	11
12q	IFNG	Ifg	10				
12	ELA1	Ela-1	15				
12	INT1	Int-1	15				
12	GPD1	Gdc-1	15				
12	PRP	Prp	8				
12	SHMT					3	
13q	ESD	Es-10	14				
14q	NP	Np	14	16,17	U23	U3	3
14q	TCRA	Tcra	14				
14q	IGH	Igh	12	IX			
14q	PI	Pre-1	12				
14q	FOS	Fos	12				
15	SORD	Sdh-1	2				
15q	B2M	B2m	2				
15q	HEXA						
15q	MPI	Mpi	9		U4	U9	3
15q	PKM2	Pk-3	9	16	U5	U4	3
15q	CYP2	P450-DX	9				
15q	IDH2	Idh-2	7				
15q	FES	Fes	7				
15q	ACTC	Actc	17				
15q	CYP2	Ah	17				
16p	HBA	Hba	11				
16p	PGP						
16	GOT2	Got-2	8				
16	CTRB	Ctrb	8				
16q	APRT	Aprt	8				
16q	MT1,2	Mt-1,2	8				
16q	HP	Hp	8				
17p	TP53	Trp53	11				
17p	MYH-1	Myh	11				
17p	MYH-2	Myh	11				
17p	MYH-3	Myh	11				
17q	TK1	Tk-1	11				
17q	GALK	Glk	11				
17q	GAA						
17q	ERBA1	Erba-1	11				
17q	HOX2	Hox-2	11				
17q	COLIA1	Colla-1	16				
17	GH						
18q	PEPA	Pep-1	18				
18	MBP	Mbp(shi)	18				
19p	INSR	Insr	7				
19	GPI	Gpi-1	7	U6	U9		15
19	PEPD	Pep-4	7				
19	LHB	Lhb	7				
19q	CYP1	P450-PB	7				
19	C3	C31	17				
19	PGK2	Pgk-2	17				
20p	ITPA	Itp	2	17	U11		
20q	ADA	Ada	2		U11		
20q	SRC1	Src	2				
21q	SOD1	Sod-1	16		U10	U10	9
21q	IFNAR	Ifrc	16	U10			
21q	PRGS	Prgs	16			U10	
21	PAIS					U10	
21	CRYA1	Crya-1	17				
22q	ARSA	As-2	15				
22q	DIA1	Dia-1	15				

Tab. 4 (continued).

Human Chr.	Human Locus	Mouse Locus	Mouse	Rabbit	Cattle	Sheep	Pig
22q	SIS	Sis	15				
22q	IGLC	Igl-1	16				
22	ACO2						
22	IDUA						
Xp	DMD	mdx	X				
Xp	STS	Sts	X				
Xp	OTC	Spf	X				
Xq	G6PD	G6pd	X	X	X	X	X
Xq	HPRT	Hprt	X	X	X	X	X
Xq	GLA	Ags	X	X	X	X	X
Xq	PGK	Pgk-1	X	X	X	X	X
Xq	XCE	Xce	X				
X	TFM	Tfm	X				
X	PYK	Phk	X				
Xp	CDPX	Bpa	X				
X	HPDR	Hyp	X				
Xq	HEMA						
X	EDA	Ta	X				
X	MNK	Mo	X				
X	HRAS2						
Xq	PLP	Plp	X				
X	IMD	Xid	X				
Y	HYA	H-Y	Y				
Ypkk	TDF	Tdy	Y				
Yp	STSY	Stsy	Y				

region that contains the 2 proximal markers. Depending on the nature of the trait of interest, different approaches may be applied to detect the cosmids that contribute to the expression of the trait. In the case of a quantitative trait, such as milk yield, one could compare the frequency of reiteration of candidate sequences in animals with high and low milk yields, assuming that quantitative differences are reflected by the frequency of reiteration of the gene. Candidate sequences might also be tested by microinjection into embryos and subsequent study of the effect of the candidate sequence on the performance of the transgenic animal. Another way of determining the sequences that contribute to the trait of interest could consist in screening cDNA libraries from various tissues that are functionally related to the trait as well as from tissues that are not related.

It is obvious that the isolation of genes contributing to animal performances is a long and tedious process. However, it is very likely that various steps involved in this procedure might have interesting applications that cannot be foreseen.

CONCLUSIONS

Gene mapping techniques are now sufficiently advanced to permit rapid progress in the establishment of detailed gene maps of domestic animals. The search for polymorphic DNA markers is expected to become an integral part of gene mapping. Restriction fragment mapping and in situ hybridization can be used to place DNA markers at random sites throughout the genome. Family studies should allow one to define linkages between unknown genes for complex traits and the references marker. The polymorphic DNA markers may then serve to predict an animal's performance. A series of DNA markers flanking one or several genes for complex polygenic traits might eventually lead to the isolation and cloning of the genes for these markers.

SUMMARY

Gene maps are constructed by the synthesis of data obtained by different methods which include family analyses, somatic cell hybridization, direct mapping of DNA segments by Southern blot analysis, and in situ hybridization to fixed metaphase chromosomes. Gene mapping has already contributed significantly to a better understanding of the mammalian genome, in particular the human genome, but the gene maps of economically important domestic species are not well-characterized. The application of somatic cell genetics and recombinant DNA methodologies now allows rapid progress to be made in the construction of detailed gene maps for domestic animals. Such gene maps will serve as tools for selection in applied animal breeding and for the analysis of polygenic traits.

ACKNOWLEDGEMENTS

This work was supported by Fellowships from the Swiss National Science Foundation to Rudolf Fries and by NIH Grant GM09966 to F.H. Ruddle. We would like to thank Suzy Pafka for excellent technical assistance.

REFERENCES

1. Andresen, E., and L.N. Baker (1964) The C-blood group system in pigs and the detection and estimation of linkage between the C and J systems. Genetics 49:379-386.
2. Andresen, E., and P. Jensen (1977) Close linkage established between HAL locus for halothane sensitivity and PHI (phosphohexose isomerase) locus in pigs of the Danish Landrace breed. Nord. Vet. Med. 29:502-504.
3. Baker, R.M., D.M. Brunette, R. Mankovitz, L.H. Thompson, G.F. Whitmore, L. Siminovitch, and J.E. Till (1974) Ouabain-resistant mutants of mouse and hamster cells in culture. Cell 1:9-21.
4. Botstein, D., R.L. White, M. Skolnick, and R.W. Davis (1980) Construction of a genetic linkage map using restriction fragment length polymorphisms. Am. J. Hum. Genet. 32:314-331.
5. Brennand, J., D.S. Konecki, and C.T. Caskey (1983) Expression of human and chinese hamster hypoxanthine-guanine phosphoribosyltransferase cDNA recombinants in cultured Lesh-Nyhan and chinese hamster fibroblasts. J. Biol. Chem. 16:9593-9596.
6. Briles, W.E., R.W. Briles, R.E. Taffs, and H.A. Stone (1983) Resistance to a malignant lymphoma in chicken is mapped to a subregion of major histocompatibility (B) complex. Science 219:977-979.
7. Caspersson, T., L. Zech, E.J. Modest, G.E. Foley, U. Wagh, and E. Simonsson (1969) Chemical differentiation with fluorescence alkylating agents in Vicia faba metaphase chromosomes. Exptl. Cell Res. 58:128-140.
8. Christensen, K. (1980) Evidence of polymorphism of the nuclear organizer region (N-Band) in pig chromosomes. Proceedings of the 4th European Colloquium on Cytogenetics of Domestic Animals, Uppsala, pp. 464-468.
9. Croce, C.M., and G. Klein (1985) Chromosome translocations and human cancer. Scientific American 252:54-60.
10. Di Berardino, D., and L. Iannuzzi (1982) Detailed description of R-banded bovine chromosomes. J. Hered. 73:434-438.
11. Dolf, G. (1984) Genkartierung beim Schwein mit Hilfe von somatischen Zellhybriden. Thesis no. 7644, Eidgenössische Technische Hochschule, Zürich.
12. Donahue, R.P., W.B. Bias, J.H. Renwick, and V.A. McKusick (1968) Probable assignment of the Duffy blood group locus to chromosome 1 in man.

Proc. Natl. Acad. Sci., USA 61:949-955.

12a. Eighth International Human Gene Mapping Workshop (1985) Cytogenet. Cell Genet. Vol. 39 (in press).

13. Förster, M., and W. Hecht (1984) Some provisional gene assignments in pig. Proceedings of the 6th European Colloquium on Cytogenetics of Domestic Animals, Zurich, pp. 351-355.

14. Förster, M., G. Stranzinger, and B. Hellkuhl (1980) X-chromosomal gene assignment of swine and cattle. Naturwissenschaften 67:48.

15. Fries, R., and F.H. Ruddle (1985) Gene mapping in domesticated animals. In Biotechnology for Solving Agricultural Problems, Vol. X, J. St. John, ed. (in press).

16. Fries, R., and G. Stranzinger (1982) Chromosomal mutations in pigs derived from X-irradiated semen. Cytogenet. Cell Genet. 34:55-66.

17. Fries, R., G. Stranzinger, and P. Vögeli (1983) Provisional assignment of the G-blood group locus to chromosome 15 in swine. J. Hered. 74:426-430.

18. Fries, R., B.A. Rasmusen, V.L. Jarrell, and R.R. Maurer (1984) Mapping of the gene for G blood antigens to chromosome 15 in swine. Anim. Blood Grps. Biochem. Genet. 15:251-258.

19. Gall. J.G., and M.L. Pardue (1969) Formation and detection of RNA-DNA hybrid molecules in cytological preparations. Proc. Natl. Acad. Sci., USA 63:378-383.

20. Geffrotin, C., C.P. Popescu, E.P. Cribiu, J. Bosher, C. Renard, P. Chardon, and M. Vaiman (1984) Assignment of MHC in swine to chromosome 7 by in situ hybridization and serological typing. Ann. Genet. 27:213-219.

21. Gellin, J., F. Benne, M.C. Hors-Cayla, and M. Gillois (1980) Gene mapping in the pig (sus-scrofa L.). Study of two syntenic groups G6PD, PGK, HPRT and PKM2, MP1. Ann. Genet. 23:15-21.

22. Gellin, J., G. Echard, F. Benne, and M. Gillois (1981) Pig gene mapping: PKM2-MP1-NP synteny. Cytogenet. Cell Genet. 30:59-62.

23. Gusella, J.F., N.S. Wexler, P.M. Conneally, S.L. Naylor, M.A. Anderson, R.E. Tanzi, P.C. Watkins, K. Ottina, M.R. Wallace, A.Y. Sakaguchi, A.B. Young, I. Shoulson, E. Bonilla, and J.B. Martin (1983) A polymorphic DNA marker genetically linked to Huntington's disease. Nature (London) 306:234-238.

24. Harding, J.D.J., J.T. Done, J.F. Harbourne, and F.R. Gilbert (1973) Congenital tremor type AIII in pigs: An hereditary sex linked cerebrospinal hypomyelinogenesis. Vet. Rec. 92:527-529.

25. Harper, M.E., and G.F. Saunders (1981) Localization of single copy DNA sequences on G-banded human chromosomes by in situ hybridization. Chromosoma 83:431-439.

26. Hruban, V., H. Simon, J. Hradecky, and F. Jilek (1976) Linkage of the pig main histocompatibility complex and the J blood group system. Tissue Antigens 7:267-271.

27. John, H.A., M.L. Birnstiel, and K.W. Jones (1969) RNA-DNA hybrids at the cytological level. Nature (London) 223:582-587.

28. Juneja, R.K., B. Gahne, I. Edfors-Lilja, and E. Andresen (1983) Genetic variation at a pig serum protein locus, Po-2, and its assignment to the Phi, Hal, S, H, Pgd linkage group. Anim. Blood Grps. Biochem. Genet. 14:27-36.

29. Kamarck, M.E., P.E. Barker, R.L. Miller, and F.H. Ruddle (1984) Somatic hybrid panels. Exptl. Cell Res. 152:1-14.

30. Klobutcher, L.A., and F.H. Ruddle (1981) Chromosome mediated gene transfer. Ann. Rev. Biochem. 50:533-554.

31. Knyazev, S.P., and V.N. Tikhonov (1984) Gene mapping of the pig chromosome no. 16. Proceedings of the 6th European Colloquium on Cytogenetics of Domestic Animals, Zürich, pp. 395-399.

32. Lande, R. (1981) The minimum number of genes contributing to quantitative variation between and within populations. Genetics 99:541-553.

33. Lax, T. (1971) Hereditary splayleg in pigs. J. Hered. 62:250-252.

34. Leong, M.M.L., C.C. Lin, and R.F. Ruth (1983) Assignment of superoxide dismutase (SOD-1) gene to chromosome no. 9 of domestic pig. Can. J. Genet. Cytol. 25:233-238.

35. Leong, M.M.L., C.C. Lin, and R.F. Ruth (1983) The localization of genes for HPRT, G6PD and -GAL onto the X-chromosome of domestic pig (sus scrofa domesticus). Can. J. Genet. Cytol. 25:239-245.

36. Littlefield, J.-W. (1964) Selection of hybrids from matings of fibroblasts in vitro and their presumed recombinants. Nature (London) 256:495-497.

37. Mayr, B., D. Schweizer, and G. Geber (1984) NOR activity, heterochromatin differentiation, and the Robertsonian polymorphism in sus scrofa L. J. Hered. 75:79-80.

38. Miller, A.D., D.J. Jolly, T. Friedmann, and I.M. Verma (1983) A transmissible retrovirus expressing human hypoxanthine phosphoribosyltransferase (HPRT): Gene transfer into cells obtained from humans deficient in HPRT. Proc. Natl. Acad. Sci., USA 80:4709-4713.

39. Morton, C.C., I.R. Kirsch, R. Taub, S.H. Orkin, and J.A. Brown (1984) Localization of the β-globin gene by chromosomal in situ hybridization. Am. J. Hum. Genet. 36:576-585.

40. Morton, N.E. (1955) Sequential tests for the detection of linkage. Am. J. Hum. Genet. 7:277-318.

41. Murray, J.M., K.E. Davies, P.S. Harper, L. Meredith, C.R. Mueller, and R. Williamson (1982) Linkage relationship of a cloned DNA sequence on the short arm of the X-chromosome to Duchenne muscular dystrophy. Nature (London) 300:69-71.

42. O'Brien, S.J., ed. (1984) Genetic Maps, Vol. 3, National Institutes of Health, Bethesda, Maryland.

43. Paris Conference, 1971 (1972) Standardization in human cytogenetics. Cytogenet. Cell Genet. 11:313-362.

44. Pontecorvo, G. (1976) Production of mammalian somatic cell hybrids by means of polyethylene glycol (PEG) treatment. Somatic Cell Genet. 1:397-400.

45. Rabin, M., R. Fries, D. Singer, and F.H. Ruddle (1985) Assignment of the porcine major histocompatibility complex to chromosome 7 by in situ hybridization. Cytogenet. Cell Genet. (in press).

46. Rabin, M., C.P. Hart, A. Ferguson-Smith, W. McGinnis, M. Levine, and F.H. Ruddle (1985) Two homeobox loci mapped in evolutionarily related mouse and human chromosomes. Nature (London) 313:175-178.

47. Rasmusen, B.A. (1981) Linkage of genes for PHI, halothane sensitivity, A-O inhibition, H red blood cell antigens, and 6-PGD variants in pigs. Anim. Blood Grps. Biochem. Genet. 12:207-209.

48. Reading Conference, 1976 (1980) Proceedings of the first international conference for the standardisation of banded karyotypes of domestic animals. Hereditas 92:145-162.

49. Renwick, J.H. (1969) Progress in mapping human autosomes. Brit. Med. Bull. 25:65-73.

50. Ruddle, F.H. (1981) A new era in mammalian gene mapping: Somatic cell genetics and recombinant DNA methodologies. Nature (London) 294:115-120.

51. Ruddle, F.H. (1984) The William Allan Memorial Award Address: Reverse genetics and beyond. Am. J. Hum. Genet. 36:944-953.

52. Seabright, M. (1971) A rapid banding technique for human chromosomes. Lancet 2:971.

53. Skolnick, M.H., H.F. Willard, and L.A. Menlove (1984) Report of the committee on human gene mapping by recombinant DNA techniques. Cytogenet. Cell Genet. 37:210-273.

54. Sumner, A.T. (1972) A simple technique for demonstrating centromeric heterochromatin. Exptl. Cell Res. 75:304-306.

55. Tikhonov, V.N., I.G. Gorelov, S.V. Nikitin, V.E. Bobovich, and N.M. Astakhova (1983) Mapping of the locus for the H-blood group system on

chromosome 15 of domestic pig. <u>Doklady Akademii Nauk S.S.S.R.</u> 272: 486–489.

56. Vögeli, P., G. Stranzinger, H. Schneebeli, C. Hagger, N. Künzi, and C. Gerwig (1984) Relationships between the H and A-O blood types, phosphohexose isomerase and 6-phosphogluconate dehydrogenase red cell enzyme systems and halothane sensitivity, and economic traits in a superior and an inferior selection line of Swiss landrace pigs. <u>J. Anim. Sci.</u> 59:1440–1450.

57. Wald, A. (1947) <u>Sequential Analysis</u>, Dover Publications, Inc., New York, 212 pp.

58. Weiss, M.C., and H. Green (1967) Human–mouse hybrid cell lines containing partial complements of human chromosomes and functioning human genes. <u>Proc. Natl. Acad. Sci., USA</u> 58:1104–1111.

59. White, R., M. Leppert, D.T. Bishop, D. Barker, J. Berkowitz, C. Brown, P. Callahan, T. Holm, and L. Jerominski (1985) Construction of linkage maps with DNA markers for human chromosomes. <u>Nature</u> (London) 313:101–105.

60. Yerganian, G., and M.B. Nell (1966) Hybridization of dwarf hamster cells by UV-inactivated Sendai virus. <u>Proc. Natl. Acad. Sci., USA</u> 55:1066–1073.

61. Yoshida, M.C., T. Ikeuchi, and M. Sasaki (1975) Differential staining of parental chromosomes in interspecific cell hybrids with a combined quinacrine and 33258 Hoecht technique. <u>Proc. Jap. Acad.</u> 51:185–187.

GENE TRANSFER INTO ANIMALS BY RETROVIRAL VECTORS

W.F. Anderson,[1] P.W. Kantoff,[1] M.A. Eglitis,[1]
and E. Gilboa[2]

[1]Laboratory of Molecular Hematology
National Heart, Lung, and Blood Institute
Bethesda, Maryland 20892

[2]Department of Molecular Biology
Princeton University
Princeton, New Jersey 08544

In a previous chapter (W.F. Anderson, this Volume), there was a discussion of why the unique life cycle of retroviruses makes them attractive candidates for use as agents for gene transfer into animals. However, a number of specific requirements need to be met (1). In order to be useful, a potential vector should be generated at a sufficiently high titer so that the infection of such rare target cells as the hematopoietic stem cells would be likely. The vector should become integrated as a stable, intact sequence, not only in cultured cells, but also in whole animals. Finally, expression of transferred genes in animals should be at or near physiological levels in order for such gene transfer to have potential practical applicability.

It has already been shown by others that recombinant retroviral vectors can be used to introduce new genetic material into the progenitor cells of the hematopoietic system of the mouse. Joyner et al. (3) were the first to report the retroviral transfer of exogenous genes into murine bone marrow stem cells. They detected expression of a transferred, bacterial, neomycin-resistance (neo^r) gene in individual colony-forming units--granulocyte, macrophage (CFU-GM) in vitro, but at a low efficiency (0.3%). Subsequently, Miller et al. (4) showed transfer of a functional human hypoxanthine-guanine phosphoribosyl transferase (HPRT) gene into hematopoietic stem cells which subsequently colonized the hematopoietic system of a whole mouse. Finally, Williams et al. (5), using the helper-free system that was described previously (W.F. Anderson, this Volume), which we used in this chapter, showed that they were able to use a retroviral vector to introduce DNA sequences containing a neo^r gene into about 15% of the stem cells [i.e., colony-forming unit--spleen (CFU-S)] of a mouse. In our study, we have characterized in vivo an efficient new retroviral vector derived from Moloney murine leukemia virus and have determined the conditions for bone marrow gene transfer such that over 85% of CFU-S are infected and the transferred gene is expressed efficiently (2).

The structure of the provirus form of the recombinant retrovirus, N2, is shown in Fig. 1. This vector was designed to produce high-titer virus

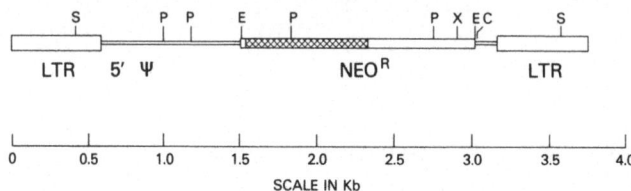

N2 PROVIRUS

SCALE IN Kb

Fig. 1.   Diagram of the integrated vector (proviral) N2. 0.0 to 1.5 and
3.0 to 3.8 kb: Moloney murine leukemia virus sequences; 1.5- to
3.0-kb box: Tn5 sequence containing the neo^{r} gene (BgII–BamHI
fragment from Tn5); the hatched area is the coding sequence; LTR,
long terminal repeat; 5', the donor splice site at the 5' end; ψ,
packaging sequence; restriction enzyme sites: S, SacI; P, PstI;
E, EcoRI; X, XhoI; C, ClaI. [Photo from "Prospects for Human
Gene Therapy" (1984); permission granted from Science.]

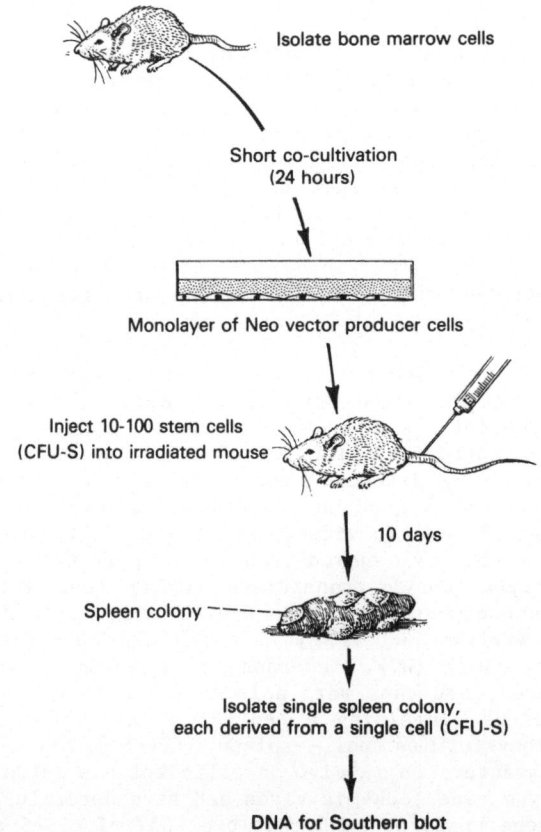

Fig. 2.   Infection and transplantation of bone marrow cells.

Tab. 1. Efficiency of gene transfer by retroviral vectors.

| | Hr co-culture | | Total |
	24	48	
+ IL-3	19/20 (95%)	14/18 (78%)	33/38 (87%)
- IL-3	4/6 (67%)	6/8 (75%)	10/14 (71%)

which would lead to a stable provirus capable of efficient expression of inserted genes. A large portion of the Moloney murine leukemia virus coding sequence has been deleted in N2 and replaced with the neo^r gene which confers upon eukaryotic cells resistance to the neomycin analog G418. After transfection into the helper-free packaging line, ψ2, and subsequent selection in G418, individual clones were isolated that produced N2 virus over a range of titers up to over 10^6 colony-forming units (CFU)/ml. The 3T3 cells infected with N2 virus produced by a high-titer clone, called F-5B, were studied by restriction enzyme analysis and Southern blotting. No evidence for deletions or rearrangements in the vector DNA was found. Furthermore, there was significant expression of the neo^r-coded phosphotransferase activity in these cells.

Having shown that the F-5B cell line was producing intact N2 virus, we next optimized the procedure for co-cultivation of bone marrow cells with the F-5B cells. The bone marrow transplantation protocol is shown in Fig. 2. This procedure was first reported by Williams et al. (5). As shown in Tab. 1, 24-hr co-cultivation in the presence of purified IL-3 (a well-characterized murine lymphokine) results in 95% of the murine stem cells (CFU-S) being infected.

Although these results indicated that vector titers of over 10^6 CFU/ml were very efficient at introducing exogenous genes into the murine hematopoietic stem cell population, the effect of lower titers on infection efficiency was unknown. To determine the efficiency of bone marrow infection

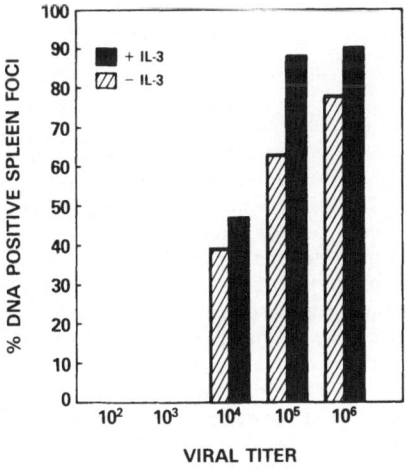

Fig. 3. Titer effects upon efficiency of bone marrow infection. Indicated titers were obtained from clonal populations of various N2-producing cells as described in the text.

with lower-titer virus, bone marrow cells were co-cultured with subconfluent plates of F-5B. In addition, bone marrow cells were co-cultivated with individual ψ2 cell clones (obtained at the same time as the high-titer clone, F-5B) having lower viral titers. No evidence of successful stem cell infection was found when cells producing virus at titers of 10^2 or 10^3 CFU/ml were used (Fig. 3). However, when the titer of the virus in the medium was 10^4 CFU/ml or greater, regardless of whether the titer was the result of diluted, high-titer cells or the total productivity of a given clone, infected CFU-S were detected. The proportion of infected stem cells increased as the viral titer increased, with efficiencies of over 80% being obtained when titers were 10^5 CFU/ml or greater. The increase in efficiency of stem cell infection with increasing viral titer was seen both in the presence and absence of IL-3 (Fig. 3). At each titer tested, the overall proportion of CFU-S infected appeared to be only slightly greater in the presence of IL-3.

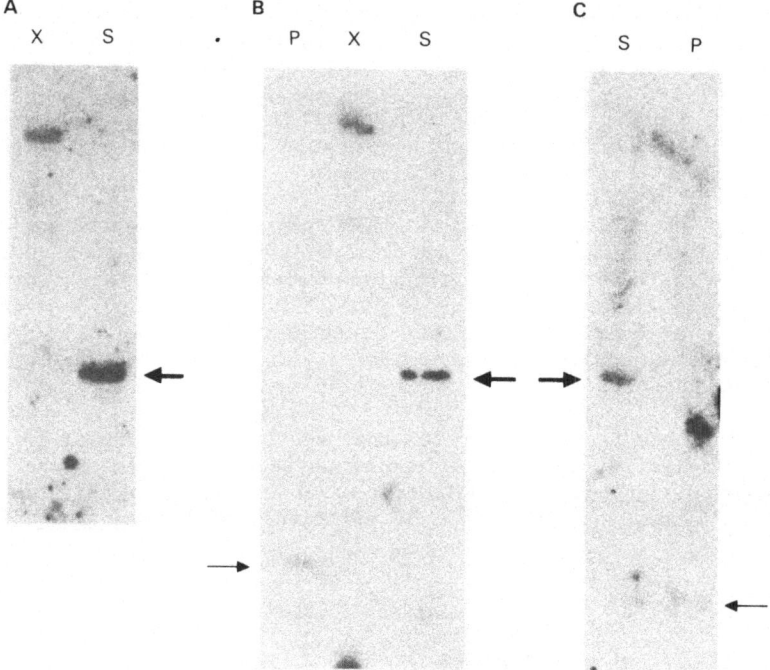

Fig. 4.   Southern blots of DNA prepared from individual primary spleen foci and a whole, reconstituted spleen. Bone marrow cells were co-cultured for 24 hr with F-5B cells producing N2 at a titer of 2×10^6 CFU/ml. DNA from 2 individually isolated, 10-day spleen foci obtained from infections done at different times (A, B) and from a long-term, reconstituted spleen (C) was prepared. The spleen DNA was prepared from a mouse 4 months after the lethally irradiated animal had received 5×10^6 infected bone marrow cells. For blots A, B, and C, equal amounts of DNA (30 μg per lane) were digested with restriction enzymes and then electrophoresed through 0.7% agarose gels. After electrophoresis, the gels were blotted onto Biotrans filter membranes. Hybridizations were as described in Ref. 2 with films exposed for 5 days. Enzyme digestions were with XhoI (X), SacI (S), and PstI (P). Large arrows indicate the position of the 3.2-kb SacI fragment (see Fig. 1); small arrows indicate the position of the 0.9-kb PstI fragment (see Fig. 1).

Although N2 was known to be stable when used to infect tissue culture cells, it was possible that, during the proliferation and differentiation of the infected hematopoietic stem cells in vivo, rearrangements or deletions of the proviral sequences could have occurred. To check the structural stability of the integrated N2 virus, Southern blots were performed using DNA obtained from individual foci (Fig. 4A and B). Using a variety of restriction endonucleases, no evidence of gross rearrangement was detected. With SacI, which cuts N2 within both long terminal repeats (LTRs), thus releasing a near unit length fragment, a band of expected length (3.2 kilobases, see Fig. 1) was always found. When the enzyme XhoI, which cuts N2 at only one internal site, was used, several bands of differing intensities were often detected on the blots when infection took place at a high viral titer. This indicated that some stem cells had been infected more than once. DNA was also digested with PstI, generating the expected 0.9-kb fragment.

Of greatest importance is the question of expression of the transferred gene. The presence of transcripts and phosphotransferase enzyme activity would provide evidence that, besides the presence of the transferred gene, the transcriptional machinery of the provirus remained functional. To this end, mice were lethally irradiated and received a portion of the cells from a DNA positive spleen focus. Spleens were removed from these secondary mice, and RNA was prepared and analyzed. RNA of the appropriate size was detected on T1 RNase gels using a neo^r probe.

The neo^r gene product was assayed by its phosphotransferase activity in extracts from individual and pooled foci. The majority of foci tested demonstrated expression of the neo^r gene. All 6 of the individual foci shown in Fig. 5 were positive but to different extents. As shown by

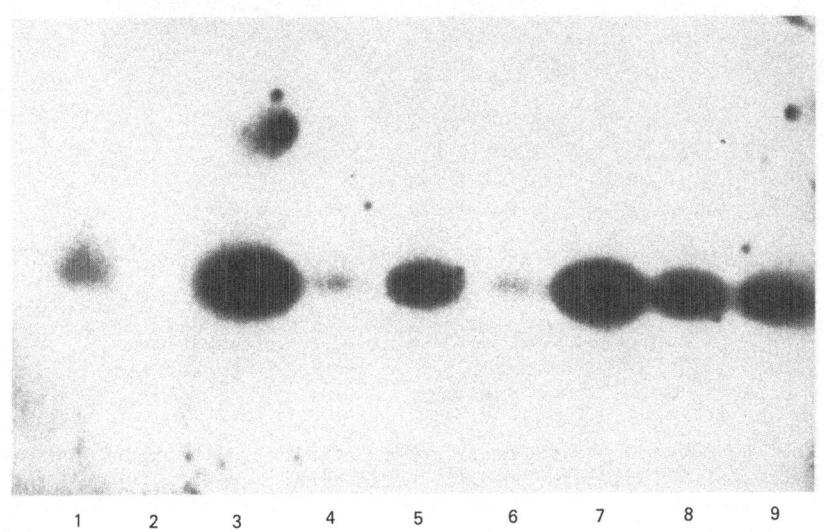

Fig. 5. Neo^r gene-coded phosphotransferase activity in extracts from spleen foci. The arrow indicates the position of neo^r gene-coded phosphotransferase activity. Lane 1, lysate of 1×10^5 F-5B cells. Lane 2, lysate of approximately 1×10^5 uninfected spleen focus cells. Lane 3, lysate of pooled foci from 1 mouse, approximately 5×10^5 cells, following bone marrow infection using F-5B cells. Lanes 4-9, lysate of 6 individual foci (approximately 0.5×10^5 to 2×10^5 cells each) following bone marrow infection using F-5B cells.

comparing lane 1 (the F-5B control) with lanes 4-9 (containing individual foci), some hematopoietic cells have less and others, more enzyme activity than a similar number of expressing, N2-containing, 3T3 tissue culture cells. The variation in activity among different foci might be partially due to varying sizes of the foci since cell numbers were not exactly equalized. More likely, however, is that the neo^r gene may be expressed at different levels in different foci, either because of multiple single-copy insertions of the N2 vector or for some other reason (e.g., a position effect due to the random chromosomal integration of each proviral sequence). Studies are underway to evaluate these possibilities.

To determine the long-term, structural stability of N2 proviral sequences, totally reconstituted animals were obtained by injecting irradiated mice with 5 x 10^6 infected bone marrow cells and then letting those cells repopulate the animal over a period of several months. Spleens, bone marrow, and blood were recovered and analyzed by Southern blot and/or phosphotransferase assays. Southern blots demonstrated that the N2 sequences remained intact even after 4 months (Fig. 4C). However, the proportion of total hematopoietic cells carrying N2 may have decreased since band intensity in spleen DNA appeared to be reduced when compared with an equal amount of DNA isolated from spleen foci (compare Fig. 4A and B with C).

Of major interest was whether or not the blood and bone marrow of long-term, reconstituted animals express the neo^r gene. Phosphotransferase was detected in the bone marrow of 3 of the 4 animals tested (Fig. 6, lanes 2, 6, and 7). In one animal (lane 5), a strong neo^r gene-coded phosphotransferase signal was found in the blood. Therefore, the neo^r gene is

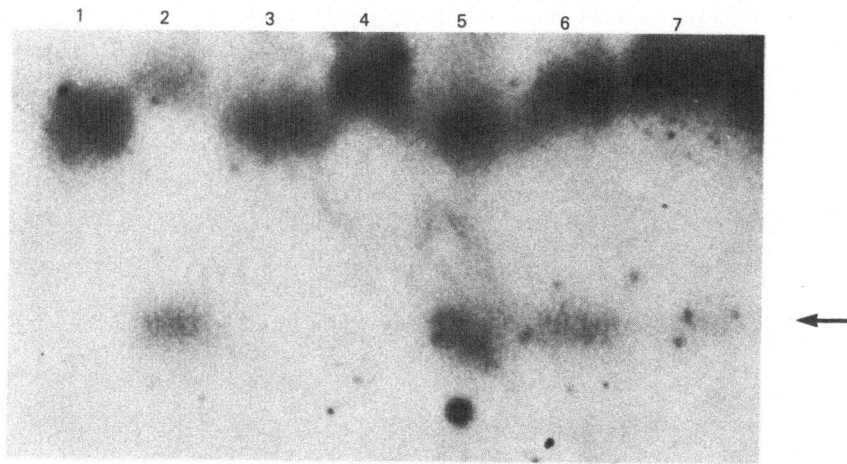

Fig. 6.   Neo^r gene-coded phosphotransferase activity in the blood and bone marrow of 4 long-term, reconstituted mice. Lane 1, lysate of 1 x 10^6 whole blood cells from mouse A. Lane 2, lysate of 1 x 10^6 bone marrow cells from mouse A. Lane 3, lysate of 1 x 10^6 whole blood cells from mouse B. Lane 4, lysate from 1 x 10^6 bone marrow cells from mouse B. Lane 5, lysate from 1 x 10^6 whole blood cells from mouse C. Lane 6, lysate from 1 x 10^6 bone marrow cells from mouse C. Lane 7, 1 x 10^6 bone marrow cells from mouse D. The positive band in lane 7 is much clearer on longer exposures. The arrow indicates the position of neo^r gene-coded phosphotransferase activity. The dark, slower-migrating band in each lane represents a phosphotransferase activity seen to various degrees in all tissues studied, and is unrelated to the presence or absence of the neo^r gene.

active in the circulating hematopoietic system of at least some of these animals 4 months after bone marrow infection and transplantation.

These data clearly indicate that retroviral vectors can be used to transfer functional genes into intact animals. We have established a similar protocol with nonhuman primates. Early results indicate that simian virus 40-promoted genes can be inserted into the XhoI site of N2 (see Fig. 1) and that low levels of gene products can be detected in the bloodstream of monkeys. It is likely that a similar procedure could be used for many different animals. Attempts to transfer genes using retroviral vectors into pig and dog bone marrow are now underway.

REFERENCES

1. Anderson, W.F. (1984) Prospects for human gene therapy. Science 226:401-409.
2. Eglitis, M.A., P. Kantoff, E. Gilboa, and W.F. Anderson (1985) Gene expression in mice following high efficiency retroviral-mediated gene transfer. Science (in press).
3. Joyner, A., G. Keller, R.A. Phillips, and A. Bernstein (1983) Retrovirus transfer of a bacterial gene into mouse haematopoietic progenitor cells. Nature 305:556-558.
4. Miller, A.D., R.J. Eckner, D.J. Jolly, T. Friedmann, and I.M. Verma (1984) Expression of a retrovirus encoding human HPRT in mice. Science 225:630-632.
5. Williams, D.A., I.R. Lemischka, D.G. Nathan, and R.C. Mulligan (1984) Introduction of new genetic material into pluripotent haematopoietic stem cells of the mouse. Nature 310:476-480.

BOOROOLA (F̲) GENE: MAJOR GENE AFFECTING

OVINE OVARIAN FUNCTION

B.M. Bindon and L.R. Piper

Division of Animal Production
Commonwealth Scientific and
Industrial Research Organization
Armidale, New South Wales 2350, Australia

INTRODUCTION

The Booroola Merino is one of a small number of sheep breeds with high genetic merit for prolificacy, defined here as the number of offspring per pregnancy. These breeds (Tab. 1) are valuable genetic resources for the genetic improvement of sheep reproductive efficiency, since they can be used to bring about by genetic substitution rapid increases (40-60%) in prolificacy of other sheep populations. Genetic progress of this magnitude would require about 30 years of within-breed selection for prolificacy to achieve the same result.

Until recently, ovine prolificacy, like most production characteristics of domestic livestock, was assumed to be a polygenic trait influenced by many genes of small effect. Although this appears to be the case for most of the prolific breeds in Tab. 1, there is now substantial evidence to show that the exceptional prolificacy of the Booroola can be traced to the action of a single gene (F̲) with a major effect on the number of eggs shed from the ovary at each estrous cycle (64). This knowledge has stimulated renewed interest in the roles of major genes in animal breeding and in the development of strategies for their recognition and utilization (79).

The Booroola gene is already being used to bring about genetic improvement of sheep reproductive efficiency as illustrated in this chapter. Using traditional genetic approaches, the F̲ gene is also being introduced into other sheep breeds by backcrossing (introgression), and elaborate mathematical models have been developed to predict the likely rate of progress of this procedure (37). Recombinant DNA technology could greatly simplify this process as well as making possible the transfer of the F̲ gene to other livestock species. A review of the Booroola gene is, therefore, especially relevant to this Volume dealing with the agricultural perspective of genetic engineering of animals.

This chapter describes the Booroola Merino and its origin, summarizes the evidence for the existence of the F̲ gene, illustrates its effect on ovine prolificacy, and reviews what is known about its physiological effects and likely mode of action. The subject has been reviewed in a number of recent publications (6,15,19,21,60,64).

Tab. 1.   Estimates of ovulation rate and litter size of the major prolific
          sheep breeds.

	Estimates of:				
	Ovulation rate		Litter size		
Breed	Mean	Range	Mean	Range	Reference(s)
Finnsheep	3.5	1-9	2.6	1-7	42
Romanov	3.4	1-7	2.6	1-5	69,70
D'Man	2.8	1-8	2.1	1-6	48
Booroola Merino	4.2	1-11	2.5	1-7	61,64
Cambridge	4.0	2-13	>2.6	1-8	41

BOOROOLA MERINO: GENETIC RESOURCE FOR OVINE PROLIFICACY

Relative Phenotypic Merit for Prolificacy

     The Booroola Merino is one of 5 breeds of sheep that are characterized
by exceptional prolificacy. Estimates of their ovulation rates and litter
sizes are presented in Tab. 1. These breeds have never been compared in
the same environment, so it is not possible to meaningfully rank them for
prolificacy. Until 1985 the Booroola was regarded as having the highest
recorded ovulation rate (i.e., 11), but recent data (Tab. 1) suggest that
the Cambridge breed (a composite derived mainly from Clun Forest and Finnish Landrace breeds) is probably the world's most prolific sheep (41).

     The extent to which the Booroola differs from other Merinos in prolif-
icacy can best be illustrated by a comparison of ovulation rates. This is
shown in Tab. 2, which contains the distributions of ovulation rates for
young (1.5 yr) and mature (2.5-7.5 yr) ewes of the Booroola (B) and randomly bred, control Merino populations compared in the same environment at
Armidale in March-April, 1985. The important difference between the genotypes is that most adult and all young control Merinos have ovulation rates
of 1 or 2, while about 90% of Booroola ewes have ovulation rates of 3 or
more. Among adult Booroola ewes, 41% have ovulation rates of 6 or more.
It will be seen later that this exceptional ovulation rate pattern is the
distinctive feature of the Booroola Merino and the major site of action of
the Booroola gene.

History of the Booroola Merino and Possible Origin of the F Gene

     The Booroola Merino was developed initially by 2 commercial sheep
breeders, the Seears Brothers of "Booroola," Cooma, New South Wales,
Australia, and later by the Commonwealth Scientific and Industrial Research
Organisation (CSIRO) Divisions of Animal Genetics and Animal Production
(84). It is a bona fide representative (84) of the medium-wooled non-
Peppin (MNP) (77) strain of Australian Merinos, and, in terms of wool and
body characteristics, is genetically similar to present day conventional
Merinos of the same strain (61). The Booroola flock maintained by the
CSIRO was formed in 1958 with triplet- and quadruplet-born ewes, and with a
quintuplet-born ram obtained from the Seears Brothers (61). The duration
and exact nature of the selection process used by these commercial sheep
breeders in developing this special multiple birth flock are not known.
However, it is established that these breeders practiced selection only on

Tab. 2.   Prolificacy of CSIRO Booroola and control Merinos.   Distribution
          (%) of ovulation rate in ewes of 2 age classes studied in March/-
          April, 1985.

Ovulation rate class	Ewes aged 1.5 yr		Ewes aged 2.5-7.5 yr	
	Booroola	Control	Booroola	Control
	n = 57	n = 54	n = 220	n = 65
1	3.5	81.5	3.2	36.9
2	7.0	18.5	7.7	61.5
3	26.3	--	9.6	1.5
4	17.5	--	17.7	--
5	28.1	--	20.9	--
6	14.0	--	18.2	--
7	3.5	--	12.3	--
8	--	--	6.8	--
9	--	--	1.8	--
10	--	--	0.9	--
11	--	--	0.9	--
Mean ovulation rate:	4.16	1.19	5.08	1.65

the ewe portion of the flock and purchased sires annually from stud flocks
with no history of multiple births (84).

     With this type of breeding program, the increase in prolificacy ob-
tained by the Seears Brothers could not have resulted from increases in the
frequency of favorable alleles at loci with small effects on litter size.
Indeed, the only reasonable genetic model to account for the observed increase (60) is that it occurred because of a gradual increase in the frequency of individuals carrying a "gene" (or duplication, deletion, or
closely linked group of genes) with a major effect on prolificacy.

     Turner (84) has examined the major gene theory in relation to the ori-
gins of the Booroola Merino.   The gene could have arisen as a mutation in
the Seears' flock.   Alternatively, it may have survived as a gene that
existed in either the (non-Merino) "Cape" or "Bengal" breeds that arrived
in Australia in 1788 and 1792, respectively.   Historical evidence favors
the "Bengal" sheep as the origin of the F gene, since this sheep was known
to be prolific and to breed twice per year (38).   In any event, both breeds
were incorporated into the early flock of Sir Samuel Marsden, to which were
added some Spanish Merinos after their arrival in Australia in 1797.   There
is further historical evidence from the Australian Merino flock register
(see Ref. 84) to show that animals from the Marsden flock were used in the
formation of the very Merino studs from which the Seears Brothers made
their annual ram purchases.

Evidence for the F Gene

     A comprehensive review of the data supporting the existence of the F
gene has recently been prepared (64).   The main lines of evidence are summarized as follows.

<u>The segregation criterion and problems in the segregation analyses.</u>
The criterion adopted to distinguish carriers of the putative allele in the CSIRO's foundation Booroolas was the occurrence of one or more sets of triplets, or higher-order litter sizes, in a ewe's lifetime litter-size record (60). Though the data were shown to be consistent with the segregation of an allele having major effects on litter size, it was recognized that there would be difficulties in confirming its existence while the effects were defined in terms of differences in the distribution of litter size. In genetic terms, the variation in litter size is largely a reflection of variation in ovulation rate, with little or no contribution from variation in embryo survival (40). Moreover, since embryo loss rate is directly proportional to ovulation rate (40), litter size becomes an increasingly inaccurate indicator of ovulation rate as the mean ovulation rate increases. For these reasons, measurements of ovulation rate have supplemented or replaced litter size records in most subsequent attempts to demonstrate the existence of the putative allele.

<u>Background genotype.</u> For Merinos in Australia where triplet births are rare (0.1%; Ref. 64), the occurrence of one or more litter-size or ovulation-rate records of 3 or more has been a useful working criterion for distinguishing carriers of the putative allele. Using this criterion, the following segregation analyses support the concept of a major gene for prolificacy in the Booroola flock.

(a) <u>Foundation CSIRO Booroola ewes and their daughters.</u> The lifetime litter-size records of these animals conform to the prediction (61) that the dams of the 13 foundation CSIRO ewes were heterozygous for the putative allele (Tab. 3).

(b) <u>Apparent segregation in crosses of Booroola (B) and conventional (MNP) Merinos in Australia.</u> Ovulation-rate and litter-size records of F_1 (B x MNP), F_2 (B x MNP)^2, and backcrosses [(B x MNP) x MNP] genotypes in Armidale provide strong support for the F-gene theory (61). As shown in Tab. 4, the observed frequency of carriers in the F_1 was 0.72, and since F_1 carriers cannot be homozygous, the estimated frequency of the allele in the F_1 was 0.36. The expected and observed frequencies in the F_2 and backcross are in good agreement (Tab. 4).

(c) <u>Apparent segregation in crosses of Booroola (B) and control Merinos (M) in New Zealand.</u> Crosses of putative heterozygous

Tab. 3. Number of ewes (mean litter size) with at least one or with no litter size records of 3 or more among foundation CSIRO Booroola ewes and their daughters (64).

Group	Total ewes	Number with litter size records >3	
		At least one $\geq$3	None >3
Foundation CSIRO Booroola ewes	14	8(2.42)	6(1.37)
Daughters from foundation ewes with at least one litter size record $\geq$3	11	6(2.52)	5(1.35)
Daughters from foundation ewes with no litter size record >3	8	0	8(1.35)

rams with F_1 (B x M) ewes (34) produced carrier frequencies (Tab. 5) that are in good agreement with expectation.

(d) <u>Apparent segregation of the allele in the progeny test records of Booroola rams.</u> Ovulation rate progeny tests of Booroola rams in New Zealand have identified rams of each of the putative F-locus genotypes (33). These data are summarized in Tab. 6. Further confirmation of the F-gene theory was provided by progeny test data on the sons of the 3 rams shown in Tab. 6 (33).

Summary of Segregation Analyses

The major gene hypothesis to account for the exceptional prolificacy of the Booroola Merino was first advanced in 1980 (60,61). Since that time a considerable research effort has been directed towards testing the theory, and most of the relevant information has been reviewed here. All of the new information is consistent with the theory, and the existence of the putative major allele should now be accepted.

Identification of F Gene Carriers

Classification of ewes with respect to the F gene is complicated by the fact that the F gene does not confer an easily recognized all-or-none characteristic on the carrier. Ewes not carrying the F gene already ovulate. The problem is to decide on an ovulation rate (i.e., number of eggs shed) that will reliably distinguish carriers from noncarriers and heterozygotes from homozygotes.

The following arbitrary classification (61,64) scheme has been adopted for identifying carriers of the F gene in Merino or Merino-Romney populations:

Genotype	Maximum ovulation rate recorded
Ewes homozygous for the F gene (i.e., $\underline{FF}$)	≥ 5
Ewes heterozygous for the F gene (i.e., $\underline{F+}$)	≥ 3 and <5
Noncarriers (i.e., $\underline{++}$)	<3

This proposed criterion to distinguish putative homozygotes may be of little value when the average prolificacy of the background genotype is

Tab. 4. Frequency ± s.e. of ewes carrying the putative allele in several Booroola (B) x medium non-Peppin (MNP) Merino genotypes (64).

Genotype	Number ewes[a]	Frequency	
		Observed	Expected[b]
F_1 (B x MNP)	136	0.72 ± 0.04	?
F_2 (B x MNP)[2]	124	0.55 ± 0.04	0.59
Backcross [(B x MNP) x MNP]	82	0.35 ± 0.05	0.36

[a] Carrier status assessed from the total litter size and ovulation rate records available for each ewe (3-14 records/ewe).
[b] Expected frequencies calculated on the basis of the estimated gene frequency of 0.36 in the F_1.

Tab. 5.   Frequency ± s.e. of ewes carrying the putative allele among the offspring of progeny-tested, heterozygous Booroola rams and F_1 (Booroola x Merino) carrier and noncarrier ewes in New Zealand (34,64).

F_1 dam putative genotype[a]	Number	Progeny Carrier frequency	
		Observed[b]	Expected
Carrier	119	0.68 ± 0.04	0.75
Noncarrier	57	0.51 ± 0.07	0.50

[a] Based on 3-6 ovulation rate records per ewe.
[b] Based on 3-5 ovulation rate records per ewe.

well above that of Merinos. For example, in a recent study involving F_1 Booroola x Border Leicester ewes (which by definition cannot be homozygous), the mean ovulation rate was 3.13 ± 0.07, and of these 14% had ovulation rates of 5 or more (16).

Identifying male carriers of the putative allele. At the present time, there is no known direct expression of the putative allele in males. The genotype of males at the hypothesized locus must therefore be determined by progeny test, except perhaps those cases in which the putative genotypes of the parents imply only one possible offspring genotype. Even in such cases, it is probably wise to progeny test, especially when the female parent, despite her Booroola ancestry, has been judged a noncarrier of the putative allele (64).

Effects of the F Gene on Ovine Prolificacy

Outcrossing studies have shown that substantial increases in prolificacy result from crosses of the Booroola with other Merinos as well as with other sheep breeds (61). Not all of the increases can be attributed to the F gene, since the remainder of the Booroola genome may also influence prolificacy. This problem has been resolved (64) by comparing animals that differ only with respect to the F locus. To do this the ovulation-rate and litter-size records of CSIRO Booroola ewes born over the period 1973-1980 have been used to assign to each of the ewes a genotype at the F locus. Ewes with less than 3 records were excluded, and genotypes were assigned according to the criteria proposed above. The data were then analyzed by least squares procedures to adjust the genotype means for the effects of age and year of measurement. The results are shown in Tab. 7. For ovulation rate, the dominance deviation [F+ - (FF + ++)/2] was not significant (-0.07 ± 0.17), and it may be concluded that the effect of the gene is additive and equal to half the homozygote difference of 3.30 ± 0.29 eggs. However, for litter size the dominance deviation was significant (0.27 ± 0.10), and the first and second copies of the gene added 0.9 and 0.4 of a lamb, respectively.

It could be argued that these estimates are biased because the data for each ewe of the FF and F+ genotypes, respectively, included by definition one record of 5 or more or one record of 3 or more but not more than 4. The analyses were therefore repeated with this genotype assignment or

Tab. 6.   Ovulation rate, progeny test data for Booroola rams of each of
          the 3 putative, major locus genotypes (33).

Sire	Number	Progeny	
		Proportion with at least one record >3*	Mean ovulation rate ± s.e.
75-1232	29	0.97	2.90 ± 0.10
Control	71	0.07	1.67 ± 0.05
77-358	58	0.52	2.48 ± 0.06
Control	54	0.06	1.51 ± 0.05
75-1492	44	0.04	1.53 ± 0.06
Control	71	0.07	1.67 ± 0.05

* Two ovulation records available for each ewe.

"ascertainment" record excluded, and the resulting estimates are shown in
Tab. 7. As before, there was no evidence of nonadditive effects on ovulation rate. However, in this analysis the dominance deviation for litter
size was also not significant (0.10 ± 0.10), and the first and second
copies of the gene added 0.7 and 0.5 of a lamb, respectively. By contrast
with the previous conclusion (61), it would appear that in breeds with
levels of prolificacy similar to or less than that of the ++ Booroola genotype (1.5), the effect of the gene on litter size may be very nearly
additive. However, if the proposed relationship between litter size and
ovulation rate (40) holds, the effects of the gene on litter size in breeds
of higher average prolificacy may well be dominant.

In practical terms it is also pertinent to ask whether the effect on
ovulation rate of a single copy of the gene is the same at all levels of
prolificacy. Data relevant to this question is provided by Piper et al.
(64), who showed that in 21 different experiments in which the ovulation
rate of noncarrier (i.e., ++) ewes ranged from 1.0 to 1.85, the effect of
the F gene (i.e., the ovulation rate difference between F+ and ++ ewes) was
independent of the ovulation rate of the base population.

Effects of the F Gene on Other Production Traits

Although there have been many studies designed to evaluate the comparative productivity of Booroola crosses with other sheep breeds, only one
has isolated the effects of the F gene on wool production and liveweight.
In that study (66) it was shown that the F gene had no undesirable pleio-
tropic effects on greasy fleece weight, clean fleece weight, fiber diameter, staple length, wool style, lamb birth weight, growth rate, or adult
(15-month) liveweight. These preliminary data confirm that the principal
effect of the F gene is to influence prolificacy resulting from a major
effect on ovulation rate (FF ewe progeny 1.9 times greater than ++ ewe
progeny in the aforementioned study).

Agricultural Significance of the Booroola Merino

The value of the Booroola Merino as a genetic resource lies in its
ability to bring about rapid increases in reproductive rate when crossed
with other Merinos or other breeds. This contrasts with the relatively

Tab. 7.   The effect of the $\underline{F}$ gene on ovulation rate (OR) and litter size
(LS) in CSIRO Booroola ewes born 1973-1980. Ewes with <3 OR or
LS records excluded (64).

Basis of estimate	Data	Genotype		
		++	F+	FF
Ascertainment record included	OR	1.40 ± 0.28	2.92 ± 0.08	4.70 ± 0.08
	LS	1.48 ± 0.14	2.38 ± 0.06	2.74 ± 0.07
Ascertainment record excluded	OR	1.40 ± 0.27	2.82 ± 0.09	4.38 ± 0.09
	LS	1.48 ± 0.13	2.17 ± 0.07	2.66 ± 0.07

slow progress achieved (i.e., annual genetic gains of 1-2 lambs born per
100 ewes joined) by within-breed selection for prolificacy (23).

Although the performance of a Booroola cross may not be ascribed entirely to the $\underline{F}$ gene, it is thought that the effects on reproductive performance are primarily due to this cause. In any event, the practical
exploitation of the $\underline{F}$ gene to date (and for the immediate future) will
depend on outcrossing procedures where the effects of the $\underline{F}$ gene are partly
confounded with the effects of the remainder of the Booroola genome. The
results described below are therefore relevant to the agricultural impact
of the $\underline{F}$ gene.

Crosses with other Merino strains. As an example of the effects observed, the results of Booroola crosses with a MNP strain are described.
In each of 3 successive years, 5 Booroola and 5 MNP rams were joined to MNP
ewes at Armidale, New South Wales, and the ewe progeny then evaluated at
ages of 2-4 yr. The results (61) are summarized in Tab. 8 and 9, which
compare ovulation rate, litter size, lamb survival, and gross reproductive
performance of MNP and Booroola x MNP ewes. The main conclusions from this
study are:

(a) Both the mean and the distribution of ovulation rates were sig-
     nificantly altered in the B x MNP ewes, so that their average
     ovulation rates were 87% higher than the MNP controls.

(b) Average litter size of B x MNP ewes increased by 56% compared
     with that of MNP ewes.

(c) Lamb survival was lower for B x MNP ewes, largely as a result of
     reduced survival among multiple births.

(d) The combined effects of fertility, prolificacy, and lamb survival
     led to a 16% advantage in lambs weaned for the B x MNP crosses.

In the above investigation (61), the B x MNP ewes had similar live-
weights, wool weight, and wool quality. In crosses with strong wool Merino
strains (e.g., Ref. 52), the Booroola-cross ewes produced less wool (about
0.5 kg clean fleece weight less per head), but 29% more lambs were weaned.
The Booroola rams used in both studies cited above were a mixture of genotypes with respect to the $\underline{F}$ locus. Further increases in performance could
be achieved if homozygous Booroola rams were used to generate the Booroola-
cross females, all of which would then carry one copy of the $\underline{F}$ gene.

74

Tab. 8.   Agricultural significance of the Booroola Merino. Comparison of
          ovulation and litter size distributions (%) in control Merino and
          Booroola x control Merino ewes. Data from Ref. 61, pooled over 3
          years and 3 ages (2-, 3-, and 4-yr-old ewes).

| Genotype | n | % Ewes with ovulation rate of: | | | | Mean ovulation rate |
		1	2	3	4	
Control Merino	179	94	6	–	–	1.06
Booroola x control	197	30	46	20	4	1.98

| Genotype | n | Litter size at birth | | | | Mean |
		1	2	3	4	
Control Merino	178	86	13	1	–	1.15
Booroola x control	191	38	46	15	1	1.79

Crosses with other breeds.   The Booroola may also be used to improve
the reproductive potential in crosses with British breeds which traditionally form the basis of the prime lamb industry in Australia. Compared with
with control Merino cross ewes, Booroola cross Border Leicester and Dorset
Horn genotypes have a 55-62% higher ovulation rate, are 46% more prolific,
and, although lamb survival is lower, have more (21-36%) lambs weaned per
ewe joined (Tab. 10), representing a 24% advantage in economic return per
ewe joined (16). These and the Merino results described earlier were
achieved without special management. More of the potential of the Booroola
crosses may be realized by using procedures to enhance lamb survival.

Although the Booroola has mainly been used in Australia and New Zealand, it could also improve sheep reproduction in other countries, particularly those countries where fine-wool production is economically important. For these, the Booroola is much more attractive than the other major
prolific breeds such as the Romanov (USSR), Finnsheep (Finland), and D'Man
(Morocco), all of which have either pigmented or poor quality fleeces. The
USSR has the world's second largest Merino population and could rapidly
disseminate the Booroola gene, since some 50 million ewes are artificially
inseminated annually (T.W. Scott, CSIRO, Prospect, pers. comm.). Benefits
of Booroola crosses could also be better exploited in that country, since
relatively intensive husbandry is practiced where ewes are wintered in-
doors.

The single gene basis of the Booroola's increased prolificacy may
limit its commercial utilization under management circumstances in which
the size of the prolificacy increase conferred by a single copy of the $\underline{F}$
gene (approximately 1 extra lamb) is considered to be too extreme. On the
other hand, this same mode of inheritance may also be an advantage in that
the $\underline{F}$ gene may be inserted into any breed of sheep without changing the
rest of that breed's genotype. For example, by backcrossing with selection
(introgression) or (in the future) by recombinant DNA technology, it will
be possible to develop a Border Leicester (or any other breed) that is
homozygous for the $\underline{F}$ gene yet retains the other desirable attributes of
that breed.

Tab. 9. Agricultural significance of the Booroola Merino. Least squares means for reproduction rate and its components in control Merino and Booroola x control Merino ewes. Data (± s.e.) are from Ref. 61, pooled across 3 years and 3 ewe ages (2-, 3-, and 4-yr-olds).

Genotype	n	Fertility (Ewes lambed/ ewes joined)	Prolificacy (Lambs born/ ewes lambed)	Survival (Lambs weaned/ lambs born)	Reproduction rate* (Lambs weaned/ ewes joined)
Control Merino	224	0.89 ± 0.02	1.13 ± 0.04	0.85 ± 0.03	0.86 ± 0.04
Booroola x control	224	0.94 ± 0.02	1.70 ± 0.04	0.64 ± 0.02	1.00 ± 0.05

* Booroola x control Merino ewes weaned annually 16% more lambs than did control ewes.

In the context of this Volume dealing with genetic engineering for agriculture, it seems reasonable also to suggest that isolation of the F gene could make possible its transfer to the genome of other livestock. We may only speculate on the likely agricultural value of, say, a beef cow carrying the F gene. Yet it seems to be a reasonable alternate approach to other nongenetic methods for increasing the ovulation rate in the cow.

REPRODUCTIVE BIOLOGY OF THE BOOROOLA MERINO AND
THE PHYSIOLOGICAL ASPECTS OF THE F GENE

Comparative Reproductive Biology of the Booroola Merino

Since the early 1970s the components of reproduction of the Booroola ewe have been studied to establish the extent to which the Booroola differs from other Merinos. To do this the Booroola has been compared with CSIRO's "T" and "O" flocks (Peppin-based Merinos selected for and against prolificacy, respectively, since 1954) and a randomly bred, control ("C") Merino flock. Complete details of the origin, design, and management of these flocks are described elsewhere (83). The results of these comparisons have been reported in detail elsewhere (6,15) and are summarized as follows.

Age at puberty, duration of estrus, and length of the estrous cycle. Following a spring lambing, less than 10% of Booroola ewes reached puberty in the first year of life. A similar proportion was observed in the "T," "O," and "C" flocks (21). Early in the second year of life Booroola ewes reached puberty significantly earlier (24 days) than control Merinos and had slightly more estrous cycles between puberty and their first joining at age 18 months.

Adult Booroola ewes have a similar duration of estrus (12.9 ± 1.1 hr, mean ± s.e.) to control Merinos (21), and this contrasts with other prolific breeds that have prolonged estrus (7,42,50). Booroola ewes have an estrous cycle of similar length (16.9 ± 0.2 days) to those of "O," "T," and "C" Merinos (21).

Seasonality of estrus and ovulation. Booroola ewes experience about 40% more estrous cycles per year than "O" ewes (6) and 60% of Booroola ewes

Tab. 10. Least squares means (± s.e.) for liveweight and the components of reproductive performance for crossbred ewes derived from Booroola and control Merinos (16).

Ewe's sire breed:	Border Leicester		Dorset Horn	
Ewe's dam breed:	Booroola	Control	Booroola	Control
Number of ewes	182	108	129	106
Liveweight (kg)	48.3 ± 0.40	52.0 ± 0.50	48.4 ± 0.50	49.8 ± 0.60
Fertility*	0.89 ± 0.02	0.83 ± 0.03	0.90 ± 0.03	0.87 ± 0.03
Ovulation rate	3.13 ± 0.07	2.02 ± 0.10	2.77 ± 0.09	1.71 ± 0.10
Prolificacy*	2.43 ± 0.06	1.66 ± 0.09	2.21 ± 0.07	1.51 ± 0.08
Lamb survival	0.72 ± 0.02	0.91 ± 0.03	0.71 ± 0.03	0.82 ± 0.04
Reproduction rate	1.51 ± 0.06	1.25 ± 0.07	1.42 ± 0.07	1.04 ± 0.08

* Definitions as for Tab. 9.

ovulate in all months of the year (12), while most "0" Merinos do not ovulate for 4 months of the year. It is not yet known if the long sexual season of the Booroola is an expression of the same gene that is responsible for the prolificacy of the Booroola.

Ovulation rate and time of ovulation. The distributions of ovulation rate in the Booroola and "C" Merino flocks in 1985 are shown in Tab. 2. Ewes with ovulation rates as high as 11 have been recorded. It is the extreme range in ovulation number (i.e., from 1 to 11) within the Booroola population that is the most distinctive feature of this genotype. As discussed above this is now known to be a reflection of the 3 separate F-locus genotypes present in the Booroola flock. The number of ovulations recorded for Booroola ewes in successive estrous cycles has a high repeatability (5,12); this is also a function of the 3 separate genotypes within a Booroola population. When separate groups of ewes are subjected to ovarian examination by laparoscopy between 20 and 32 hr after the onset of estrus, it is estimated that the Booroola ewe ovulates about 7.5 hr earlier than comparable "C" Merinos (22).

Litter size at birth. The following data for the Booroola and "C" Merino flocks have recently been reported (61). Booroola ewes (n=522) had an average litter size of 2.29 (range 1-7), with 40% of litters being >3, while "C" Merinos (n=835) had a mean litter size of 1.22 (range 1-2). Booroola survival rates for lambs born in litters of 1, 2, 3, 4, 5, and 6 were 90, 77, 55, 37, 30, and 28%, respectively. For "C" Merinos comparable figures were 88% for singles and 79% for twins.

Fertilization and uterine capacity. Estimates (± s.e.) of fertilization rates for Booroola, "T," and "C" Merinos were 0.93 ± 0.08 (n=16), 0.83 ± 0.05 (n=41), and 0.83 ± 0.06 (n=35), respectively (62). These rates are not significantly different, indicating that failure of fertilization is unlikely to account for reproductive wastage in the Booroola. Within the Booroola population fertilization rate was also independent of ovulation rate (62).

Uterine capacity of Booroola ewes has been compared with that of "C" Merinos by transferring one or three Day 3 embryos of a third Merino genotype to Booroola and "C" ewes (17). For ewes given 3 embryos, Booroola recipients had a mean litter size of 2.6 and "C" ewes, a mean of 2.4. The results show that at this level of prolificacy the Booroola ewe does not seem to have any advantage over "C" ewes with respect to uterine capacity.

Wastage of potential embryos. When ovulation rates of Booroola and "C" Merinos are related to the number of lambs subsequently born (18), there is a progressive increase in the wastage of potential embryos as ovulation rate increases. When ovulation rate exceeds 3 there is little additional increase in mean litter size. Within the Booroola population the distribution of ovulations between the left and right ovaries had no influence on embryo survival. For example, for ewes with 4 ovulations, mean litter sizes were 0.95, 1.05, and 0.86 for ewes with ovulations distributed 2:2 (n=16), 3:1 (n=23), and 4:0 (n=7), respectively, between ovaries. The embryo wastage pattern relative to ovulation rate in the Booroola is similar to that seen in other prolific breeds (40).

Ovarian activity postpartum. Finnsheep, Romanov, and D'Man breeds resume ovarian activity postpartum earlier than nonprolific breeds in the same environment (15). The Booroola does not share this attribute; at 40 days following parturition in the spring, few (<2%) Booroola ewes or "C" Merinos had begun ovulating (21).

Freemartins in Booroola populations. It is established that the freemartin condition, although rare, does exist in sheep and that, as in cattle, its occurrence depends on there being a pregnancy of mixed sexes and an early placental anastomosis between opposed sexes. It is not yet understood why the frequency of freemartins should be so low in sheep populations in which the frequency of twins is high relative to that seen in cattle.

Sheep freemartins do seem to occur more frequently in litter sizes greater than 2 (78), perhaps as a result of uterine crowding which may promote placental anastomosis. If this is true the freemartin syndrome should be more common in the prolific sheep breeds. Preliminary evidence from the Booroola flocks supports this idea. The frequency of freemartins may be as high as 5% of all females born in Booroola populations, and these are observed in ewes with high ovulation rates (4-9) and litter sizes greater than 2 (20).

Physiological Characteristics of the Booroola Ewe and Effects of the F Gene

Examination of the reproductive biology of the Booroola leaves little doubt that the principal role of the F gene is to cause major alteration of the ewe's ovarian physiology. It can be argued that the Booroola's regulatory control of ovarian function has been impaired to an undesirable degree since ovulation rates of the magnitude seen in FF ewes (up to 11) would not be compatible with survival of this genotype under normal Australian environmental conditions. The extent of the abnormality can be gauged from comparisons with the normal Merino in Australia, where the ovulation rate rarely exceeds 2 and the predominant litter size is one.

What is the basis of the Booroola's ovarian abnormality and what is the site of action of the F gene? These questions have stimulated much interesting research in ewes carrying the F gene. Not all the studies have been between animals that differ only at the F locus, since these are difficult to identify reliably. Contrasts of Booroola and Booroola-cross ewes with randomly bred, control Merinos in the same environment do not exclusively measure effects of the F gene. However, such studies should largely reflect the presence of the F gene since the difference in prolificacy between noncarrier Booroolas (++) and ordinary Merinos in the same environment is only 0.3 lambs born (see Tab. 7 and 8).

Ovarian follicle number, morphology, and development. Booroola ewes have a similar number of growing follicles to control Merinos (36; see Tab. 11), as do Romney-cross ewes with and without the F gene (54). This is in contrast to the prolific Romanov breed which has been shown to have twice as many growing follicles as the Ile-de-France breed of low prolificacy (27). This also contrasts with results obtained in a comparative study of the D'Man and Timahdite breeds (47). Within the Booroola population (ovulation rates of 3 to 9), there was no correlation between ovulation rate at the previous cycle and the number of antral follicles (36).

These data suggest that the Booroola ovary has different follicular characteristics to 2 other prolific breeds, and, secondly, that the F gene does not act by increasing the number of growing follicles in the ovary. Rates of atresia among growing follicles also appear to be similar in ewes with the F gene and noncarriers (36,54).

The Booroola is characterized by having follicles that are significantly smaller at the time of ovulation and that contain only half the number of granulosa cells per growing follicle (36); this can be attributed directly to the action of the F gene (4,54; Tab. 12). Among the other prolific breeds, the Finnsheep at least appears to share this characteristic of small follicle size (87).

Differences in follicle development as well as morphology appear to be involved in the high ovulation rate of the Booroola. This has been demonstrated clearly by identifying the developmental history of follicles that ovulate in Booroola and control Merinos (36); in the latter, recruitment of follicles >2 mm diameter occurs between Days 13 and 15 (estrus = Day 0), and selection of the ovulatory follicle has occurred by Day 15. In the Booroola, follicles less than 2 mm diameter are recruited between Days 13 and 15 and they grow to ovulatory size (3-4 mm) by Day 15. However, recruitment continues during Days 15-17, and additional Booroola follicles reach ovulable size by the time of the luteinizing hormone (LH) discharge. This phenomenon is not seen in the Romanov breed (36) and may be peculiar to the Booroola.

Tab. 11. Estimates of ovarian, antral follicle populations in ewes with and without the F gene.

Estimate	No. of animals per genotype	Age (yr)	Basis of estimate	Follicle diameter	Booroola or F+	Control or ++
1	5-8	8	per ewe	>0.35 mm	27 ± 4	64 ± 5
2	6	2 & 8	per ovary	>3 cell layers	50 ± 6	45 ± 6
3	6	2 & 8	per ovary	>3 cell layers	78.5	46.7
4	17-18	3-4	per ewe	>1 mm	43.3 ± 4.1	50.7 ± 4.6
5	21	6-7	per ewe	>1 mm	40 ± 3	44 ± 3

References (for estimates):

1 and 2: K.J. Turnbull, pers. comm.; 3: Ref. 26; 4: Ref. 36; 5: Ref. 54

Tab. 12. Granulosa cell populations in preovulatory follicles of ewes with and without the F gene.

Reference	Contrast	No. of ewes per genotype	Booroola	Non-Booroola
4	F+ vs ++ (Merino)	8	0.68 ± 0.05 (20)*	1.36 ± 0.19 (9)
36	Booroola vs control Merino	7	1.09 ± 0.46 (27)	2.78 ± 0.57 (7)
54	F+ vs ++ (Merino x Romney)	9	2.00 ± 0.10 (31)	5.30 ± 0.70 (11)

* Total number of follicles per genotype.

So the two Merino genotypes (i.e., Booroola and control) begin at luteolysis with a similar number of "candidates" (i.e., antral follicles) for ovulation. In the control Merino, one "dominant" follicle is soon selected and the others cease to develop and become atretic. In the Booroola, these extra "candidates" are not deleted, but their development and that of the follicles selected as late as Day 17 is sustained until terminal follicle growth and ovulation are induced by increased LH-pulse frequency and LH discharge. What is it that sustains the integrity of these Booroola follicles? This may be follicle-stimulating hormone (FSH) whose peripheral concentrations and ovarian uptake are influenced by intraovarian protein hormones dealt with below.

Gonadotrophin studies. Pituitary gonadotrophic hormones may be involved in the mode of action of the F gene. There is compelling evidence from hypophysectomy experiments that the pituitary is indispensable for ovulation and that treatment with exogenous gonadotrophin will artificially increase ovulation rate in all species examined.

(a) The role of luteinizing hormone and luteinizing hormone-pulse frequency. It is clear that follicle maturation and ovulation in the sheep are dependent upon an increase in basal LH and LH-pulse frequency during the period following luteolysis (2). The increased LH activity induces progressively increasing estradiol secretion which results in estrus and preovulatory LH discharge. Ovulation occurs about 24 hr after this discharge.

The question of interest here is whether carriers of the F gene have different LH plasma concentrations or LH-pulse characteristics compared to noncarriers. The direct evidence on this point (6,75) shows that plasma LH profiles are similar in the 2 genotypes. This conclusion is in agreement with studies of other prolific sheep breeds (49,86).

Indirect evidence for the involvement of LH-pulse frequency in control of ovulation rate should come from experiments in which pulse frequency is artificially altered by exogenous LH treatment in the follicular phase of the estrous cycle. The most comprehensive study of this type (53) showed that ovulation and normal luteal function could be achieved with LH levels (in the follicular phase) ranging from 0.7 to 4 times that seen in the normal estrous cycle. Similar ovarian function was achieved with as few as 16 or with as many as 70

LH peaks of comparable size. Since exceptional ovulation rates (eg., >2) were not observed in the ewes subjected to these manipulations, one must again conclude that ovulation rate is independent of the amount and the secretory pattern of LH in the preovulatory period.

The timing of the preovulatory LH discharge in prolific sheep may be important. In 6 separate investigations, prolific sheep geno-types other than the Booroola were shown to have a LH discharge that occurred significantly later after the onset of estrus (and luteoly-sis) than breeds of low prolificacy (13). The Booroola did not con-form to this pattern since both it and "C" Merinos had a LH discharge about 4.5 hr after the onset of estrus. The effects of the preovula-tory LH surge on the ovaries of all genotypes may not be identical, however, since ovulation occurs, on average, 7.5 hr earlier after on-set of estrus in Booroola ewes than in "C" Merinos, despite a similar timing of the LH discharge (22). The F gene appears to render preovu-latory follicles more sensitive to endogenous LH.

(b) The role of follicle-stimulating hormone. Most authors (39) agree that in sheep there is a discharge of FSH coincident with the preovulatory LH discharge, and that this is followed about 24 hr later by a further FSH discharge at a time (Day 2) when LH concentration re-mains low. During the luteal phase of the cycle, plasma FSH profiles of individual ewes (73) show waves of FSH. These are normally ob-scured when values from different ewes are pooled for a particular day of the cycle, as recently confirmed (49,55). In the follicular phase of the cycle when progesterone levels are declining, average FSH con-centrations do not increase to the same relative extent as basal LH (39). Concentrations may decline from the time of luteolysis. There is general agreement, however, that FSH concentrations decline signif-icantly in the 24 hr preceding the preovulatory FSH and LH discharges.

Until recently there has been conflicting evidence for elevated plasma FSH concentrations in the Booroola and other prolific sheep breeds. This has been reviewed in detail elsewhere (6,15,19).

The most convincing between-breed data relating plasma FSH con-centrations to prolificacy (49) show that the prolific D'Man ewe had higher plasma FSH concentrations than a nonprolific breed during the follicular phase, at the time of the FSH discharge that coincides with the LH discharge, and at the time of the so-called "Day 2 FSH peak." Present evidence favors the idea that it is FSH stimulation of ovarian follicles during the follicular phase that is important for increasing ovulation rate (6,15,19). Studies of the prolific Romanov breed have shown either no difference in plasma FSH concentration (7) or a dif-ference only on Day 2 of the estrous cycle (28).

There are some data (Tab. 13) to show that both prepubertal and adult Booroola ewes have higher plasma concentrations of FSH than con-trol Merinos. This difference has also been confirmed in the pitui-tary gland, by both radioreceptor and radioimmunoassay techniques, and in the urine (Tab. 13). However, when FSH was measured in FF, F+, and ++ Booroola ewes by heterologous FSH radioimmunoassay, there was no significant difference between the genotypes (9). It is not, there-fore, possible to conclude at present that carriers of the F gene are characterized by elevated tissue FSH concentrations.

Indirect evidence for involvement of FSH in ovine prolificacy comes from attempts to artificially increase ovulation rate with exo-genous gonadotrophin. It has been shown that superovulation can be readily induced with porcine FSH and that ovulation rate is not increased by adding porcine LH to the treatment (88). A more recent study (3) confirmed that ovine FSH, infused at a dose that doubled the peripheral plasma FSH concentrations for periods of up to 60 hr from just before the time of luteolysis, caused an approximately 6-fold increase in the mean ovulation rate of a nonprolific breed. A "rebound" increase (4-fold) in plasma FSH following luteal phase suppression of FSH by bovine follicular fluid is thought to be responsible for the significant increases in ovulation rate (85).

Based on present knowledge, it is difficult not to conclude that multiple ovulation in the sheep results from excessive FSH stimulation of the ovary during the late luteal and early follicular phases of the estrous cycle. However, there are no convincing quantitative data relating adult plasma FSH concentrations to the ovulation rate of ewes within any of the prolific breeds. Repeated plasma measurements may be made only on a few individual animals and these provide too few data for meaningful correlations to be derived. Significant correlations have been recorded between prepubertal plasma FSH concentrations and subsequent ovulation rates (68).

(c) <u>Ovarian sensitivity to gonadotrophin.</u> The ovarian activity of Booroola ewes may be independent of plasma gonadotrophin concentrations and may simply reflect increased sensitivity to gonadotrophin by ovarian follicles of these animals. Evidence for this comes from the fact that Booroola ewes have been shown to have a higher ovulatory response to Pregnant Mare Serum Gonadotrophin (PMSG) than "C" Merinos (14), and from the fact that ewes with the $\underline{F}$ gene are more sensitive to PMSG than ewes without the $\underline{F}$ gene (46,63). The difference in

Tab. 13. Tissue concentrations of follicle-stimulating hormone (FSH) in Booroola ewes relative to control Merinos.

Tissue	FSH concentrations* in:		% Advantage for Booroola	Reference
	Booroola ewe	Control ewe		
Pituitary gland				
RRA method	1,234	764	+62	71
RIA method	79	55	+44	19
Plasma				
Preovulatory	330	201	+64	19
Postovulatory (Day 2)	335	202	+66	19
Prepubertal ewe lambs (Day 30)	145	39	+272	10
Urine				
Estrous cycle (Day 9)	564	354	+59	19
Estrous cycle (Day 15)	567	312	+82	19

* Pituitary gland concentrations are µg/g wet weight pituitary tissue. Plasma concentrations are ng NIAMDD-oFSH-RP$_1$/ml or ng NIH-FSH-S6/ml. Urine excretion rate is ng NIAMDD-oFSH-RP$_1$/3 hr.

sensitivity is evident at as early as 5 months of age (57). These data are in accord with a similar difference in sensitivity to PMSG between the moderately prolific "T" and lowly prolific "O" Merinos (8), and has also been confirmed in Romney ewes selected for prolificacy (80) and, more recently, in the Finnsheep and D'Man breeds (J.F. Quirke, pers. comm.). It may also be a general feature of prolific genotypes since genetic associations between prolificacy and sensitivity to PMSG have also been described in mice (11) and cattle (81).

It is not yet established that ovaries or individual follicles of prolific breeds are more sensitive to FSH as distinct from exogenous PMSG. Such evidence could come from the in vitro response by isolated follicles to added ovine FSH or the ovarian response by hypophysectomized ewes to injected ovine FSH. Such experiments are in progress.

If Booroola follicles had more gonadotrophin receptors, then they might acquire more gonadotrophin from the circulation than follicles of other Merinos. This might explain differences in ovarian sensitivity between the genotypes. It is confirmed (86) that the acquisition of LH receptors in the mural granulosa is an essential requirement for growing sheep follicles to ovulate, and that Finnsheep have more such follicles in the follicular phase than the Suffolk breed. However, there is no evidence that prolific sheep have more LH receptors per follicle. Since LH receptors result from the action of estradiol and FSH on granulosa cells (see Ref. 87), there seem to be good grounds for suspecting that the extra follicles that develop in Booroola ewes do so because they have FSH receptors (or more FSH receptors), thus enabling them to respond to this gonadotrophin. This possibility has not yet been explored.

The Role of Inhibin

It has been concluded (39) that an ovarian factor other than estradiol and progesterone must contribute to the feedback regulation of FSH during the ovine estrous cycle. It seems likely that this substance is "inhibin," a nonsteroidal, gonadal hormone capable of selective suppression of FSH release from the pituitary gland (82). Although inhibin has now been purified to homogeneity (72), the evidence for its involvement in ovine prolificacy is based mainly on experiments that used steroid-free follicular fluid (FF). It is recognized that other protein hormones exist in follicular fluid, and these may act directly on the ovary.

Steroid-free, ovine FF was shown to suppress estrus and ovulation in Booroola and "C" Merinos in a dose-dependent manner (30). Repeated injections of ovine FF for 5 days delayed ovulation for at least 12 days. Partially purified ovine FF [i.e., the peak-2 fraction from a red gel agarose column (45), containing high inhibin activity] was as effective as unpurified ovine FF in suppressing ovulation (29). This effect is most likely the result of the inhibin in ovine FF, since ovine FF selectively suppresses FSH in the ewe (31).

Direct evidence that inhibin may be involved in the action of the $\underline{F}$ gene comes from measurements of inhibin bioactivity (Tab. 14) in Booroola and "C" Merino ovarian cytosols using the in vitro, pituitary, cell culture assay (76). Booroola ewes with a mean ovulation rate of 2.8 had only one-third the ovarian inhibin content of "C" Merino ewes which had a mean of 1.2 ovulations (31). Inhibin measurements in peripheral blood should advance knowledge of the physiological significance of this substance. The recent purification of inhibin (72) is an important step towards this objective.

Tab. 14. Ovarian inhibin content, ovarian weight, and ovulation rate of 10 Booroola and 10 "C" Merinos (means ± s.e. for single ovaries). [Reproduced with the permission of the Journal of Reproduction and Fertility (31).]

Measurement	Genotype Booroola	Control	Significance of differences between genotypes
Ovarian weight (g)	1.21 ± 0.09	1.23 ± 0.12	not significant
Ovulations/ovary	1.40 ± 0.30	0.60 ± 0.10	P < 0.010
Inhibin content (units/ovary)	400 ± 60	1,230.00 ± 130.00	P < 0.001

The studies of inhibin in the Booroola ovary provide a link between this ovarian hormone and prolificacy. Indirectly, this link is substantiated by the increased ovulation rates seen in "C" Merinos following active immunization against partially purified inhibin. This was first described in "C" Merinos in 1982 (59) and has recently been confirmed in another breed (44). Further results on larger groups of ewes have since been reported (58).

As Tab. 15 shows, there are some spectacular increases in the ovulation rate in the immunized groups. These animals appear to have lost regulatory control of their ovarian function in a way not unlike the Booroola. An extreme example of this is seen in the recent demonstration (1) that Merino lambs immunized against bovine FF from the third week of life showed advancement of puberty and up to 8 ovulations.

Present evidence would suggest, therefore, that a deficiency of ovarian inhibin may be a key to the Booroola's high ovulation rate, probably mediated by elevated concentration of FSH during the estrous cycle. The abnormalities of the Booroola follicle, including the reduced granulosa cell population reviewed earlier, are compatible with inhibin deficiency, since these cells are the principal site of inhibin synthesis (43). The advancement of puberty and increased ovulation rates of normal Merinos immunized against an inhibin-enriched fraction of bovine FF provide additional circumstantial evidence for this conclusion.

Local Ovarian Follicle Growth Inhibitor

While inhibin is, by definition, the ovarian protein that selectively regulates FSH secretion by feedback action on the hypothalamo-pituitary axis, there is increasing evidence for an intraovarian regulatory substance that may be involved in ovine prolificacy. It has recently been shown (25) that steroid-free FF contains a substance ("follicle growth-inhibiting factor") that inhibits the follicular response to 2,250 i.u. PMSG in chronically hypophysectomized ewes. These data are summarized in Tab. 16. This follicle growth inhibitor cannot be inhibin, but rather is thought to act locally within the ovarian follicle by preventing mitotic division in granulosa cells (25).

This attractive theory could explain how, in the sheep, a large or "dominant" follicle among the follicle "candidates" for ovulation could secrete a substance that locally suppresses the development of other candidates. A most important objective is to establish if the Booroola ovary is deficient in follicle growth inhibitor as well as in inhibin.

84

Tab. 15. Ovulation rates of control Merinos immunized against either adjuvant or an inhibin-enriched fraction of bovine follicular fluid (58).

Year of examination	Immunization	n	No. of ewes with ovulation rates of:						Mean ovulation rate
			1	2	3	4-8	>8	18	
1982[a]	Adjuvant (C)	147	60	84	3	0	0	0	1.61
1982	bovine FF	147	71	56	10	9	1	0	1.87
1983	C	126	75	51	0	0	0	0	1.40
1983	bovine FF[b] (X1)	63	31	21	4	5	2	0	2.05
1983	bovine FF[b] (X2)	64	28	17	4	12	2	1	2.87

[a] Same ewes studied in both years.
[b] Ewes reimmunized either once [X1 (i.e., 10 days before)] or twice [X2 (i.e., 30 and 10 days)] before joining in 1983. Vaccination details as in Ref. 59.

Perhaps the sheep has developed a 2-tiered security system to ensure that the number of eggs shed is always maintained within closely defined, low limits, this number being optimum for survival of the breed. Thus, all breeds enter the follicular phase of a new estrous cycle with an excess of follicles capable of ovulating (superovulation of any sheep with PMSG or FSH confirms this). After luteolysis in the normal sheep, when follicle growth commences, inhibin production reduces the FSH supply so that new candidates cannot begin development. At the same time, as a fail-safe precaution, the dominant follicle is producing follicle growth inhibitor which eliminates the possibility of further follicles beginning growth by granulosa cell division. This would also ensure that whatever FSH is being produced is preserved for the dominant follicle.

What enables the Booroola Merino to escape both tiers of ovarian regulation? Further studies of granulosa cell function in ewes carrying the F gene may provide the answer to this question.

Plasma Concentrations of Estrogen and Androgen and Pituitary Sensitivity to Steroid Feedback

Peripheral concentrations of estradiol and androstenedione are not significantly different in Booroola and "O" Merinos during the estrous cycle (75). Ovarian vein concentrations of estradiol are also similar in ewes with and without the F gene (54). Evidently the extra follicles that mature and ovulate in Booroola ewes do not secrete additional estrogen. It has been proposed (51) that differences in sensitivity to negative feedback by steroids may be responsible for the high ovulation rates of prolific Finnsheep. Negative feedback depression of LH by estradiol in ovariectomized Booroola and "C" Merinos has been studied (29), but the two genotypes had similar sensitivity.

Luteal Function

A curious feature of luteal function in the Booroola Merino is that plasma concentrations of progesterone on Days 9-11 of the estrous cycle do not increase with increasing ovulation rate (6), which is in sharp contrast to the pattern seen when ovulation rates are increased by PMSG (8). The basis for this is not known, although it appears to result from reduced weights of individual Booroola corpora lutea (75).

Tab. 16. Effects of Pregnant Mare Serum Gonadotrophin (2,250 i.u.) and steroid-free, ovine follicular fluid (4 ml/day) on the number of large antral follicles. [Reproduced with the permission of the Australian Academy of Science (25).]

Treatment of hypox ewes	Number of follicles per size					
	Visible on ovaries (per ewe)*			Histological exam (per ovary)*		
	<2 mm	2-4 mm	>4 mm	<0.5 mm	0.5-2.0 mm	>2.0 mm
Control	29.30	0	0	53.80	8.8	0
PMSG	10.80	4.80	15.80	53.80	16.3	7.80
PMSG + ovine FF	4.80	5.30	2.80	51.20	9.3	1.30
P (Anova)	<0.010	<0.05	<0.01	NS[†]	<0.1	<0.01

* Histological preparation reduces fresh ovarian dimension by X 0.5-0.6.
† NS = not significant.

REPRODUCTIVE BIOLOGY AND ENDOCRINOLOGY OF THE BOOROOLA RAM

Although prolificacy can only be expressed in the female, there are good reasons for trying to establish the existence of correlated variables in the male. In the case of the Booroola, studies of the ram are even more important, since they may lead to methods of direct identification of male carriers of the F gene, presently only possible by progeny test. Although progeny test procedures are exacting and expensive, they have been simplified by the recent demonstration (57) that the ovulatory response to PMSG of prepubertal ewe lambs (approximately 5 months of age) may be used to differentiate heterozygous (F+) rams from those not carrying the Booroola gene (++). Subsequent experiments in New Zealand (32) involving Booroola-cross ewe lambs of known genotype (FF, F+, ++) suggest that this technique may also distinguish between FF and F+ sires. Initial results from progeny tests of such rams in Australia (65) are in agreement with this expectation.

Testicular Growth and Production of Spermatozoa

Testis growth has been studied during the period 4-18 months of age in rams of Booroola, "T," and "O" flocks (12). When expressed relative to liveweight, there was no evidence that Booroola rams had faster testis growth. This has been confirmed in crossbred rams with and without the F gene (67). The lack of any significant difference in total daily sperm production between Booroola and "C" Merino rams (21) is consistent with the testis size results.

Endocrinology of Booroola Rams

There are no convincing data to show that adult Booroola rams have higher plasma concentrations of androgens or other steroids (15). Plasma concentrations of FSH and LH are similar in adult Booroola and "C" Merino rams (19). In fact, the concentrations of FSH in the plasma of the adult ram are the lowest seen in sheep of any physiological state. Perhaps this reflects the constant feedback suppression of pituitary FSH secretion by testicular androgens and inhibin. This leads then to the possibility that in the adult the presence of the testis masks potential differences in gonadotrophin concentrations between the Booroola and rams of other genotypes. Indirect support for this idea comes from the recent demonstration (74) that prepubertal Finnsheep rams had significantly (P < 0.01) higher plasma FSH than Suffolk rams, yet there was no difference when the rams were compared as adults. Manipulation of the light environment has also exposed some differences in plasma FSH concentrations between Finnsheep and Suffolk rams (35).

When Booroola and control Merino rams were surgically castrated and the postcastration rise of plasma FSH and LH concentrations was measured, the rate of rise in FSH and LH was similar for both genotypes in the period up to 192 hr after castration (19). While there is good evidence that pre-pubertal, female lamb progeny of prolific sires (68) and Booroola female lambs (10) have elevated plasma FSH, this is not seen in male lambs of these breeds.

What does the F gene do in male carriers of the gene? It remains a paradox that the F gene, which has such a deregulating effect on the ovulatory physiology of the ewe, has no apparent impact on the gonad of the ram or, evidently, the gonadotrophic hormones that regulate it. It is not yet possible to say if the Booroola testis is deficient in inhibin, but the absence of any differences in plasma or pituitary FSH concentrations or spermatogenesis suggests that this is unlikely. The recent purification and characterization of inhibin (72) should lead to the development of radioimmunoassays that can be used for measurement of inhibin in plasma or testicular fluids of Booroola rams. Based on present evidence, however, the effects of the F gene must be regarded as sex-limited. Until further knowledge of the mode of action of the F gene becomes available, it is not possible to speculate on how the expression of this gene may be masked in the ram.

SIGNIFICANCE OF THE BOOROOLA GENE AND ITS POSSIBLE MODE OF ACTION

It is interesting to speculate on the mode of action of the Booroola gene. The evidence presented in this chapter indicates that the primary effect of the F gene is on the ovary, where it causes more than the normal (i.e., for Merino sheep) number of antral follicles to mature and ovulate in each estrous cycle. What is the basis of this phenomenon? The most plausible theory that complies with the evidence available is that the Booroola gene causes a deficiency of the ovarian peptide hormone "inhibin" (31), resulting in reduced feedback inhibition of FSH secretion or alteration of follicle sensitivity to FSH. The data in this chapter confirm that the prepubertal Booroola ewe lamb and adult ewe have elevated plasma FSH, and this is substantiated by the higher concentrations of bioactive and immunoactive FSH in the Booroola ewe pituitary gland and by the elevated immunoactive FSH in the urine. Is FSH the key regulator of ovulation rate? Based on present knowledge it is difficult not to conclude that multiple ovulation in the sheep results from excessive FSH stimulation of the ovary. The fact that superovulation can be induced in sheep by infusion of FSH, but not LH, supports this view.

The Booroola ovarian follicle and possibly the granulosa cells within the follicle are the most likely sites of the "lesion" caused by the F gene. Such follicles begin maturation at a smaller size and with fewer granulosa cells than those of other sheep (4,54) and retain the ability to enter into the follicle selection process until later in the estrous cycle than follicles of other sheep (36). This may simply mean that the F gene causes granulosa cells to become more sensitive to FSH (54). This is compatible with earlier studies showing that follicles of prolific genotypes are more sensitive to exogenous gonadotrophin (8,11), which was recently confirmed in ewes carrying the F gene (14,46,63). This picture contrasts with that of the Romanov, whose prolificacy seems to result from having more antral follicles rather than from any differences in follicle characteristics.

Other possible explanations for the effects of the Booroola gene, such as altered FSH metabolism resulting from lysosomal enzyme deficiencies, have been discussed elsewhere (6). Definitive evidence to support these theories is not yet available.

There is no precedent for major genes affecting prolificacy of sheep. Nor indeed is there any precedent in other domestic livestock for a single gene that influences ovarian function in the way described here for the Booroola. It has been suggested that dizygotic twinning in man is controlled by a completely recessive gene (24), but this is in contrast to the Booroola $\underline{F}$ gene, whose effects at background levels of prolificacy similar to the Merino are additive for both ovulation rate and litter size (64). An interesting endocrine analogy is provided by the report (56) that women with high frequencies of dizygotic twinning have elevated FSH levels during the periovulatory period of the estrous cycle.

CONCLUSION

The Booroola Merino is a most interesting genetic model of ovine prolificacy. This review illustrates that the animal has an apparently unique genetic basis and an obscure background that may go back to the earliest introductions of sheep into Australia. The agricultural significance of the $\underline{F}$ gene is that it has a major effect on prolificacy and may be used to bring about rapid improvement in sheep reproductive efficiency. The effect of a single copy of the $\underline{F}$ gene is comparable to that achieved only by many years of within-breed selection for litter size or ovulation rate. Significant economic gains have been shown to accrue from $\underline{F}$-gene substitution in Merinos and other sheep populations.

Although the exceptional ovulatory physiology of the Booroola has provided the basis for much interesting reproductive biology, it is clear that a complete understanding of its physiological basis has yet to be achieved. Of particular interest is the fact that no evidence has been found for the expression of the $\underline{F}$ gene in the Booroola male. Further research may lead to identification of the primary product of this gene, making novel manipulations possible in sheep and other species.

Studies of the Booroola and other prolific sheep have stimulated many exchanges of genetic material that make it possible to begin comparisons in one environment of at least 2 of the prolific breeds. There are programs involving the Booroola and the Romanov in France, the Booroola and the Finnsheep in Scotland, and possibly the Booroola, Finnsheep, and Romanov in the United States. This research should lead to further advances in our understanding of the control of ovine prolificacy.

ACKNOWLEDGEMENTS

The authors acknowledge, with gratitude, the technical assistance of R.D. Nethery and Y.M. Curtis, and the typing staff of this laboratory. The work was supported by the Australian Wool Corporation and the Australian Meat Research Committee.

REFERENCES

1. Al-Obaidi, S.A.R., B.M. Bindon, T. O'Shea, M.A. Hillard, and M. Cheers (1983) Advancement of puberty in ewe lambs vaccinated with an inhibin-enriched fraction from bovine follicular fluid. Proc. Aust. Soc. Reprod. Biol. Vol. 15 (Abstr. 80).

2. Baird, D.T., and A.S. McNeilly (1981) Gonadotrophic control of follicular development and function during the oestrous cycle of the ewe. J. Reprod. Fert. Suppl. 30, pp. 119-133.

3. Baird, D.T., A.S. McNeilly, J. Wallace, and R. Webb (1984) Infusion of FSH increases ovulation rate in Welsh Mountain ewes. Vth Reinier de Graaf Symposium, 1984.

4. Baird, D.T., M.M. Ralph, R.F. Seamark, F. Amato, and B.M. Bindon (1982) Pre-ovulatory follicular activity and oestrogen secretion of high (Booroola) and low fecundity Merino ewes. Proc. Aust. Soc. Reprod. Biol. Vol. 14 (Abstr. 84).

5. Bindon, B.M. (1975) Ovulation in ewes selected for fecundity. Effect of synthetic GnRH injected on the day of oestrus. J. Reprod. Fert. 44:325-328.

6. Bindon, B.M. (1984) Reproductive biology of the Booroola Merino sheep. Aust. J. Biol. Sci. 37:163-189.

7. Bindon, B.M., M.R. Blanc, J. Pelletier, M. Terqui, and J. Thimonier (1979) Periovulatory gonadotrophin and ovarian steroid patterns in sheep of breeds with differing fecundity. J. Reprod. Fert. 55:15-25.

8. Bindon, B.M., T.S. Ch'ang, and H.N. Turner (1971) Ovarian response to gonadotrophin by Merino ewes selected for fecundity. Aust. J. Agric. Res. 22:809-820.

9. Bindon, B.M., J.K. Findlay, and L.R. Piper (1982) Preovulatory plasma FSH in high fecundity Booroola ewes. Proc. Aust. Soc. Reprod. Biol. Vol. 14 (Abstr. 84).

10. Bindon, B.M., J.K. Findlay, and L.R. Piper (1985) Plasma FSH and LH in prepubertal Booroola ewe lambs. Aust. J. Biol. Sci. 38:215-220.

11. Bindon, B.M., and Pamela R. Pennycuick (1974) Differences in ovarian sensitivity of mice selected for fecundity. J. Reprod. Fert. 36:221-224.

12. Bindon, B.M., and L.R. Piper (1976) Assessment of new and traditional techniques of selection for reproductive rate. In Sheep Breeding, G.J. Tomes, D.E. Robertson, and R.J. Lightfoot, eds. West Australian Institute of Technology, Perth, pp. 357-371.

13. Bindon, B.M., and L.R. Piper (1982) Physiological characteristics of high fecundity sheep and cattle. In Proceedings of the World Congress on Sheep and Beef Cattle Breeding, Vol. 1: Technical, R.A. Barton and W.C. Smith, eds. The Dunmore Press, Palmerston North, pp. 315-331.

14. Bindon, B.M., and L.R. Piper (1982) Physiological basis of the ovarian response to PMSG in sheep and cattle. In Embryo Transfer in Cattle, Sheep and Goats, J.N. Shelton, A.O. Trounson, N.W. Moore, and J.W. James, eds. Australian Society for Reproductive Biology, Canberra, pp. 1-5.

15. Bindon, B.M., and L.R. Piper (1986) The reproductive biology of prolific sheep breeds. In Oxford Reviews of Reproductive Biology, Vol. 8, J.R. Clarke, ed. Oxford University Press, Oxford (in press).

16. Bindon, B.M., L.R. Piper, and T.S. Ch'ang (1984) Reproductive performance of crossbred ewes derived from Booroola and Control Merinos and joined to rams of two terminal sire breeds. In Reproduction in Sheep, D.R. Lindsay and D.T. Pearce, eds. Australian Academy of Science, Canberra, pp. 243-246.

17. Bindon, B.M., L.R. Piper, M.A. Cheers, and Y.M. Curtis (1978) Uterine capacity of low and high fecundity Merinos. Proc. Aust. Soc. Reprod. Biol. Vol. 10 (Abstr. 83).

18. Bindon, B.M., L.R. Piper, M.A. Cheers, Y.M. Curtis, and R.D. Nethery (1980) Reproductive wastage in control and Booroola Merinos. Proc.

Aust. Soc. Reprod. Biol. Vol. 12 (Abstr. 67).

19. Bindon, B.M., L.R. Piper, L.J. Cummins, T. O'Shea, M.A. Hillard, J.K. Findlay, and D.M. Robertson (1985) Reproductive endocrinology of prolific sheep: Studies of the Booroola Merino. In Genetics of Reproduction in Sheep, R.B. Land and D.W. Robinson, eds. Butterworths, London, pp. 217-235.

20. Bindon, B.M., L.R. Piper, Y.M. Curtis, and R.D. Nethery (1985) Freemartins in Booroola Merino populations. Proc. Aust. Soc. Reprod. Biol. Vol. 17 (Abstr. 108).

21. Bindon, B.M., L.R. Piper, and R. Evans (1982) Reproductive biology of the Booroola Merino. In The Booroola Merino, L.R. Piper, B.M. Bindon, and R.D. Nethery, eds. CSIRO, Melbourne, pp. 21-34.

22. Bindon, B.M., L.R. Piper, and J. Thimonier (1984) Preovulatory LH characteristics and time of ovulation in the prolific Booroola Merino ewe. J. Reprod. Fert. 71:519-523.

23. Bradford, G.E. (1985) Selection for litter size. In Genetics of Reproduction in Sheep, R.B. Land and D.W. Robinson, eds. Butterworths, London, pp. 3-18.

24. Bulmer, M.G. (1970) The Biology of Twinning in Man, Clarendon Press, Oxford.

25. Cahill, L.P. (1984) Folliculogenesis and ovulation rate in sheep. In Reproduction in Sheep, D.R. Lindsay and D.T. Pearce, eds. Australian Academy of Science, Canberra, pp. 92-98.

26. Cahill, L.P., T.A. Loel, K.E. Turnbull, L.R. Piper, B.M. Bindon, and R.J. Scaramuzzi (1982) Follicle populations in strains of Merino ewes with high and low ovulation rates. Proc. Aust. Soc. Reprod. Biol. Vol. 14 (Abstr. 76).

27. Cahill, L.P., J.C. Mariana, and P. Mauleon (1979) Total ovarian follicular populations in ewes of high and low ovulation rates. J. Reprod. Fert. 55:27-36.

28. Cahill, L.P., J. Saumande, J.P. Ravault, M. Blanc, J. Thimonier, J. Mariana, and P. Mauleon (1981) Hormonal and follicular relationships in ewes of high and low ovulation rates. J. Reprod. Fert. 62:141-150.

29. Cummins, L.J. (1983) Ovarian function in the Booroola Merino. The role of ovarian inhibiting factors in the regulation of ovulation rate. Ph.D. Thesis, University of New England, Armidale, New South Wales.

30. Cummins, L.J., B.M. Bindon, T. O'Shea, and L.R. Piper (1980) Effects of ovine follicular fluid given during induced luteolysis in control and Booroola Merinos. Proc. Aust. Soc. Reprod. Biol. Vol. 12 (Abstr. 50).

31. Cummins, L.J., T. O'Shea, B.M. Bindon, V.W.K. Lee, and J.K. Findlay (1983) Ovarian inhibin content and sensitivity to inhibin in Booroola and control strain Merino ewes. J. Reprod. Fert. 67:1-7.

32. Davis, G.H., and P.D. Johnstone (1985) Ovulation response to pregnant mares serum gonadotrophin in prepubertal ewe lambs of different Booroola genotypes. Anim. Reprod. Sci. Vol. 8 (in press).

33. Davis, G.H., and R.W. Kelly (1983) Segregation of a major gene influencing ovulation rate in progeny of Booroola sheep in commercial and research flocks. Proc. N.Z. Soc. Anim. Prod. 43:197-199.

34. Davis, G.H., G.W. Montgomery, A.J. Allison, and R.W. Kelly (1982) Segregation of a major gene influencing fecundity in progeny of Booroola sheep. N.Z. J. Agric. Res. 25:525-529.

35. D'Occhio, M.J., B.D. Schanbacher, and J.E. Kinder (1984) Profiles of luteinizing hormone, follicle stimulating hormone, testosterone and prolactin in rams of diverse breeds: Effects of contrasting short (8L:16D) and long (16L:8D) photoperiods. Biol. Reprod. 30:1039-1054.

36. Driancourt, M.A., L.P. Cahill, and B.M. Bindon (1985) Ovarian follicular populations and preovulatory enlargement in Booroola and control Merino ewes. J. Reprod. Fert. 73:93-107.

37. Elsen, J.M., J. Vu Tien, J. Bouix, and G. Ricordeau (1985) Linear programming model for incorporating the Booroola gene into another breed. In Genetics of Reproduction in Sheep, R.B. Land and D.W. Robinson, eds. Butterworths, London, pp. 175-181.

38. Garran, J.C., and L. White (1985) Merinos, Myths and Macarthurs, Australian National University Press, Sydney.

39. Goodman, R.L., S.M. Pickover, and F.J. Karsch (1981) Ovarian feedback control of follicle stimulating hormone in the ewe: Evidence for selective suppression. Endocrinology 108:772-777.

40. Hanrahan, J.P. (1980) Ovulation rate as the selection criterion for litter size in sheep. Proc. Aust. Soc. Anim. Prod. 13:405-408.

41. Hanrahan, J.P., and J.B. Owen (1985) Variation and repeatability of ovulation rate in Cambridge ewes. Brit. Soc. Anim. Prod. (Abstr.).

42. Hanrahan, J.P., and J.F. Quirke (1975) Repeatability of the duration of oestrus and breed differences in the relationship between duration of oestrus and ovulation rate of sheep. J. Reprod. Fert. 45:29-36.

43. Henderson, K.M., and P. Franchimont (1983) Inhibin production by bovine ovarian tissues in vitro and its regulation by androgens. J. Reprod. Fert. 67:291-298.

44. Henderson, K.M., P. Franchimont, M.J. Lecomte-Yerna, N. Hudson, and K. Ball (1984) Increase in ovulation rate after active immunisation of sheep with inhibin partially purified from bovine follicular fluid. J. Endocrinol. 102:305-309.

45. Jansen, E.H., J. Steenbergen, F.H. de Jong, and H.J. van der Molen (1981) The use of affinity matrices in the purification of inhibin from bovine follicular fluid. Molec. Cell. Endocrinol. 21:109-117.

46. Kelly, R.W., J.L. Owens, S.F. Crosbie, K.P. McNatty, and N. Hudson (1983) Influence of Booroola Merino genotype on the responsiveness of ewes to pregnant mares' serum gonadotrophin, luteal weights and peripheral progesterone concentrations. Anim. Reprod. Sci. 6:199-207.

47. Lahlou-Kassi, A., and J.C. Mariana (1984) Ovarian follicular growth during the oestrus cycle in two breeds of ewe of different ovulation rate. J. Reprod. Fert. 72:301-310.

48. Lahlou-Kassi, A., and M. Marie (1985) Sexual and ovarian function in the D'Man ewe. In Genetics of Reproduction in Sheep, R.B. Land and D.W. Robinson, eds. Butterworths, London, pp. 245-260.

49. Lahlou-Kassi, A., D. Schams, and P. Glatzel (1984) Plasma gonadotrophin concentrations during the oestrous cycle and after ovariectomy in two breeds of sheep with high and low fecundity. J. Reprod. Fert. 70:165-173.

50. Land, R.B. (1970) A relationship between the duration of oestrus, ovulation rate and litter size of sheep. J. Reprod. Fert. 23:49-53.

51. Land, R.B., A.G. Wheeler, and W.R. Carr (1976) Seasonal variation in the oestrogen-induced LH discharge of ovariectomized Finnish Landrace and Scottish Blackface ewes. Annls. Biol. Anim. Biochim. Biophys. 16:521-528.

52. McGuirk, B.J., I.D. Killeen, L.R. Piper, B.M. Bindon, G. Caffery, and C. Langford (1982) Evaluation of the Collinsville Merino and its crosses with the Booroola and the Border Leicester in Southern N.S.W. In The Booroola Merino: Proceedings of a Workshop, L.R. Piper, B.M. Bindon, and R.D. Nethery, eds. CSIRO, Melbourne, pp. 69-75.

53. McNatty, K.P., M. Gibb, C. Dobson, and D.C. Thurley (1981) Evidence that changes in luteinizing hormone secretion regulate the growth of the pre-ovulatory follicle in the ewe. J. Endocrinol. 90:375-389.

54. McNatty, K.P., K.M. Henderson, S. Lun, D. Heath, L. Kieboom, N. Hudson, J. Fannin, K. Ball, and P. Smith (1985) Ovarian activity in Booroola x Romney ewes which have a major gene influencing their ovulation rate. J. Reprod. Fert. 73:109-120.

55. Miller, K.F., E.V. Nordheim, and O.J. Ginther (1981) Periodic fluctuations in FSH concentrations during the ovine estrous cycle. Theriogenology 16:669-679.

56. Nylander, P.P.S. (1978) Causes of high twinning frequencies in Nigeria. In Twin Research: Biology and Epidemiology, W.E. Nance et al., eds. Alan R. Liss, Inc., New York, pp. 35-43.

57. Oldham, C.M., S.J. Gray, P. Poindron, and B.M. Bindon (1984) Progeny testing for the F gene using prepubertal ewe lambs. In Reproduction in Sheep, D.R. Lindsay and D.T. Pearce, eds. Australian Academy of Science, Canberra, pp. 260-261.

58. O'Shea, T., S.A.R. Al-Obaidi, B.M. Bindon, M.A. Hillard, and J.K. Findlay (1983) Ovarian activity in ewes vaccinated with an inhibin-enriched fraction from bovine follicular fluid. In Reproduction in Sheep, D.R. Lindsay and D.T. Pearce, eds. Australian Academy of Science, Canberra, pp. 335-337.

59. O'Shea, T., L.J. Cummins, B.M. Bindon, and J.K. Findlay (1982) Increased ovulation rate in ewes vaccinated with an inhibin-enriched fraction from bovine follicular fluid. Proc. Aust. Soc. Reprod. Biol. Vol. 14 (Abstr. 85).

60. Piper, L.R., and B.M. Bindon (1982) Genetic segregation for fecundity in Booroola Merino sheep. In Proceedings of the World Congress on Sheep and Beef Cattle Breeding. Vol. 1. Technical, R.A. Barton and W.C. Smith, eds. The Dunmore Press, Palmerston North, pp. 395-400.

61. Piper, L.R., and B.M. Bindon (1982) The Booroola Merino and the performance of medium non-Peppin crosses at Armidale. In The Booroola Merino, L.R. Piper, B.M. Bindon, and R.D. Nethery, eds. CSIRO, Melbourne, pp. 9-20.

62. Piper, L.R., B.M. Bindon, M.A. Cheers, and Y.M. Curtis (1980) Relation between ovulation rate and fertilization in sheep. Proc. Aust. Soc. Reprod. Biol. Vol. 12 (Abstr. 69).

63. Piper, L.R., B.M. Bindon, Y.M. Curtis, M.A. Cheers, and R.D. Nethery (1982) Response to PMSG in Merino and Booroola Merino crosses. Proc. Aust. Soc. Reprod. Biol. Vol. 14 (Abstr. 82).

64. Piper, L.R., B.M. Bindon, and G.H. Davis (1985) The single gene inheritance of the prolificacy of the Booroola Merino. In The Genetics of Reproduction in Sheep, R.B. Land and D.W. Robinson, eds. Butterworths, London, pp. 115-125.

65. Piper, L.R. B.M. Bindon, S.K. Walker, J.R. Walkley, and D. Phillips (1985) Further observations on F gene progeny testing using the ovarian response to PMSG in pre-pubertal ewe lambs. Proc. Aust. Soc. Reprod. Biol. Vol. 17 (Abstr. 14).

66. Ponzoni, R.W., M.R. Fleet, J.R. Walkley, and S.K. Walker (1985) A note on the effect of the F gene on wool production and live weight of Booroola x South Australian Merino rams. Animal Production 40:367-369.

67. Purvis, I.W., L.R. Piper, B.M. Bindon, T.N. Edey, Y.M. Curtis, and R.D. Nethery (1983) Further studies of testis size in crossbred Booroola rams. Proc. Aust. Soc. Reprod. Biol. Vol. 15 (Abstr. 34).

68. Ricordeau, G., M.R. Blanc, and L. Bodin (1984) Teneurs plasmatiques en FSH des agneaux males et femelles issus de beliers Lacaune prolifiques et non prolifiques. Genet. Sel. Evol. 16:195-210.

69. Ricordeau, G., J. Razungles, and D. Lajous (1982) Heritability of ovulation rate and level of embryonic losses in Romanov breed. Proc. 2nd Wld. Congr. Genet. Appld. Livest. Prod. 7:591-595.

70. Ricordeau, G., L. Tchamitchian, J. Thimonier, C. Flamant, and M. Theriez (1978) First survey results obtained in France on reproductive and maternal performance in sheep with particular reference to the Romanov and crosses with it. Livest. Prod. Sci. 5:181-201.

71. Robertson, D.M., S. Ellis, L.M. Foulds, J.K. Findlay, and B.M. Bindon (1984) Pituitary gonadotrophin in Booroola and Control Merino sheep. J. Reprod. Fert. 71:189-197.

72. Robertson, D.M., L.M. Foulds, L. Leversha, F.J. Morgan, M.P.W. Hearn, H.G. Burger, R.E.H. Wettenhall, and D.M. De Kretser (1985) Isolation

of inhibin from bovine follicular fluid. Biochem. Biophys. Res. Commun. 126:1220-1226.

73. Salamonsen, L.A., H.A. Jonas, H.G. Burger, J.M. Buckmaster, W.A. Chamley, I.A. Cumming, J.K. Findlay, and J.R. Goding (1973) A heterologous radioimmunoassay for follicle stimulating hormone: Application to measurement of FSH in the ovine oestrous cycle and in several other species including man. Endocrinology 93:610-618.

74. Sandford, L.M., W.M. Palmer, and B.E. Howland (1982) Influence of age and breed on circulating LH, FSH and testosterone levels in the ram. Can. J. Anim. Sci. 62:767-776.

75. Scaramuzzi, R.J., and H.M. Radford (1983) Factors regulating ovulation rate in the ewe. J. Reprod. Fert. Suppl. 69:353-367.

76. Scott, R.S., H.G. Burger, and H. Quigg (1980) A simple and rapid in vitro bioassay for inhibin. Endocrinology 107:1536-1542.

77. Short, B.F., and H.B. Carter (1955) Bulletin No. 276, CSIRO, Melbourne, pp. 1-35.

78. Slee, J. (1963) Immunological tolerance between litter mates in sheep. Nature 200:654-656.

79. Smith, C. (1985) Utilization of major genes. In Genetics of Reproduction in Sheep, R.B. Land and D.W. Robinson, eds. Butterworths, London, pp. 151-158.

80. Smith, J.F. (1976) Selection for fertility and response to PMSG in Romney ewes. Proc. N.Z. Soc. Anim. Prod. 36:247-251.

81. Thimonier, J.T., B.M. Bindon, and L.R. Piper (1979) Ovarian response to PMSG in cattle selected for twinning. Proc. Aust. Soc. Reprod. Biol. Vol. 11 (Abstr. 5).

82. Tsonis, C.G., H. Quigg, V.W.K. Lee, L. Leversha, A.O. Trounson, and J.K. Findlay (1983) Inhibin in individual ovine follicles in relation to diameter and atresia. J. Reprod. Fert. 67:83-90.

83. Turner, Helen Newton (1978) Selection for reproduction rate in Australian Merino Sheep: Direct responses. Aust. J. Agric. Res. 29:327-350.

84. Turner, Helen Newton (1982) Origins of the CSIRO Booroola. In The Booroola Merino, L.R. Piper, B.M. Bindon, and R.D. Nethery, eds. CSIRO, Melbourne, pp. 1-7.

85. Wallace, Jacqueline M., and A.S. McNeilly (1984) Increase in ovulation rate following treatment of ewes with bovine follicular fluid (bFF) in the luteal phase of the oestrous cycle. Society for the Study of Fertility Annual Conference (Abstr. 81).

86. Webb, R., and B.G. England (1982) Identification of the ovulatory follicle in the ewe: Associated changes in follicle size, thecal and granulosa cell luteinizing hormone receptors, antral fluid steroids and circulating hormones during the preovulatory period. Endocrinology 110:873-881.

87. Webb, R., and I.K. Gauld (1985) Folliculogenesis in sheep: Control of ovulation rate. In Genetics of Reproduction in Sheep, R.B. Land and D.W. Robinson, eds. Butterworths, London, pp. 261-274.

88. Wright, R.W., K. Bondioli, J. Grammer, F. Kuzan, and A. Menino (1981) FSH or FSH plus LH superovulation in ewes following oestrus synchronization with medioxyprogesterone acetate pessaries. J. Anim. Sci. 52:115-118.

CASEIN GENES AND GENETIC ENGINEERING OF THE CASEINS

Young Kang, Rafael Jimenez-Flores, and Tom Richardson

Department of Food Science and Technology
University of California
Davis, California 95616

INTRODUCTION

The caseins are the major proteins of milk providing needed amino acids to the suckling infant. The bovine caseins have been the most thoroughly studied and serve as the principal basis for the dairy industry. The bovine caseins are comprised of 4 major polypeptide families: α_{s1}-, α_{s2}-, β-, and κ-caseins (52). They are phosphorylated to varying degrees generally containing 8, 10-13, 5, and one seryl phosphate residues per monomer, respectively. The fact that some molecules of κ-casein are also glycosylated while others do not contain carbohydrate residues gives rise to electrophoretic heterogeneity.

The distribution of polar and apolar amino acids within each casein tends to be asymmetric (52), yielding protein molecules with hydrophobic-hydrophilic polarities, giving the caseins excellent functional properties (e.g., emulsifiers). Moreover, the seryl phosphate groups of α_{s1}-, α_{s2}-, and β-caseins associate with calcium phosphate complexes in milk to form loosely ordered aggregates termed casein micelles. These aggregates are stabilized in colloidal suspension by a surface coating of κ-casein. Thus, the caseinate system of milk is thought to be comprised of proteinaceous submicellar particles which associate in the presence of calcium phosphate to form larger micelles stabilized by κ-casein on the surface (49). These structures tend to maximize the amount of calcium phosphate and protein available in a limited volume for effective nutrition of the young.

The physical stability of the colloidal casein micelle is important in defining the characteristics of dairy products. For example, association of casein micelles induced by thermal treatments on the one hand (18) or by enzymatic action on the other (12) can lead to unwanted gelation or to the formation of cheese curd, respectively. The physical interactions of the milk proteins important in dairy processing are determined essentially by their structural properties. Secondary and 3-dimensional structures of caseins (and other proteins) to a large extent are dictated by their primary sequence of amino acids (28,52). The physico-chemical and functional properties of the proteins in foods thus reflect the primary sequence of protein amino acids. It is now possible to systematically alter the primary sequence of protein amino acids using oligodeoxynucleotide site-directed mutagenesis in order to modify the functional properties of proteins (7,

61). In the long term, such basic studies on the structure-function of the bovine caseins will eventually pave the way to genetic engineering of the dairy cow's genome resulting in overproduction of selected caseins or the production of completely novel caseins by the cow.

This chapter focuses on where genetic engineering, particularly with regard to the dairy cow, can possibly be used in designing more nutritious and more functional food proteins. Eventually, the food scientist can interact with the molecular biologist and geneticist to help design more desirable food proteins. Currently available literature on bovine casein genes and rat casein genes will be surveyed and the prospects for genetic engineering of the caseins will be discussed.

STRUCTURE AND EXPRESSION OF RAT CASEIN GENES

Hormonal Regulation

The casein genes encode a family of milk proteins whose synthesis is regulated by peptide and steroid hormones (54). Genetic analysis shows that the bovine caseins occur as a gene cluster (32). Chromosomal mapping studies, using mouse-hamster somatic cell hybrids, indicate that the caseins genes are located on mouse chromosome 5 (19).

Both prolactin and hydrocortisone induce casein synthesis by stimulating the accumulation of casein messenger RNA (mRNA) (31,35). Prolactin increases both the rate of casein gene transcription and stability of casein mRNAs (45). Hydrocortisone is necessary for maximum induction of casein mRNA by prolactin (20). Progesterone is known to antagonize the stimulatory effect of prolactin in the mammary gland (31).

Complementary DNAs (cDNAs) coding for rat α-, β-, and γ-casein have been isolated and sequenced (1,24). These cDNAs have been used to measure the level of each of the casein mRNAs in total RNA extracts, both during normal mammary gland development and in organ culture in response to prolactin and hydrocortisone (23). In the developed mammary gland, the levels of the α-, β-, and γ-casein mRNA increased 10-, 26-, and 160-fold, respectively, over the levels found in virgin gland.

In organ culture, the differential effect of both hormones on the rate of accumulation of each of the mRNAs was observed. By 24 hours after the addition of prolactin to organ culture that was incubated in the presence of insulin and hydrocortisone, the concentrations of α-, β-, and γ-casein mRNAs were increased 8-, 35-, and 250-fold, respectively, over the control levels. The addition of hydrocortisone to the organ culture in the absence of prolactin increased the level of α- and β-casein mRNAs only 2- to 3-fold, while there was no change in the level of γ-casein mRNA. These data indicate that the largest increase in casein mRNA concentration in vivo and in vitro is essentially a response to prolactin.

Rat γ- and β-Casein Structural Genes

The entire γ-casein structural gene has been isolated from Charon 4A rat genomic libraries (60). R-loop and restriction enzyme mapping indicate that the γ-casein gene is 15 kilobases in length and contains a minimum of 9 small exons separated by large intervening sequences (Fig. 1). The 15-kb γ-casein gene is an unusually large and complex gene which is 17.4-fold larger than the mature γ-casein mRNA. DNA sequencing of the 5' end of the γ-casein gene showed the presence of an unusual Goldberg-Hogness sequence

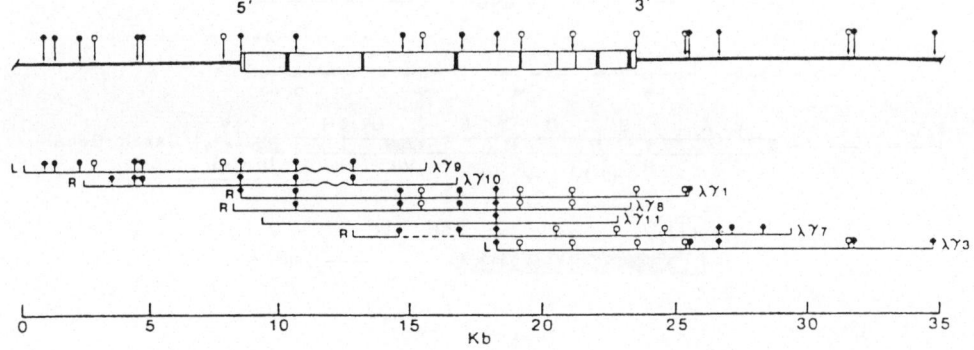

Fig. 1. Map of the rat γ-casein region. Overlapping γ-casein-specific phage clones are shown in the 5' to 3' orientation. L and R indicate the orientation of the left and right arms, respectively, of the Charon 4A vector. Solid boxes represent exons, open boxes represent intervening sequences, and the solid line represents flanking sequences. The dashed line denotes a 2.2-kb deletion in clone λγ7. The wavy line denotes a 2.3-kb insertion in clones λγ9 and λγ10. Clone λγ10 has been mapped with only EcoRI and BamHI. Clone λγ11 has only been characterized with EcoRI and MspI (data not shown). Clones λγ1 and λγ3 have true EcoRI ends; all of the other clones have artificial EcoRI linkers as ends. ♦ , EcoRI; ●, BamHI; o, SstI. The scale is shown in kb. (From Ref. 60.)

TTTAAAT.* Assignment of the Cap site was based on comparisons with the nucleotide sequence around the mRNA start site in casein mRNAs (24).

The β-casein structural gene and 5'- and 3'-flanking regions have been isolated and characterized (27). The β-casein gene is 7.2-kb long and contains 9 exons (Fig. 2). The TATA box of the β-casein gene TATATAT shows a greater homology to the Goldberg-Hogness sequence TATA$^{A}_{T}$A$^{A}_{T}$ (2) than the more divergent TATA box (TTTAAAT) of γ-casein, α-casein, and whey acidic proteins (3,27,60).

In other genes, a T, rather than an A, at the second position in the Goldberg-Hogness sequence is known to reduce its in vitro promoter efficiency (4). This difference in the TATA sequence may be in part responsible for the higher level of β-casein gene expression in comparison to other caseins (27).

Casein proteins are among the most divergent protein families studied (14). However, caseins, which are major nutritional proteins to the newborn animal, should have 3 characteristic functions: (a) secretion, (b) formation of casein micelles, and (c) phosphorylation for calcium binding. Calcium-sensitive caseins (rat α- and γ-caseins) show considerable homology at the nucleotide level between (a) the 5'-noncoding regions, (b) the regions encoding the sites of phosphorylation, and (c) the signal peptide-coding regions (24). These conserved sequences might be involved in (a) hormonal regulation of casein expression, (b) casein phosphorylation with attendant calcium binding and micelle formation, and (c) casein secretion, respectively.

---

* A, adenine; C, cytosine; G, guanine, T, thymine.

97

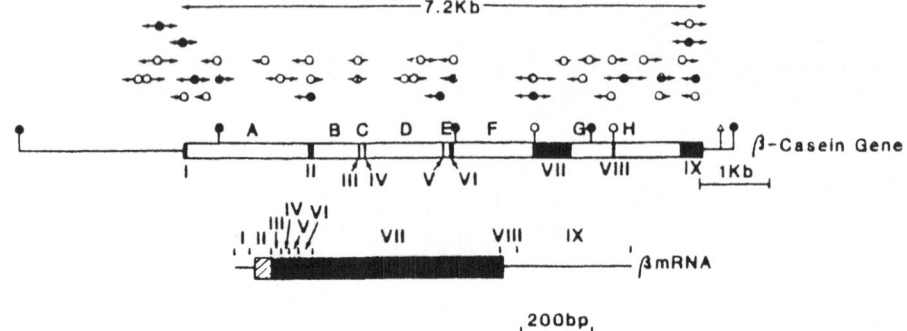

Fig. 2.   Beta-casein gene and sequencing strategy. A map of the β-casein
          gene is shown in the middle.   Closed boxes represent exons (I-IX)
          and open boxes represent introns (A-H).   Each arrow above repre-
          sents one sequencing gel. Open circles represent 3'-labeling and
          closed circles represent 5'-labeling. The region of the β-mRNA
          which each exon encodes is indicated at the bottom of the figure.
          (From Ref. 27.)

Flanking Regions

The 5'-flanking regions of the rat β- and γ-casein genes were compared
and examined for sequences with greater than 70% homology (Fig. 3) (27).
Three regions of homologous sequences were observed. In the β-casein gene,
the proximal region of homology lies between -48 and -63, the medial between -106 and -119, and the distal between -130 and -165. In the γ-casein
gene these regions are at the same positions ± 3 base pairs (Fig. 3).

In the β-casein gene, one sequence between -157 and -143 (TGTCCCCCAGA-
ATT) shows 86% homology to the sequence between -184 and -171 of the chick-
en ovalbumin gene (TGTTTACCCAGAATT) which lies within the progesterone receptor binding site defined by DNase I footprinting (4). This region was
also shown to be involved in progesterone and estrogen regulation by deletion analysis (15). It is interesting that this sequence is found within
the distal conserved region of the β-casein gene. The β-casein gene does
not have a CAAT box at -80 bp from the Cap site. The sequence CAAAT is
located near -58 bp from the Cap site and within the proximal conserved region.

The 3 conserved 5'-flanking regions of casein genes (Fig. 3) are candidates for regulatory elements (27). The similarity of the proximal
region to the CAAT sequence and of the distal region to the progesterone
receptor binding site suggests possible functions for these 2 regions.
More research is required to find out the functions of these regions.

Phosphorylation Sites

Casein kinase phosphorylates the amino-terminal serine of the sequence
Ser-X-Y where Y is either a seryl phosphate or an acidic, usually glutamic,
residue (36). Minor phosphorylation sites contain only one Glu residue,
whereas the more common major phosphorylation sites have 2 Glu residues,
Ser-Ser-Glu-Glu (27). The coding region for this series of residues, one
of the most highly conserved in casein mRNA, is as follows (27):

TCN-AGN-GAG-GAA

Ser-Ser-Glu-Glu

98

```
γ                                      CAC CACACATCTT

β                                      TTC CTGACAAGTT


        -193
γ       CCATTATCTT   ATCATGGCCT   CAATCAAACG   GTTAAGAAC  TCCCTAGAA-
                                               |||||  |   ||| ||||
β       CCTTCACCAG   CTTCTGAATT   G-CTGC-CTT   GTTAATGTC  CCCC-AGAAT
                                                 '    '  '
        -146                                   ──────────Progesterone?
γ       -TCTGTGGAA   -CAAAATCCA   GAG-AGACAA   TTTCTAATGA  TATTGCTTTC
        ||| ||||      |||||                ||  ||||||||| |   |
β       TTCTTGGGAA   AGAAAATGA   AAGAAACCAT   TTTCTAATCA  TGTGAACTTC


        -96
γ       TTAGAATTCG   AATGT-CTTT   TT-AGGTATT   TGAAACCACA  GAATTAGCAT
                                               |||||||    ||||||||||
β       TTGGAATTAA   AAGGAACTTT   TGAATATCTT   ACGAACCACA  -AATTAGCAT


        -47                    TATA                              -1
γ       ATGATCGTAG   AACCTGGTTT  AAATAGTGCG   GGAGCTACCC  ACTGCT----
                              | |  | |||
β       GTCATT-AAG   --TATGGTAT  ATATACAGTC   ACAGAGTCTG  ATAGACCATC
```

Fig. 3.   Comparison of the first 200 bp of 5'-flanking DNA of rat γ- and β-caseins. The sequences were aligned to give maximum homology. The sequence is arranged 5' to 3' with the -1 nucleotide in the lower right hand corner. The Goldberg-Hogness sequences and the 3 conserved regions are underlined. (From Ref. 27.)

The conserved major phosphorylation site for rat β-casein is not encoded by a single exon but rather is formed by an RNA splicing event involving the 2 terminal Glu residues (27). The conserved sequence of the major phosphorylation site is thus split with Ser-Ser-Glu residues encoded by the 3' region of an exon separated by an intron from the 5' end of an exon encoding for the C-terminal Glu residue. In this way, a minor phosphorylation site is converted to a major phosphorylation site by a splicing event which juxtaposes a glutamate codon with the minor phosphorylation site to form a major site.

It has been hypothesized that exon/intron junctions often encode peptides found at the surface of the native proteins (8). Since nucleotides encoding the major phosphorylation site for rat β-casein exist at an exon/-intron junction, the phosphorylation site may exist at the protein surface. This would allow access of casein kinase to surface polar residues for phosphorylation. Moreover, it would facilitate calcium binding and associated interactions of the caseins.

CLONING OF BOVINE CASEIN COMPLEMENTARY DNAs

Cloning of bovine casein cDNAs has been stimulated by the knowledge of the amino acid sequence of all the major bovine caseins (16,37) and the great abundance of casein-coding mRNA in lactating mammary gland (46).

A common methodology to isolate mRNA, used to synthesize bovine casein cDNAs, is its extraction from frozen mammary gland, separation of poly(A)-RNA by chromatography on oligo(dt)-cellulose, and sucrose gradient purification of the fraction richest in casein-coding mRNA. Generally, the cDNA fragments obtained are inserted into a plasmid (commonly pBR322) and Escherichia coli is transformed with it. However, screening procedures of the desired clones have been different in many cases. For example, radioactive (^{32}P) single-stranded cDNAs, synthesized from total poly(A)-mRNA as templates, were used as probes (59) to screen for α_{s1}-casein cDNAs. Also, direct immunological screening of cDNA products has been used to detect fusion proteins of β-lactamase:casein that may be produced in <u>E.</u> <u>coli</u> transformed with pBR322 containing cDNA inserts in the ampicillin site (25).

Research has focused mainly on bovine α-, β-, and κ-caseins and β-lactoglobulin. The following is a report on the cDNA cloning of the major bovine caseins.

Alpha$_{s1}$-Casein

The nucleotide sequence coding for this casein has been reported independently by Stewart et al. (50) and Nagao et al. (39). Both groups cloned a full-length cDNA fragment, each of 1,194 and 1,161 bases, respectively. Both fragments include the coding region for the signal peptide of the pre-α_{s1}-casein signal peptide (38), and the mature protein sequence agrees with the published α_{s1}-casein (16). A restriction enzyme map for α_{s1}-casein cDNA is shown in Fig. 4.

Comparison of the nucleotide sequences between rat α- and bovine α_{s1}-caseins showed 87% homology according to Nagao et al. (39). The constraints for genetic divergence operating upon the α_{s1}-casein amino acid sequence, which include retention of phosphorylation sites, secondary structure, protein interactions, and amino acid composition, can be approximately equated with the constraints operating upon the silent sites. The silent sites include mRNA secondary structure, codon usage, and maintenance of splice junctions (50).

Beta-Casein

Willis et al. (59) and Ivanov et al. (25) cloned different partial sequences of β-casein. Willis et al. (59) found a segment that represents

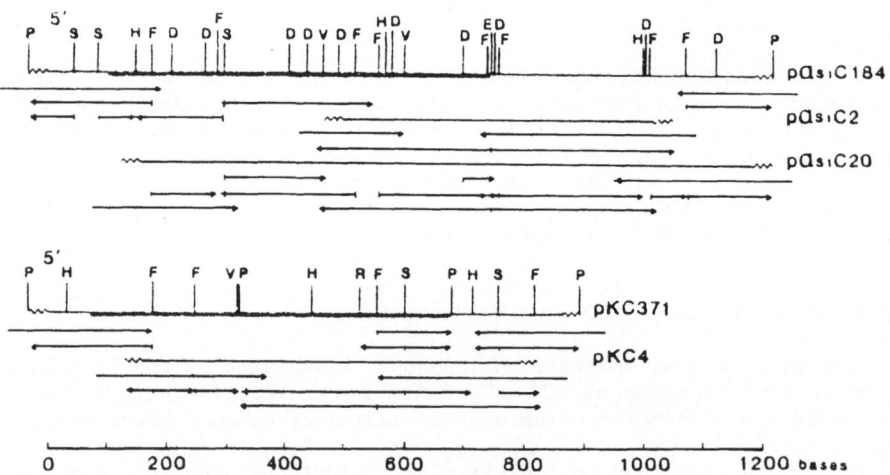

Fig. 4.   Restriction maps and sequencing strategies of bovine casein cDNA clones. The solid lines represent the cDNA inserts, the wavy lines represent the GC tails, and the arrows represent the extent and direction of sequence obtained from the restriction sites indicated by the short vertical lines. Those arrows that have their ends beyond the GC tails represent sequencing runs made from external <u>HpaII</u> sites. The thickened lines correspond to the protein coding regions. Key: D, <u>DdeI</u>; E, <u>EcoRI</u>; F, <u>HinfI</u>; H, <u>HaeIII</u>; P, <u>PstI</u>; R, <u>RsaI</u>; S, <u>Sau3A</u>; V, <u>PvuII</u>.   (From Ref. 50.)

50% of the mRNA that encodes for β-casein. In this fragment the carboxy terminus is included and is 600 bp long. The coding region of the fragment starts at amino acid 151 of β-casein.

Ivanov et al. (25) cloned a fragment 330 bp long, and the coding region starts at amino acid 53 of the bovine β-casein. Partial restriction enzyme maps are reported in both cases.

Kappa Casein

A restriction enzyme map (Fig. 4) and the full-length sequence of κ-casein published by Stewart et al. (50) show the complete coding region for mature κ-casein and the leader signal peptide sequence, which differs from the published ovine sequence by one amino acid. The cDNA fragment is 869 bp long and has a poly(A) tail 20 bases long. Bovine κ-casein mRNA displays little homology to other caseins in the 5'-noncoding region and in the signal peptide, which is highly conserved in α- and β-caseins. Furthermore, the 3'-noncoding region is shorter than that of other casein mRNAs. These observations support the contention that κ-casein proceeds from a different ancestral gene than the other caseins. However, heritability studies show that the relative amounts of the different caseins remain constant (29).

Beta-Lactoglobulin

Beta-lactoglobulin is not a casein. It is a globular protein soluble in the aqueous portion of the milk; whereas, the caseins are not globular in nature and are associated primarily in the aforementioned micellar structures. Willis et al. (59) published the restriction map of a cDNA fragment of β-lactoglobulin, which contains the carboxy-terminus code and represents 60% of the length of the mRNA. The 3-dimensional structure of β-lactoglobulin has recently been published (48).

CLONING OF CASEIN COMPLEMENTARY DNAs OF OTHER SPECIES

Mouse

Mehta et al. (35) isolated a 400-bp cDNA fragment coding for mouse β-casein. Hennighausen and Sippel (22) characterized the mouse milk protein mRNA and obtained cDNAs for α-, β-, γ-, and ε-casein, and also for a novel acidic whey protein. They also described a partial homology between mouse and rat caseins. Mouse ε-casein possesses an amino acid sequence homologous to bovine $α_{s2}$-casein in the N-terminal region of the mature protein (22).

Guinea Pig

The cDNAs of A, B, and C caseins (6) and α-lactalbumin of guinea pig were isolated and inserted in the plasmid pAT153 by Craig et al. (5). Fragments coding for casein A and α-lactalbumin appear to be full-length with respect to the mRNA from which they were synthesized. The casein B cDNA insert represents only 80%, whereas the casein C fragment contains only 55-60% of the mRNA sequence.

Rabbit

Caseins of rabbit milk were separated and identified by Dayal et al. (13). Comparison of their proteinase digest and electrophoretic mobilities showed that the rabbit casein with α-mobility corresponded to bovine $α_{s1}$; the casein with γ-mobility exhibits properties of cow's κ-casein, including chymosin susceptibility. However, the rabbit casein with β-mobility has little homology to the bovine β-casein. Therefore, it was termed casein X (13).

Suard et al. (51) reported the restriction map of cDNA fragments coding for α- and γ-caseins and α-lactalbumin; their lengths were 100, 60, and 65%, respectively, according to their mRNA lengths.

Sheep

Studies on sheep's casein have been based on in vitro translation of ovine mammary mRNAs (38). This system has allowed the study of the amino acid sequence of the signal peptide of ovine caseins, and its comparison to the peptides of other species' caseins.

Signal Peptides

The signal peptide sequences of ovine, rat, and mouse caseins (22,27, 36) are shown in Fig. 5. Note the intraspecies homology among ovine α_{s1}-, α_{s2}-, and β-caseins. However, the signal peptide for ovine κ-casein diverges substantially from the former caseins, perhaps supporting the idea of the origin of κ-casein from a different ancestral gene. Also, there is substantial interspecies homology in the signal peptides between ovine, rat β-, and mouse ε-caseins. These homologies apparently reflect the functional involvement of the signal peptides in the secretory process.

PROSPECTS

The following is an attempt to focus on where genetic engineering, particularly with regard to the dairy cow, may apply to the design of more nutritious and more functional food proteins. It is meant to anticipate what genetic engineering of food proteins may mean to the food scientist and how the food scientist can interact with the molecular biologist and geneticist to help design more desirable proteins.

Transgenic Animals

A restrictive definition of transgenic animals relates to those animals that have integrated foreign DNA into their germline as a consequence of experimental introduction of DNA, usually by microinjecting recombinant DNA directly into pronuclei of fertilized eggs (42).

Most of the work on transgenic animals has been with mice. In the case of mice, many of the injected genes are expressed and are carried in

```
                                                                        -1  +1
Ovine αs1-Casein  Met-Lys-Leu-Leu-Ile-Leu-Thr-Cys-Leu-Val-Ala-Val-Ala-Leu-Ala--Arg

Ovine αs2-Casein  Met-Lys-Val-Leu-Met-Lys-Ala-Cys-Leu-Val-Ala-Val-Ala-Leu-Ala--Lys

Ovine β-Casein    Met-Lys-Val-Leu-Ile-Leu-Ala-Cys-Leu-Val-Ala-Leu-Ala-Leu-Ala--Arg

                                    Ile-Leu-
Ovine κ-Casein    Met-Arg-Lys-Ser-Phe-Phe-Leu-Val-Val-Thr-Ile-Leu-Ala-Leu-Thr-Leu-Pro-Phe-Leu-Ile-Ala--Gln

Rat β-Casein      Met-Lys-Val-Phe-Ile-Leu-Ala-Cys-Leu-Val-Ala-Ala-Leu-Ala-Arg

Mouse ε-Casein    Met-Lys-Phe-Ile-Ile-Leu-Thr-Cys-Leu-Leu-Ala-Val-Ala-Leu-Ala--Lys
```

Fig. 5. Amino acid homology among signal peptides for caseins from various species. Note intra- and interspecies conservation of amino acid sequences. (From Ref. 22, 27, and 36.)

the germ cells as well as somatic cells. However, the level of expression from one transgenic mouse to another varies and usually does not correlate with gene copy number (42). Although microinjected genes can be expressed, techniques must be developed for targeting integration of the DNA into the genome at predetermined chromosomal sites.

As with many genes, the casein genes are normally expressed in a tissue-specific manner, possibly because cis-acting DNA elements are involved in developmental programming of gene expression. Transgenic animals offer a stringent means for assessing whether a particular DNA sequence is sufficient for appropriate developmental regulation since that sequence can be tested in every possible cell type in the animal throughout development. It has been shown that several intact genes are expressed in a tissue-specific manner in transgenic mice. For example, expression of the rat elastase gene in the acinar cells of the pancreas is 10,000-fold greater than in any other tissue examined (41,53).

The exact localization of cis-acting, regulatory DNA sequences necessary for tissue-specific gene expression has only just begun for genes introduced into mice. Thus, virtually nothing is known about tissue-specific expression of foreign genes in other species, particularly domestic farm animals. However, research is beginning on the production of transgenic rabbits, sheep, and pigs by microinjection of foreign DNA (21). The fusion gene, mouse metallothionein-human growth hormone (MT-hGH), was microinjected into the pronuclei or nuclei of eggs from superovulated rabbits, sheep, and pigs. The gene was retained or integrated in the rabbit (12.8%) and the pig (11.0%) but only to a low extent in the sheep (1.3%) compared to about 27% for mice. Four of the 16 rabbits and 11 of the 20 pigs elaborated detectable MT-hGH mRNA, but the levels of hGH expressed were at ng/ml levels in the serum. However, as more is learned about the hormonal, developmental, and regulatory elements involved in controlling levels of protein expression in animals, it should become possible to enhance the quantities of protein elaborated.

In a sense the bovine-casein system is not truly transgenic because initially, at least, it is likely that homologous structural genes coding for the bovine caseins (and not foreign DNA) will probably be microinjected into bovine ova to cause overproduction of selected homologous caseins. The bovine casein genes are tightly linked and behave somewhat as a super-locus. The heritabilities of the milk proteins (Tab. 1) indicate there is little likelihood that it will be possible to alter the relative concentrations of the caseins using conventional breeding techniques (29). Thus, a change in the proportion of one casein will probably require microinjection

Tab. 1. Heritabilities of relative percentages of bovine milk protein fractions (29).

Protein fraction	Heritability
α_s-Casein	0.02
β-Casein	0.03
κ-Casein	0.05
β-Lactoglobulin	0.24
α-Lactalbumin	0.14
Bovine serum albumin	0.15
Immunoglobulins	0.02

of the structural gene and associated control elements into bovine ova. From a food technological point of view, there may be good reasons for wanting to do this.

For example, the overproduction of κ-casein in bovine milk by microinjection of the κ-casein structural gene and flanking control regions into bovine ova may enhance the stability of milk to certain thermal treatments. In general, a high molar ratio of κ-casein to β-lactoglobulin might lead to a milk product that would not be subject to unwanted gelation problems in a thermally processed fluid milk (18). In essence, the physical stability of many concentrated and unconcentrated milk products as a function of processing is related to the molar ratios of certain milk proteins. Since these molar ratios seem to be under relatively tight genetic and hormonal control, microinjection of additional genes coding for selected milk proteins could lead to a more desirable ratio of milk proteins to maintain the physical stability and shelf-life of certain dairy products.

Insertional Mutagenesis

Evidently the integration of microinjected, foreign DNA into the host's genome is currently more or less a random process. The insertion of foreign DNA into endogenous genes can destroy the reading frame and result in insertional mutagenesis (42). In transgenic mice foreign DNA has been shown to disrupt genes essential for embryonic development. If insertional mutagenesis could be controlled so that selected genes could be mutated, it might prove useful in manipulating the composition of milk in desirable ways. For example, a major milk whey protein, α-lactalbumin, is involved in the biosynthesis of lactose or milk sugar (26,56), which exists at a concentration in milk of ∿5% weight/volume. Lactose is responsible for a number of defects in dairy products because of its limited solubility. Lactose crystals, for example, lead to a gritty or sandiness defect in ice cream. Moreover, many people in the world cannot hydrolyze lactose in their gut because of a deficiency of the enzyme β-galactosidase. This leads to a condition called "lactose intolerance" wherein the lactose that reaches the colon is fermented, leading to an uncomfortable and explosive diarrhea (56). Insertional mutation of the α-lactalbumin gene would inhibit the biosynthesis of lactose, leading to an accumulation of its precursors galactose and glucose. Since these 2 sugars would have twice the osmotic pressure as the resultant lactose on a molar basis, perhaps half as much of the monosaccharides would be needed to control the osmotic pressure of milk (isotonic with blood), thereby decreasing the total sugar content of milk. This would obviate technological problems as well as lactose intolerance.

Site-Directed Mutagenesis

In analogy with the microinjection of foreign DNA into the ova of animals to derive transgenic animals, it may be possible to microinject homologous DNA which has been engineered to yield novel proteins with desirable functional properties in a food. Functionalities of food proteins are related to the physico-chemical properties of the proteins that define their behavior in the manufacture of food products. It is the superior functionalities of caseins that dictate their use in cheesemaking and in egg whites for whipping meringues. Thus, functional behavior of proteins can be crucial in the development of novel and improved food products.

The secondary structures of the caseins are largely defined by the primary amino acid sequences along with the hydrophobic-hydrophilic balance of the proteins and the sites of phosphorylation and glycosylation (10,52). Consequently, the properties of the caseins (and other proteins) and their aggregates are dictated to a great extent by the primary sequence of amino acids (28). Since it is possible to alter the primary amino acid sequence of a protein using genetic engineering techniques (7,11,61), it should be possible to systematically change the physical properties of resultant products to yield proteins with more desirable functional properties. Such alterations in the primary sequences in proteins have come to be known as protein engineering (33,55) and may eventually prove useful in designing novel enzymes and food proteins.

In the shorter term, it is now feasible to produce engineered enzymes and proteins in microorganisms (7,11,58). Systematic alterations in the primary sequences of proteins allow detailed basic studies of the structure-function relationships of the isolated proteins. In the case of the caseins, this information will provide background needed to eventually engineer the dairy cow or its mammary gland (43) to produce milk proteins with more desirable properties. Although this latter possibility is a long-term prospect (possibly 15-20 years), research on the genetic engineering of bovine and other caseins is progressing and serves as the basis for the following discussion.

Molecular biologists and synthetic oligonucleotide chemists have provided us with the tools for systematic genetic manipulation of the primary structures of food proteins. Using recombinant DNA techniques, molecular biologists have modified microorganisms such as E. coli and Saccharomyces cerevisiae so that they will synthesize animal proteins in vivo. In general, ∿5% (17,44) of the cellular proteins are the cloned protein. These levels of biosynthesis can provide sufficient quantities of the desired protein for research on protein functionality.

Techniques are now available to systematically change the primary structure of a protein and to correlate these changes with alterations in protein functionality. It is possible to prepare semisynthetic genes by substituting a synthetic oligodeoxynucleotide segment containing desired changes in its nucleotide sequence into the total DNA gene coding for the cloned protein. The change in the DNA sequence dictates changes in the primary sequence of amino acids in the protein. Using this technique of oligodeoxynucleotide site-directed mutagenesis, one can alter a single amino acid in the primary sequence of a protein. Thus, it becomes possible to engineer proteins and enzymes to modify their behavior in food systems.

Essentially, there are 6 steps (11) involved in the site-directed mutagenesis of a protein (Tab. 2). Limitations of space prevent a detailed discussion of the first 4 steps. The following remarks will largely be restricted to steps 5 and 6, the actual events for protein modifications. To start with, one should know as much as possible about the protein in question, including its 3-dimensional structure if available.

Principles for the cloning of relevant genes into a desirable vector and the expression of the cloned gene to produce proteins in an appropriate

Tab. 2. Six steps to restructure proteins (11).

1. Clone relevant gene in appropriate vector.
2. Gain expression of cloned gene.
3. Determine DNA sequence of gene.
4. Chemical synthesis of an oligodeoxynucleotide.
5. Oligodeoxynucleotide-directed in vitro mutagenesis.
6. Identification of mutant colonies.

organism have been reviewed extensively (40,57). Suffice it to indicate that the sequence of foreign DNA spliced into the vector acting as a vehicle codes for the primary sequence of the protein as mediated by the intervening complementary mRNA and transfer RNA (tRNA). These latter nucleic acids act upon the instructions of the parent DNA which dictates the primary sequence of the resultant protein. Alterations (mutations) of the nucleotide bases in the parent DNA are reflected in a change in the primary sequence of the protein. To achieve production of the foreign protein in the host cell, specific DNA sequences that control expression or biosynthesis of the foreign protein must also be inserted into the cloning vector (plasmid) preceding the DNA coding for the proteins. Once this expression vector (containing the added control elements) is inserted into competent microbial cells such as E. coli, Bacillus subtilis, or S. cerevisiae, the stage is set for the cell to start producing the foreign protein (step 2). The DNA sequences inserted into various vectors can be removed or replaced at will with the aid of commercially available enzymes. It is now possible to accurately sequence the bases in replicated cloned DNA removed from host cells using the method of Sanger et al. (47) or Maxam and Gilbert (34) (step 3). The sequence of the DNA can be used to verify that it indeed codes for the primary sequence of the protein in question. Moreover, once the sequence of the DNA insert is known, it can be systematically altered to change the primary sequence of the protein using site-directed oligodeoxynucleotide mutagenesis. This technique relies on the synthesis of an oligodeoxynucleotide fragment of the DNA coding stand, but containing 1 or 2 base substitutions resulting in desired mutations. There are several strategies available for manipulating the primary sequence of proteins using site-directed mutagenesis (7,11,57,61). In one strategy shown in Fig. 6, a plasmid vector is nicked with a restriction enzyme. Exonuclease III is used to digest away a portion of the coding strand at the 3' end. A synthetic oligonucleotide, 16 to 19 nucleotides in length, is made suffi-

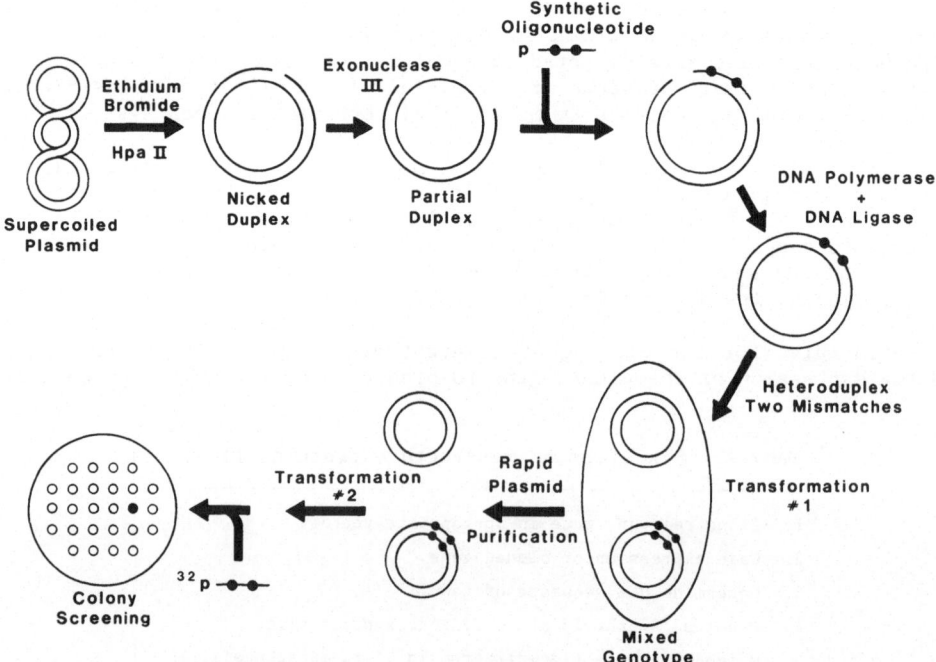

Fig. 6. A strategy for oligodeoxynucleotide site-directed mutagenesis of a protein involving a plasmid vector. (From Ref. 11.)

ciently complementary to the template so as to hybridize with the appropriate sequence. A number of techniques are available for oligodeoxynucleotide synthesis (57), and synthetic oligodeoxynucleotides can be obtained from custom suppliers. Because DNA polymerase enzymes require a double-strand segment for initiation of DNA replication, the synthetic oligodeoxynucleotide primes DNA synthesis and is, itself, incorporated into the resulting heteroduplex molecule. Upon transformation of E. coli, semiconservative, in vivo replication of this heteroduplex gives rise to homoduplexes whose sequences are either that of the original wild-type DNA or that of the synthetic oligodeoxynucleotide containing desired mutations. Colonies are then screened using the synthetic oligodeoxynucleotide primer labeled with ^{32}P as a hybridization probe to allow easy detection of the desired mutant. With this method it is possible to identify mutants containing single-base mismatches; however, differences of 2 bases between mutant and wild-type (as shown in Fig. 6) greatly increase the ability to identify mutants unambiguously. Thus, it is possible to make single amino acid substitutions in the primary sequence of proteins.

Oligodeoxynucleotide-directed mutagenesis opens up numerous possibilities for the genetic engineering of proteins and enzymes (33,55). We are now entering the era of genetic engineering of proteins and enzymes to tailor them to fit our requirements.

Potential for Engineering Caseins

In the case of casein cDNAs, for all species studied thus far, none of them have been specifically inserted into expression vectors to produce caseins for further examination. Initially it will be possible for the primary sequences of the caseins to be produced by microorganisms once the appropriate cDNA in an expression vector has been incorporated into a host cell such as E. coli. One can then obtain sufficient protein to examine the effects of structural changes, elicited using the aforementioned oligodeoxynucleotide site-directed mutagenesis.

In the longer term, reliance must be placed on the possibility of stable incorporation of casein structural genes into the bovine genome followed by appropriate development and expression of the gene. For example, it may be possible to inject the structural gene along with its controlling elements into the bovine embryo and have the gene stably integrated into the bovine genome for expression in the adult under appropriate circumstances (21,30,42,57). Once the gene is stably integrated into the genomic DNA, it is possible that it will be transmitted to the progeny via the germ cells (42,57). However, much research remains to be done before this potential becomes a reality.

There are numerous modifications possible in altering the caseins to be more desirable from a functional point of view. Based on current knowledge of the caseinate system in milk, it is possible to propose changes in casein structure which may eventually prove useful. Production of modified caseins superimposed on normal casein biosynthesis may alter behavior of the system to physical, chemical, or enzymic treatments, yielding novel or improved products.

For example, the ripening of cheese involves the proteolytic breakdown of the protein matrix to yield a more desirable texture. Conversion of α_{s1}- to α_{s1}-I-casein and an amino terminal peptide by cleavage of a Phe_{23}-Phe_{24} or Phe_{24}-Val_{25} bond by residual clotting enzyme (chymosin or other acid protease) is thought to be one of the primary changes in the caseins during maturation of cheddar cheese (9). It might be possible to introduce additional bonds with enhanced chymosin sensitivity to accelerate the rate of textural development in cheese. Conversion of Ile_{71}-Val_{72} to Phe_{71}-

Phe$_{72}$ (by changing 2 nucleotide bases) in α_{s1}-casein would generate an additional Phe$_{71}$-Phe$_{72}$ bond in a region of α_{s1}-casein disordered by a series of seryl phosphates, other charged amino acid residues, and an adjacent prolyl residue (52). This bond should be readily available for cleavage by acid proteases with a propensity for cleaving bonds adjacent to aromatic amino acids. Cleavage of a bond more toward the middle of the α_{s1}-casein (as at position 71) should maximize the rheological effects of proteolysis in promoting faster ripening of the cheese. More rapid textural development in cheese would be highly desirable from an economic, storage point of view.

The foregoing is a rather speculative attempt to introduce the reader to the potential that genetic engineering of the bovine caseins holds forth upon successful application to food science. Much additional research is necessary before successful genetic manipulation of plants and animals to tailor-make more functional proteins becomes a reality. As more is learned about the molecular basis for protein functionality in foods, it becomes increasingly possible to design proteins for particular uses. Consequently, the future holds the possibility of the food scientist working with the molecular biologist and geneticist to design food proteins with more desired functionalities.

ACKNOWLEDGEMENT

This review was made possible by support from the College of Agricultural and Environmental Sciences, University of California, Davis, California.

REFERENCES

1. Blackburn, D.E., A.A. Hobbs, and J.M. Rosen (1982) Rat β-casein DNA: Sequence analysis and evolutionary comparisons. Nucl. Acids Res. 10:2295-2307.
2. Breathnach, R., and P. Chambon (1981) Organization and expression of eukaryotic split genes coding for proteins. Ann. Rev. Biochem. 50: 349-383.
3. Campbell, S.M., and J.M. Rosen (1984) Comparison of the whey acidic protein genes of the rat and mouse. Nucl. Acids Res. 12:8685-8697.
4. Compton, J.G., W.T. Schrader, and B.W. O'Malley (1983) DNA sequence preference of the progesterone receptor. Proc. Natl. Acad. Sci., USA 80:16-20.
5. Craig, R.K., L. Hall, D. Parker, and P. Campbell (1981) The construction, identification and partial characterization of plasmids containing guinea-pig milk protein complementary DNA sequences. Biochem. J. 194:989-998.
6. Craig, R.K., D. McIlreavy, and R.L. Hall (1978) Separation and partial characterization of guinea-pig caseins. Biochem. J. 173:633-641.
7. Craik, C.S. (1985) Use of oligonucleotides for site-specific mutagenesis. BioTechniques 3:12-16.
8. Craik, C.S., W.J. Rutler, and R. Fletlerick (1983) Splice junctions: Association with variation in protein structure. Science 220:1125-1129.
9. Creamer, L.K., and N.F. Olson (1982) Rheological evaluation of cheddar cheese. J. Food Sci. 57:631-636.
10. Creamer, L.K., T. Richardson, and D.A.D. Parry (1981) Secondary structure of bovine α_{s1}- and β-casein in solution. Arch. Biochem. Biophys. 211:689-696.
11. Dalbadie-McFarland, G.D., L.W. Cohen, A.D. Riggs, C. Morin, K. Itakura, and J.H. Richards (1982) Oligonnucleotide-directed mutagenesis

as a general and powerful method for studies of protein function. Proc. Natl. Acad. Sci., USA 79:6409-6413.

12. Dalgleish, D.G. (1982) The enzymatic coagulation of milk. In Developments in Dairy Chemistry, Vol. 1, P.F. Fox, ed. Applied Sci. Publ., New York, pp. 157-188.

13. Dayal, R., J. Huslmann, Y.M.L. Suard, and J.-P. Kraebenbuhl (1982) Chemical and immunochemical characterization of caseins and the major whey proteins of rabbit milk. Biochem. J. 201:71-79.

14. Dayhoff, M.O. (1976) Miscellaneous proteins. In Atlas of Protein Sequence and Structure, Vol. 5 (Suppl. 2), M.O. Dayhoff, ed. National Biomedical Research Foundation, Bethesda, Maryland, pp. 261-263.

15. Dean, D.C., R. Gope, B.J. Knoll, M.E. Riser, and B.W. O'Malley (1984) A similar 5'-flanking region is responsible for estrogen and progesterone induction of ovalbumin gene expression. J. Biol. Chem. 259: 9967-9970.

16. Eigel, W.N., J.E. Butler, C.A. Ernstrom, H.M. Farrell, Jr., V.R. Harwalkar, R. Jenness, and R. McL. Whitney (1984) Nomenclature of proteins of cow's milk. Fifth revision. J. Dairy Sci. 67:1599.

17. Emtage, J.S., S. Angal, M.T. Doel, T.J.R. Harris, B. Jenkins, G. Lalley, and P.A. Lome (1983) Synthesis of calf prochymosin (prorennin) in Escherichia coli. Proc. Natl. Acad. Sci., USA 80:3671-3676.

18. Fox, P.F. (1982) Heat-induced coagulation of milk. In Developments in Dairy Chemistry, Vol. 1, P.F. Fox, ed. Applied Sci. Publ., New York, pp. 189-228.

19. Gupta, P., J.M. Rosen, P. D'Eustachio, and F.M. Ruddle (1982) Localization of the casein gene family to a single mouse chromosome. J. Cell. Biol. 93:199-204.

20. Guyette, W.A., R.J. Matusik, and J.M. Rosen (1979) Prolactin-mediated transcriptional and posttranscriptional control of casein gene expression. Cell 17:1013-1023.

21. Hammer, R.E., V.G. Pursel, C.E. Rexroad, Jr., R.J. Wall, D.J. Bolt, K.M. Ebert, R.D. Palmiter, and R.L. Brinster (1985) Production of transgenic rabbits, sheep and pigs by microinjection. Nature 315:680-683.

22. Hennighausen, L.G., and A.E. Sippel (1982) Characterization and cloning of the mRNAs specific for the lactating mouse mammary gland. FEBS 125:131-141.

23. Hobbs, A.A., D.A. Richards, D.J. Kessler, and J.M. Rosen (1982) Complex hormonal regulation of rat casein gene expression. J. Biol. Chem. 257:3598-3605.

24. Hobbs, A.A., and J.M. Rosen (1982) Sequence of rat α- and γ-casein in mRNA: Evolutionary comparison of the calcium-dependent rat casein multigene family. Nucl. Acids Res. 10:8079-8098.

25. Ivanov, V.N., D.R. Kersholite, A.A. Bayer, A.A. Akhondora, G.E. Solimova, E.S. Judin-Kora, and S.I. Gorodetsky (1984) Identification of bacterial clones encoding bovine caseins by direct immunological screening of the cDNA library. Gene 32:381-388.

26. Jenness, R. (1982) Interspecies comparisons of milk proteins. In Developments in Dairy Chemistry, P.F. Fox, ed. Applied Sci. Publ., New York, pp. 87-114.

27. Jones, W.K., L.-Y. Yu-Lee, S.M. Clift, T.L. Brown, and J.M. Rosen (1985) The rat casein multigene family. III. Fine structure and evolution of the β-casein gene. J. Biol. Chem. 260:7042-7050.

28. Kinsella, J.E. (1982) Relationships between structure and functional properties of food proteins. In Food Proteins, P.F. Fox and J.J. Condon, eds. Applied Sci. Publ., New York, pp. 51-104.

29. Kroeker, E.M., K.F. Ng-Kwai-Hang, J.F. Hayes, and J.E. Moxeley (1985) Heritabilities of relative percentages of major bovine casein and serum proteins in test-day milk samples. J. Dairy Sci. 68:1346-1348.

30. Kurcherlapati, R., and A.I. Skoultchi (1984) Introduction of purified genes into mammalian cells. CRC Crit. Rev. Biochem. 16:349-382.

31. Matusik, R.J., and J.M. Rosen (1978) Prolactin induction of casein mRNA in organ culture. J. Biol. Chem. 253:2343-2347.

32. Matyukov, V.S., and A.P. Urmysher (1980) Linkage of α_{s1}-, β- and κ-casein loci in cattle. Genetika 16:884-886.

33. Maugh, II, T.H. (1984) Need a catalyst? Design an enzyme. Science 223:269-271.

34. Maxam, A.M., and W. Gilbert (1977) A new method of sequencing DNA. Proc. Natl. Acad. Sci., USA 74:560-567.

35. Mehta, N.M., M.R. El-Gewely, J. Joshi, R.B. Helling, and M.R. Banerjee (1981) Cloning of various β-casein gene sequences. Gene 15:285-288.

36. Mephan, T.B., P. Gage, and J.C. Mercier (1982) Biosynthesis of milk proteins. In Developments in Dairy Chemistry, Vol. 1, P.F. Fox, ed. Applied Sci. Publ., New York, pp. 115-156.

37. Mercier, J.-C., and P. Gaye (1982) Early events in secretion of main milk proteins: Occurrence of precursors. J. Dairy Sci. 65:299.

38. Mercier, J.-C., G. Haze, P. Gaye, and D. Hue (1978) Amino terminal sequence of the precursor of ovine β-lactoglobulin. Biochem. and Biophys. Res. Comm. 82(4):1236-1245.

39. Nagao, M., M. Maki, R. Sasaki, and H. Chiba (1984) Isolation and sequence analysis of bovine α_{s1}-casein cDNA clone. Agric. Biol. Chem. 48(6):1663-1667.

40. Old, R.W., and S.B. Primrose (1981) Principles of Gene Manipulation: An Introduction to Genetic Engineering, 2nd ed., University of California Press, Berkeley, California, 214 pp.

41. Ornitz, D.M., R.D. Palmiter, R.E. Hammer, R.L. Brinster, G.H. Swift, and R.J. MacDonald (1985) Specific expression of an elastase-human growth hormone fusion gene in pancreatic acinar cells of transgenic mice. Nature 313:600-602.

42. Palmiter, R.D., and R.L. Brinster (1985) Transgenic mice. Cell 41: 343-345.

43. Patton, S., U. Welsch, and S. Singh (1984) Genetic engineering of the mammary gland by intramammary infusion technique. J. Dairy Sci. 67: 1323-1328.

44. Queen, C. (1983) A vector that uses phage signals for efficient synthesis of proteins in Escherichia coli. J. Mol. Appl. Gen. 2:1-10.

45. Rosen, J.M., R.J. Matusik, D.A. Richards, P. Gupta, and J.R. Rodgers (1980) Multihormonal regulation of casein gene expression at the transcriptional and posttranscriptional levels in the mammary gland. Recent Prog. Horm. Res. 36:157-193.

46. Rosen, J.M., S.L.C. Woo, and J.P. Comstock (1975) Regulation of casein messenger RNA during the development of the rat mammary gland. Biochemistry 14(13):2895-2903.

47. Sanger, F., S. Nicklen, and A.R. Coulson (1977) DNA sequencing with chain terminating inhibitors. Proc. Natl. Acad. Sci., USA 74:5463-5468.

48. Sawyer, L., M.Z. Papiz, C.T. North, and E.E. Eliopoulos (1985) Structure and function of bovine β-lactoglobulin. Biochem. Soc. Transac. 13:265-266.

49. Schmidt, D.G. (1982) Association of caseins and casein micelle structure. In Developments in Dairy Chemistry, Vol. 1, P.F. Fox, ed. Applied Sci. Publ., New York, pp. 61-86.

50. Stewart, A.F., I.M. Willis, and A.G. MacKinlay (1984) Nucleotide sequences of bovine α_{s1}- and κ-casein cDNAs. Nucl. Acids Res. 12(9): 3895-3907.

51. Suard, Y.M.L., M. Tosi, and J.P. Kraehenbuhl (1982) Characterization of the translation products of the major mRNA species from rabbit lactating mammary glands and construction of bacterial recombinants containing casein and α-lactalbumin complementary DNA. Biochem. J. 201:81-90.

52. Swaisgood, H.E. (1982) Chemistry of milk protein. In Developments in

Dairy Chemistry, Vol. 1, P.F. Fox, ed. Applied Sci. Publ., New York, pp. 1-60.

53. Swift, G.H., R.E. Hammer, R.J. MacDonald, and R.L. Brinster (1984) Tissue-specific expression of the rat pancreatic elastase I gene in transgenic mice. _Cell_ 38:639-646.

54. Topper, Y.S. (1970) Multiple hormone interactions in the development of mammary gland _in vitro_. _Recent Prog. Horm. Res._ 26:287-308.

55. Ulmer, K. (1983) Protein engineering. _Science_ 219:666-669.

56. Walstra, P., and R. Jenness (1984) _Dairy Chemistry and Physics_, John Wiley and Sons, New York, pp. 359-361.

57. Watson, J.D., J. Tooze, and D.T. Kurtz (1983) _Recombinant DNA: A Short Course_, W.H. Freeman and Co., New York, 260 pp.

58. Wilkinson, A.J., A.F. Fersht, D.M. Blow, P. Carter, and G. Winter (1984) A large increase in enzyme-substrate affinity by protein engineering. _Nature_ 307:187-188.

59. Willis, I.M., A.F. Stewart, A. Caputo, A.R. Thompson, and A.G. Mackinlay (1982) Construction and identification by partial nucleotide sequence analysis of bovine casein and β-lactoglobulin cDNA clones. _DNA_ 1(4):375-386.

60. Yu-Lee, L.-Y., and J.M. Rosen (1983) The rat casein multigene family. I. Fine structure of the γ-casein gene. _J. Biol Chem._ 258:10794-10804.

61. Zoller, M.J., and M. Smith (1983) Oligonucleotide-directed mutagenesis of DNA fragments cloned into M13 vectors. _Methods in Enzymology_ 100:468-478.

AVIAN HORMONES

Frederick C. Leung

Animal Physiology
Merck, Sharp & Dohme Research Laboratories
Rahway, New Jersey   07065

INTRODUCTION

During recent years, genetic engineering technology has come into routine use at basic research laboratories throughout the world (5). It is expected that this technology will find practical application in the agricultural industries in the near future (16,64). With the aid of recombinant DNA technology, scientists have been able to express exogenous genes of interest in both prokaryotic and eukaryotic cell systems (5,37). Such expression systems could conceivably provide large quantities of gene products for the study of biological phenomena and presumably for use as commercial products. Scientists have been able to insert exogenous genes of interest into animal genomes, either by direct injection into the pronuclei of fertilized eggs or by using a retroviral vector as carrier (1,11,17,21, 51). Thus, with the rapid advance of recombinant technology, the techniques for producing a large (unlimited) quantity of gene products for agricultural use are available, as are the techniques for inserting desirable genes into agricultural animals. The question of which are the most desirable genes remains open. However, the sites of integration and the fine control of expression of the exogenous genes that are necessary to make such technology applicable to the agricultural industry require further research and experimentation. This chapter will discuss the hormonal regulation of growth in chickens. Due to space limitations, the author will limit the discussion to those hormones that have been reported to influence growth in the chicken. These include growth hormone, thyroid hormones, insulin, and glucocorticoids.

GROWTH HORMONE

Growth hormone (GH) was first isolated and characterized from bovine pituitary glands in 1945 (45), but chicken GH (cGH) was not purified and characterized until 1977 (23). Partial sequence of the cGH was obtained from a highly purified preparation, and it was found to show high homology (79%) with the bovine GH (bGH) sequence (43). Recently, Souza et al. (61), using bovine and rat GH complementary DNA (cDNA) as a probe, successfully isolated cGH cDNA. They also successfully expressed the recombinant cGH from <u>Escherichia coli</u> and found it was essentially the same as the GH preparation isolated from chicken pituitary glands. As the name implies, the primary biological activity of GH is to control somatic growth. Nevertheless, there are many unresolved questions regarding the biological activities and the chemistry of GH. Critical questions remain concerning the mechanism of GH actions at the cellular level and the biological implications of the heterogeneity of GH. For example, is the growth-promoting activity due to a direct action of GH, is it mediated by somatomedin, or are both pathways involved? Do different forms of GH possess different biological activities? Where are the effectors for different biological activities located in the GH molecule? Are there different GH receptors in different target tissues? Is there more than one gene coding for the GH molecule?

It is evident that GH controls and regulates growth in normal animals (63,65), but it has not been proven that an increased concentration of GH results in an increase of growth. A deficiency of GH produced either through a genetic defect or by surgical removal of the pituitary gland (hypophysectomy) results in reduction of growth. Replacement with purified preparations of GH reverses the process (60). Due to the unavailability of avian growth hormone, and the technical difficulty of hypophysectomizing chickens, GH replacement experiments using hypophysectomized chickens have not been performed. With the aid of radioimmunoassays (RIAs), Burke and Marks (8) reported that GH concentration at various ages (between 1-8 weeks) was significantly correlated with relative growth rate in selected broiler strain chickens. However, when they compared the circulating GH concentrations in faster-growing, selected lines and slower-growing, nonselected lines, they observed that the selected lines of chickens, which have faster growth rates and greater body weights, have consistently lower GH levels than the smaller, nonselected lines. In contrast to those observations, Harvey et al. (25) reported a positive correlation between GH concentration and growth rates in chickens, and Proudman and Wentworth (53) found a positive correlation between GH and growth rates in turkeys. Thus, the positive correlation of GH and growth rates within a selected strain suggests that GH is a major factor in regulating the growth of avian species. But the higher GH concentration in the smaller, nonselected chickens contradicts this hypothesis unless the selection of rapid growth rates results in an increase in GH receptors in target tissues. Recently, we isolated a highly purified and biologically active preparation of cGH from chicken pituitary glands (43) and determined its effect on body weight gain in 4-week-old cockerels. The results of the experiment are shown in Tab. 1. Birds receiving daily injections of 5 μg or 10 μg of cGH via the brachial vein showed a transient, but significant, increase in body weight gain over the control birds. Birds receiving 50 μg of cGH daily showed no significant increase in body weight gain. These results indicate that exogenous GH injection can stimulate body weight gain in 4-week-old cockerels, but the increase is transient, which suggests that the exogenous GH may cause down-regulation of its receptor.

Recently we reported that a synthetic, human pancreatic, growth hormone-releasing factor (hpGRF), which has been shown to be a specific GH-releaser in mammals, also stimulates cGH release in chickens, both in vivo and in vitro (41). We also determined the effect of hpGRF on body weight gain in 4-week-old cockerels, and the results are shown in Tab. 2. Birds receiving a daily injection of 0.1 μg and 1.0 μg of hpGRF showed a transient stimulation in body weight gain over the control birds. Birds receiving a daily injection of 10 μg hpGRF showed no significant increase in body weight gain as compared with controls. These data suggest that the transient stimulation in body weight gain by hpGRF could be mediated via pituitary GH since both hpGRF and cGH caused a similar biological response when injected into 4-week-old chickens. If daily injection of exogenous GH down-regulates its receptor, one can expect only a temporary increase in rate of gain by supplying exogenous GH.

Tab. 1. Effects of chicken growth hormone (cGH) on body weight gain in 4-week-old cockerels.

Treatment[+]	Body weight gain (g): Day of treatment			
	3	6	10	14
Control	102 ± 6	318 ± 12	486 ± 17	716 ± 28
cGH, 5 µg/bird daily	123 ± 7* (20.6)	361 ± 12* (13.5)	531 ± 21 (9.2)	774 ± 23 (8.1)
cGH, 10 µg/bird daily	122 ± 6* (19.6)	354 ± 11 (11.3)	531 ± 12 (9.2)	771 ± 17 (7.7)
cGH, 50 µg/bird daily	118 ± 5 (15.6)	349 ± 15 (9.7)	518 ± 22 (6.6)	763 ± 33 (6.5)

[+] Chicken GH was given via the brachial vein daily for 2 weeks and control birds received saline vehicle injections.
* $P < 0.05$ as compared with control.
( ) = Percent difference as compared with control.

We have compared circulating GH levels and growth rates in different strains of chickens (46,47). In general, circulating GH concentrations are consistently inversely related to growth. For example, chickens that carry a dwarfing gene (sex-linked dwarf) grow 30-50% less than fast-growing broilers, and have significantly higher circulating GH concentrations than do normal fast-growing broilers (Tab. 3). Layer strain chickens, which grow slower than meat strain chickens, also have higher circulating GH concentrations than the latter birds (Tab. 4). Our findings as to the relationship between circulating GH concentrations and growth in chickens differ from those reported in mammals (65). Using gene insertion techniques, Palmiter et al. (52) produced transgenic mice by direct injection of a cloned rat GH (rGH) or human GH (hGH) gene into the pronuclei of fertilized eggs. This resulted in transgenic mice that had higher than normal circulating levels of rGH or hGH, and that grew to twice the size of their controls. Hammer et al. (19) have used the gene transfer technique successfully in correcting the dwarfism of a strain of "little" mice that is growth hormone-deficient. In fact, the "little" mice bearing the extra GH gene grow even larger than normal mice. Recently, Hammer et al. (20) have generated transgenic mice that carry and express the human growth hormone-releasing factor (hGRF) gene. These mice have hGRF measurable by RIA, increased levels of circulating mouse GH, and accelerated growth rates relative to control littermates. In all of these experiments, higher circulating GH concentrations have resulted in higher growth rates for the mice. Since we reported that the circulating GH concentration is inversely related to growth of chickens, it would be of interest to determine what would happen to the growth rate of a chicken if a GH gene were inserted into the genome.

The biological activities of GH were mainly demonstrated and established in mammals by using mammalian GH preparations. The main biological activity of mammalian GH is concerned with regulating and controlling growth. Growth hormone also influences a wide range of metabolic activities as follows: (a) stimulates protein synthesis and amino acid transport, (b) stimulates lipolysis, and (c) inhibits insulin action on glucose metabolism. It is generally accepted that most of the anabolic effects of GH are mediated in part, if not wholly, by somatomedin, but the lipolytic effect and the anti-insulin-like activity on blood glucose levels are due

Tab. 2. Effects of hpGRF on body weight gain in 4-week-old cockerels.

Treatment[+]	Body weight gain (g): Day of treatment			
	3	7	10	14
Control	64 ± 7	264 ± 12	437 ± 16	658 ± 21
hpGRF, 0.1 µg/bird	83 ± 5 (29.6)	307 ± 10* (16.2)	487 ± 13* (11.4)	718 ± 18* (9.1)
hpGRF, 1.0 µg/bird	86 ± 7 (34.3)	308 ± 16* (16.6)	486 ± 19 (11.2)	716 ± 28 (8.8)
hpGRF, 10.0 µg/bird	74 ± 7 (15.6)	281 ± 9 (6.4)	462 ± 12 (5.7)	697 ± 18 (5.9)

[+] hpGRF was given via the brachial vein daily for 2 weeks and control birds received saline vehicle injections.
* $P < 0.05$ as compared with control.
( ) = Percent difference as compared with control.

directly to GH itself (10,63). Data concerning the biological activity of cGH are scarce. We have shown that purified cGH stimulates body weight gain and epiphysial width of hypophysectomized rats in a dose-dependent manner and is parallel with the bGH standard. We also reported that purified cGH failed to stimulate weight gain and [^{3}H]-proline and $^{35}SO_4$ incorporation into 9- to 10-day-old chick pelvic cartilage cultured in serum-free media; but a purified, human somatomedin-C is very potent under the same culture conditions (7). These results suggest that the growth-promoting effect of cGH is most likely mediated through a somatomedin-like peptide in the chicken as has been proposed for mammals. To explore such a hypothesis, chicken somatomedin-C has first to be purified and characterized. There is some indirect evidence that chickens also possess somatomedin-C-like peptides. Using a human somatomedin-C RIA, we have been able to show that immunoreactive, somatomedin-C concentrations increase after injection of purified cGH (39). Huybrechts et al. (29) have also reported that circulating, immunoreactive somatomedin-C concentrations increase between 6 and 12 weeks of age in chickens, and decrease following hypophysectomy. Sex-linked dwarf chickens also have lower circulating somatomedin-C concentrations than faster-growing birds (29). Purified chicken somatomedin-C/IGF-I and/or IGF-II should facilitate progress in understanding the physiology of the somatomedin-C-like peptides in avian species.

Limited data are also available that indicate that GH is lipolytic in chickens. Harvey et al. (24) reported that GH decreases lipogenesis in chicken hepatocytes in the presence of insulin and stimulates the rates of lipolysis in chicken adipocytes in vitro. These preliminary data suggest that cGH could act directly to modulate lipid metabolism in the chicken.

The hypothesis concerning the mediation of somatomedin in the biological activity of GH has been challenged by Isaksson et al. (31). They have injected purified hGH directly into the proximal tibia epiphyseal plate and observed stimulation of cartilage growth. Recently, using homologous hormone preparations, Russell and Spencer (56) reported that purified rGH can stimulate growth when injected locally into the tibia plate. In addition, they obtained similar results with human somatomedin-C. Thus, it appears that GH and somatomedin-C can act directly on cartilage to promote growth. The hypothesis that GH and growth factors interact in promoting growth has been put forward by 2 groups of investigators. Green et al. (18) proposed

Tab. 3.   Effects of genotype and age on body weight and serum growth hor-
          mone (GH) concentrations in 2 genotypes of broiler chickens.

Age (weeks)	Strain	Body weight (g)	Serum GH (ng/ml)*
0[†]	F-DW/DW	44.0 ± 0.8	414.49 ± 35.02
	dw/dw	37.0 ± 0.5	1,584.80 ± 346.70
2	F-DW/DW	299.0 ± 8.0	107.75 ± 5.84
	dw/dw	192.0 ± 9.0	309.45 ± 31.87
4	F-DW/DW	904.0 ± 14.0	59.98 ± 6.66
	dw/dw	531.0 ± 15.0	329.55 ± 42.37
6	F-DW/DW	1,699.0 ± 37.0	30.44 ± 2.49
	dw/dw	1,076.0 ± 29.0	168.05 ± 25.69
8	F-DW/DW	2,610.0 ± 38.0	28.79 ± 2.83
	dw/dw	1,693.0 ± 73.0	180.98 ± 13.45

[†] Day-old chickens.
* Serum GH concentrations were measured by a homologous radio-
immunoassay developed in our laboratory.
N = 17-25 per strain; each value is presented as mean ± s.e.m.
F-DW/DW = Fast-growing broiler; dw/dw = Sex-linked dwarf broiler.

that GH stimulates the conversion of preadipocytes of 3T3 cells into adipocytes, and that somatomedin-C (growth factors) acts as a mitogen after the
differentiation.   In the 3T3 cell-culture system, GH has dual functions.
Growth hormone promotes differentiation of the 3T3 cells into adipocytes
and also acts as a lipolytic agent for the differentiated 3T3 adipocytes.
Isaksson et al. (30) suggest that GH acts directly to promote prechondro-
cyte differentiation. It then activates a growth factor gene(s) to promote
growth in the chondrocyte.

THYROID HORMONES

     Thyroid hormones are necessary for growth and development in birds
(2,3,34). Growth is severely reduced if chickens are made hypothyroid by
treatment with propylthiouracil (PTU) or methimazole (33,34). This occurs
in the embryo and also in hatched chickens (33). The reduction of growth
is particularly apparent in skeleton and muscle (36). Excess thyroid
hormone levels produced in chickens by feeding T3 or T4 in the diet are also detrimental to growth and feed efficiency (44).

     The effect of thyroid hormone on skeleton and bone growth is apparent-
ly direct. King and King (35) reported that gastrocnemius muscle weight,
actin content, and total DNA content were significantly lower in PTU-treated birds than in control birds, and that growth returned to near normal
following T4 replacement. Burch and Lebowitz (6), using pelvic cartilage
from 9-day-old chicks, reported that T3 increased weight gain and [^{14}C]-
leucine and $^{35}SO_4$ incorporation. They also demonstrated that T3 acts
directly to stimulate cartilage growth and maturation. Figures 1 and 2
summarize the relationship between circulating T3 and T4 concentrations and
body weights. Optimal body weight is obtained within a narrow range of T3
and T4 concentrations. Thus, if one raises the serum T3 or T4 concentration by feeding thyroid hormones or decreases the T3 or T4 concentration
by treatment with PTU or methimazole, one would anticipate a reduction in
body weight. Considerable attention has been given to genetic variation in

Tab. 4. Comparison of serum growth hormone (GH) concentrations and body weight in 6-week-old broiler and layer strains of chickens.

Strain	N	Body weight (g)	Serum GH (ng/ml)*
Broiler	57	1,630	18.23 ± 2.01
Layer	55	410	130.09 ± 14.57

* Serum growth hormone concentrations were measured by a homologous radioimmunoassay.

thyroid function between avian species, but no consistent relationship between growth rate and circulating thyroid hormone concentration has been established (34).

The hormonal profile of the sex-linked dwarf chickens is well-characterized. It has been found that the sex-linked dwarf chickens have a higher circulating, immunoreactive GH concentration and a significantly reduced T3 concentration (47,58). Scanes et al. (58) proposed that the reduction is due to the decreased conversion of T4 to T3 since liver 5'-monodiodinase activity was reduced. We reported that T3 or T4 supplementation did not improve body weight gain in sex-linked dwarf chickens and suggested that the sex-linked dwarf could have low or no T3 receptors in the target tissue (42). Stewart et al. (62) reported that the depressed malic enzyme activities associated with the dwarfing genes are not simply attributable to the availability of circulating T3, and suggested that the effect of the dwarfing gene could be either on the number or the affinity of the T3 receptors of the target tissue.

The interaction between thyroid hormones and GH in chickens is of interest. In mammals, it is evident that thyroid hormones stimulate GH synthesis and release in vitro (57). Marked reductions in both pituitary GH content and growth rate in hypothyroid animals are also well-documented (59). Thus, the decrease of body weight gain in hypothyroid animals is thought to reflect low pituitary GH. Harvey et al. (22,26) reported that serum GH was elevated in thyroidectomized chickens, and injection of T3 and T4 significantly lowered circulating GH concentration. We also showed that dietary T3 and/or T4 lowered circulating concentrations of GH in both normal and PTU-treated cockerels and pullets (44). The site where T3 and T4 inhibit GH release could be the pituitary gland (Leung et al., unpubl. observ.). It seems possible that the influence of thyroid hormones on growth is not mediated through pituitary GH in chickens as has been suggested for mammals.

Thyroid hormone action on skeletal and bone growth may also be directed at the target level. Thyroid hormones may play a permissive role in allowing other hormones, e.g., GH and/or somatomedin, to act. Froesch et al. (15) showed that T3 is necessary for a maximum response to somatomedin in an in vitro chicken chondrocyte assay, and Burnstein et al. (9) showed that the thyroid hormones are necessary for normal production of somatomedin-C in rats. Thus, a physiological interaction may exist between thyroid hormones and somatomedin-C that involves growth in chickens.

In general, thyroid hormone is important for normal skeletal and bone growth and development, but its relationship to growth requires further investigation.

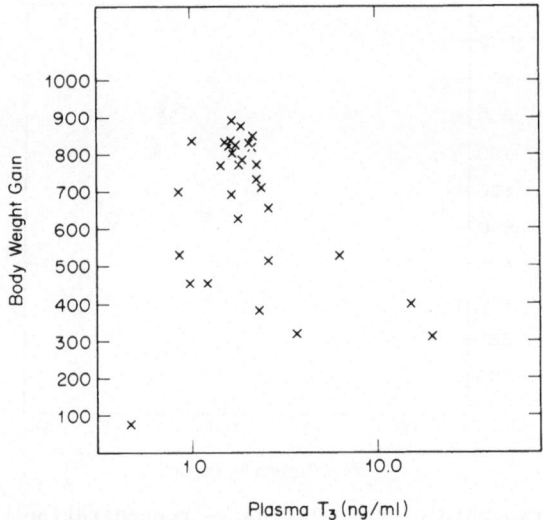

Fig. 1.   The relationship of circulating T3 concentrations and body weight gain from 4- to 6-week-old cockerels.

INSULIN

Chicken insulin has been isolated and characterized and its amino acid sequence found to be very similar to that of human insulin (32). Chicken insulin differs from human insulin by substitution of His, Asn, Thr at positions 8, 9, and 10, respectively, on the A chain, and Ala and Ala substitution at positions 1 and 30 on the B chain. The amino acid sequence of turkey insulin was found to be identical to that of the chicken. Chicken insulin genes have been cloned and shown to be highly conserved throughout the vertebrate kingdom. Biologically, chicken insulin is as potent as mammalian insulin in promoting glucose uptake in an in vitro rat diaphragm assay, but is more potent than mammalian insulin in promoting hypoglycemia in intact chickens (27). In fact, chicken insulin has 2-3 times the affinity for the mammalian receptor than does mammalian insulin, and its relative affinity correlated well with its biological potency. Chicken insulin is 2-3 times more potent than rat insulin in stimulating glucose oxidation in fat cells in the rat (55). Chicken insulin also stimulates glycogen synthesis in the liver, and promotes glucose uptake and incorporation into muscle glycogen (27). McMurtry et al. (49) have developed a highly sensitive, homologous RIA for chicken insulin that recognizes chicken insulin about 10-fold better than it does mammalian insulin. With the aid of the RIA, de Pablo et al. (14) showed that immunoreactive insulin could be detected in chicken embryos before beta-cells were recognizable. Since immunoreactive insulin was also detected in yolks and whites of unfertilized eggs, the possibility exists that it is derived from material in the eggs that is transferred from the hens. Specific binding of ^{125}I-insulin to 3- to 4-day-old embryos was also reported. This early detection of insulin and insulin receptors suggests that insulin may participate in regulating metabolic processes, growth, and differentiation during embryonic development of the chicken. Indeed, de Pablo et al. (13) showed a reduction in growth when anti-insulin antibody was applied to 2-day-old chicken embryos when compared with embryos receiving normal IgG. In addition, they also reported stimulation of growth in 4-day-old chicken embryos by treating them with purified insulin at 100 ng/dose. Thus, it appears that there are different hormonal requirements for growth in early embryonic development.

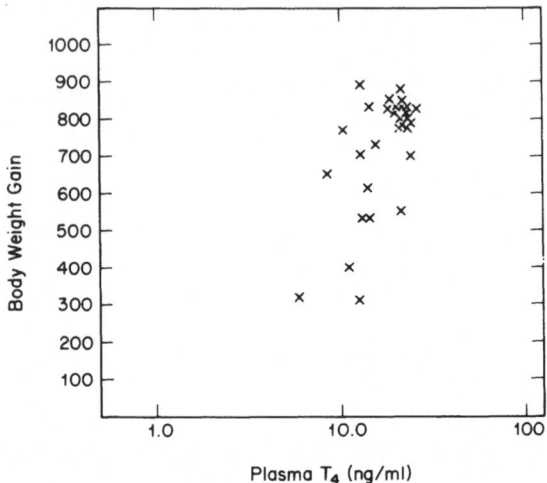

Fig. 2.   The relationship of circulating T4 concentrations and body weight
gain from 4- to 6-week-old cockerels.

GLUCOCORTICOIDS

    Corticosterone is the major glucocorticoid in birds (54).   Injection
of corticosterone stimulates hepatic glucogenesis in chickens; corticoid
injection also increases fat content of the liver (4,12,38).   Numerous investigators have reported that corticosterone, either by injection or by
feeding, had a general anti-anabolic effect (12,50).   A marked reduction in
growth and development is evident.   An excess of glucocorticoid decreases
linear growth, increases fat deposition, and causes a reduction in feed
efficiency.

HORMONE AND GENOTYPE

    Poultry breeders have applied the principles of genetics to develop
strains or breeds of chickens best suited for production of meat and/or
eggs.   Poultry breeders are actually applied geneticists, and they have
markedly improved strains of chickens bred for either meat or egg production (48).   For example, modern day broilers bred for meat production can
grow to market weight in 6-8 weeks, and modern day Leghorns can lay an average of 274 eggs over a 54-week production period.   Unfortunately, traits
such as egg production, egg size, growth rate, body conformation, feed efficiency, and disease resistance do not have a simple inheritance pattern
as do traits such as comb type or plumage color.

    At the molecular level, poultry breeders have either amplified desir-
able traits or suppressed undesirable ones through phenotypic selection to
generate the best meat- or layer-type birds.   Since hormones mediate the
physiological processes, understanding the physiology of different genotypes may contribute to the understanding of how hormones and genes interact to cause a manifestation of those traits for which poultry breeders
have selected.   In the section on growth hormones above, we stated that
slower-growing broilers and Leghorns that carry the sex-linked dwarfing
genes have higher circulating levels of GH than do normal birds.   When we
examined the hepatic GH receptors in these chickens, we observed that these
receptors in the sex-linked dwarf broiler and Leghorn chickens were virtually nonexistent (40).   This suggests that sex-linked dwarf chickens grow
slower because of an absence or lack of GH receptors.   The question of this

120

reduced number or lack of GH receptors in the sex-linked dwarf requires further investigation.

There are tremendous differences in the growth rates of meat-type (broiler) versus layer-type (Leghorn) birds. It takes an average of 20 weeks for a Leghorn to reach 3 lb; broilers require only 4–5 weeks. When comparing the circulating GH and thyroid hormones between the two types of birds, T3 and T4 do not seem to be different, but circulating GH concentration is significantly higher in the Leghorn. Again, when hepatic GH receptor-binding is examined, the Leghorns have significantly fewer GH receptors than broilers (Leung et al., unpubl. observ.). Burke and Marks (8) have also reported that the circulating concentration of GH in a slower-growing, nonselective line of chickens is significantly higher than in a faster-growing, selected line. In general, a faster rate of growth in chickens is associated with a low level of circulating GH and high GH receptor-binding. Huybrechts et al. (29) recently reported that immunoreactive somatomedin-C is significantly lower in the sex-linked dwarf chicken. Thus, it is possible that the physiological marker that the poultry breeder has selected for is sensitivity of target tissue to GH. One must keep in mind that the lack of GH receptors in the sex-linked dwarf is also reflected in a slower than normal growth rate.

Circulating thyroid hormone levels (T3 and T4) are not different between broiler- and layer-type birds, but T3 is significantly lower in both sex-linked dwarf broilers and Leghorns (58). Scanes et al. (58) suggested that the lower circulating T3 may be due to depressed liver monodiodinase activities. Our results with dietary T3 and/or T4 supplements in the sex-linked dwarfs revealed that raising circulating T3 and T4 to normal levels did not correct the reduction in body weight gain typical of sex-linked dwarf chickens (42). In addition to the data of Stewart et al. (62) suggesting a relationship between hepatic malic enzyme and T3 affinity for its receptor in the sex-linked dwarf, we suggest that dwarfism in the sex-linked dwarf could be caused by a reduction in the number of T3 receptors or in the affinity of T3 for those receptors. Such hypotheses will have to be further examined by direct measurement of thyroid hormone receptors in the sex-linked dwarf.

CONCLUSION

Growth is a highly complex physiological phenomenon. This process is regulated and mediated by a multitude of hormones and their interactions. Hormonal mediation of the growth process is evidenced by the fact that reduction of body weight gain is the consequence of removal of any one of the hormones that were mentioned in this chapter. Other hormones that have been reported to influence growth include various gonadal steroids, growth factor(s), and prolactin. It is clear that an excess of any one of the hormones that influence growth may not be beneficial to the animal. Since circulating levels of hormones only reflect one aspect of hormone action, affinity for the target tissue receptors may be very important for the manifestation of a biological response to the ligand. Hormone receptors may represent another level of regulation of biological response and may well be a controlling and/or limiting factor for such processes. Thus, information concerning hormone-receptor interactions and the identity of the limiting factor(s) in the growth process will definitely aid scientists in choosing the appropriate gene(s) for the desirable phenotypes in chickens. Growth is a developmental process and there is information suggesting that different stages may be under different hormonal requirements. Therefore, it is also important to study the involvement of hormones and their receptors in growth at different stages of development. A recent review by Hunton (28) suggests that, even after 40 years of relatively intense selection by poultry breeders, genetic variation does not appear to be exhausted due to a large reservoir of genetic diversity. Thus, with the application of recombinant technology, one might hope to duplicate what poultry geneticists have done in the last 50 years to improve the growth rates of broilers in the laboratory setting.

ACKNOWLEDGEMENTS

I would like to acknowledge the collaboration of Drs. Mike Lilburn and Jim Smith, and also thank J. Taylor, A. Van Iderstine, C.A. Ball, K.N. Ngiam-Rilling, B. Goggins, and C.I. Rosenblum for their expert assistance. Also, I thank Ms. Barbara Prieto and Ms. Karen Rosa for their expert secretarial assistance, and Drs. J. Brooks and H. Chen for reviewing this manuscript.

REFERENCES

1. Anderson, W.F. (1984) Prospects for human gene therapy. _Science_ 226: 401-409.
2. Assenmacher, I. (1973) The peripheral endocrine glands. In _Avian Biology_, Vol. 3, D.S. Farner and J.R. Kings, eds. Academic Press, New York, pp. 183-286.
3. Barker, S.B. (1955) Thyroid. _Ann. Rev. Physiol._ 17:417-442.
4. Bartov, I. (1982) Corticosterone and fat deposition in broiler chicks: Effect of injection time, breed, sex and age. _Brit. Poultry Sci._ 23: 161-170.
5. Bollon, A.P., E.A. Barron, S.L. Berent, P.W. Bragg, D. Dixon, M. Fuke, C. Hendrix, M. Mahmoudi, R.S. Sidhu, and R.M. Torczynski (1984) Recombinant DNA techniques: Isolation, cloning and expression of genes. In _Recombinant DNA Products: Insulin, Interferon and Growth Hormone_, A.P. Bollon, ed. CRC Press, Boca Raton, Florida.
6. Burch, W.M., and H.E. Lebovitz (1982) Triiodothyronine stimulation of _in vitro_ growth and maturation of embryonic cartilage. _Endocrinology_ 111:462-468.
7. Burch, W.M., G. Corda, J.J. Kopchick, and F.C. Leung (1985) Homologous and heterologous growth hormones fail to stimulate avian cartilage growth _in vitro_. _J.Clin. Endocrinol. Metab._ 60:747-750.
8. Burke, W.H., and H.L. Marks (1982) Growth hormone and prolactin levels in nonselected and selected broiler lines of chickens from hatch to eight weeks of age. _Growth_ 46:283-295.
9. Burnstein, P.J., B. Draznin, C.J. Johnson, and D.S. Schalch (1979) The effect of hypothyroidism on growth, serum growth hormone, the growth hormone-dependent somatomedin insulin-like growth factor and its carrier protein in rats. _Endocrinology_ 104:1107-1111.
10. Chawla, R.K., J.S. Parks, and D. Rudman (1983) Structural variants of human growth hormone: Biochemical, genetic and clinical respects. _Ann. Rev. Med._ 34:519-547.
11. Constantini, F., and E. Lacy (1981) Introduction of a rabbit B-globin gene into the mouse germ line. _Nature_ 294:92-94.
12. Davison, T.F., J. Rea, and J.G. Rowell (1983) Effects of dietary corticosterone on the growth and metabolism of immature _Gallus domesticus_. _Gen. Comp. Endocrinology_ 50:463-468.
13. de Pablo, F., M. Girbau, J.A. Gomez, E. Hernandez, and J. Roth (1985) Does endogenous insulin regulate growth and differentiation of early embryos? _Proc. 67th Annual Meeting of the Endocrine Society_, Endocrine Society, p. 91.
14. de Pablo, F., J. Roth, E. Hernandez, and R.M. Pruss (1982) Insulin is present in chicken eggs and early chick embryos. _Endocrinology_ 111: 1909-1916.

15. Froesch, E.R., J. Żafp, T.K. Audhya, E. Ben-Porath, B.J. Segen, and K.D. Gibson (1976) Non-suppressible insulin-like activity and thyroid hormones: Major pituitary-dependent sulfation factors for chick embryo cartilage. <u>Proc. Natl. Acad. Sci., USA</u> 73:2904-2908.

16. Glick, J.L., M.V. Peirce, D.M. Anderson, C.A. Vaslet, and H.Y. Hsiao (1983) Utilization of genetically engineered microorganisms for the manufacture of agricultural products. In <u>Genetic Engineering: Application to Agriculture</u>, L.D. Owens, ed. Rowman & Allanheld, Totowa, pp. 67-87.

17. Gordon, J.W., and F.H. Ruddle (1981) Integration and stable germ line transmission of genes injected into mouse pronuclei. <u>Science</u> 214: 1244-1246.

18. Green, H., M. Morikawa, and T. Nixon (1985) A dual effector theory of growth hormone action. In <u>Human Growth Hormone Symposium</u>, S. Raiti, ed. Plenum Press, New York (in press).

19. Hammer, R.E., R.D. Palmiter, and R.L. Brinster (1984) Partial correction of murine hereditary growth disorder by germ-line incorporation of a new gene. <u>Nature</u> 311:65-67.

20. Hammer, R.E., R.L. Brinster, M.G. Rosenfeld, R.M. Evans, and K.E. Mayo (1985) Expression of human growth hormone-releasing factor in transgenic mice results in increased somatic growth. <u>Nature</u> 315:413-416.

21. Hammer, R.E., V.G. Pursel, C.E. Rexroad, Jr., R.J. Wall, D.J. Bolt, K.M. Ebert, R.D. Palmiter, and R.L. Brinster (1985) Production of transgenic rabbits, sheep and pigs by microinjection. <u>Nature</u> 315: 680-683.

22. Harvey, S. (1983) Thyroid hormones inhibit growth hormone secretion in domestic fowl (<u>Gallus</u> <u>domesticus</u>). <u>J. Endocrinol.</u> 96:329-334.

23. Harvey, S., and C.G. Scanes (1977) Purification and radioimmunoassay of chicken growth hormone. <u>J. Endocrinol.</u> 73:321-329.

24. Harvey, S., C.G. Scanes, and T. Howe (1977) Growth hormone effects on <u>in vitro</u> metabolism of avian adipose and liver tissue. <u>Gen. Comp. Endocrin.</u> 33:322-328.

25. Harvey, S., C.G. Scanes, A. Chadwick, and N.J. Bolton (1979) Growth hormone and prolactin secretion in growing domestic fowl: Influence of sex and breed. <u>Brit. Poultry Sci.</u> 20:9-17.

26. Harvey, S., R.J. Sterling, and H. Klandorf (1983) Concentrations of triiodothyronine, growth hormone, and luteinizing hormone in the plasma of thyroidectomized fowl (<u>Gallus</u> <u>domesticus</u>). <u>Gen. Comp. Endocrinol.</u> 50:275-281.

27. Hazelwood, R.L., J.R. Kimmel, and H.G. Pollock (1968) Biological characterization of chicken insulin activity in rats and domestic fowl. <u>Endocrinology</u> 83:1331-1336.

28. Hunton, P. (1984) Selection limits: Have they been reached in the poultry industry? <u>Can. J. Anim. Sci.</u> 64:217-221.

29. Huybrechts, L.M., D.B. King, T.J. Lauterio, J. Marsh, and C.G. Scanes (1985) Plasma concentrations of somatomedin-C in hypophysectomized, dwarf and intact growing domestic fowl as determined by heterologous radioimmunoassay. <u>J. Endocrinol.</u> 104:233-239.

30. Isaksson, O.G.P., S. Eden, and J.A. Jansson (1985) Mode of action of pituitary growth hormone on target cells. <u>Ann. Rev. Physiol.</u> 47:483-499.

31. Isaksson, O.G.P., J.O. Jansson, and I.A.M. Gause (1982) Growth hormone stimulates longitudinal bone growth directly. <u>Science</u> 216:1237-1239.

32. Kimmel, J.R., H.G. Pollack, and R.L. Hazelwood (1968) Isolation and characterization of chicken insulin. <u>Endocrinology</u> 83:1323-1330.

33. King, D.M. (1969) Effect of hypophysectomy of young cockerels with particular reference to body growth, liver weight and liver glycogen level. <u>Gen. Comp. Endocrinology</u> 12:242-255.

34. King, D.B., and J.D. May (1984) Thyroidal influence on body growth. <u>J. Exp. Zool.</u> 232:455-460.

35. King, D.B., and C.R. King (1973) Thyroidal influence on early muscle growth of chickens. <u>Gen. Comp. Endocrinol.</u> 21:517-529.
36. King, D.B., and C.R. King (1976) Thyroidal influence on gastrocnemius and sartorius muscle growth in young white leghorn cockerels. <u>Gen. Comp. Endocrinol.</u> 29:473-479.
37. Kopchick, J.J., R. Malavarca, T. Livelli, and F.C. Leung (1985) Use of avian retroviral-bovine growth hormone DNA recombinants to direct expression of bovine growth hormone by cultured fibroblasts. <u>DNA</u> 4:23-31.
38. Langslow, D.R., and C.N. Hales (1969) Lipolysis in chicken adipose tissue <u>in vitro</u>. <u>J. Endocrinol.</u> 43:285-294.
39. Leung, F.C. (1985) Hormonal regulation of growth in chickens. In <u>Control and Manipulation of Animal Growth</u>, P.J. Buttery, N.B. Haynes, and D.B. Lindsay, eds. Butterworth & Co., United Kingdom (in press).
40. Leung, F.C., W.J. Styles, C.I. Rosenblum, M.S. Lilburn, J.A. Marsh, and J.J. Kopchick (1985) Diminished hepatic growth hormone receptor bindings in sex-linked dwarf broiler and leghorn chickens. <u>Proc. Soc. Exp. Bio. Med.</u> (submitted for publication).
41. Leung, F.C., and J.E. Taylor (1983) <u>In vivo</u> and <u>in vitro</u> stimulation of growth hormone release in chickens by synthetic human pancreatic growth-hormone-releasing factor (hpGRFs). <u>Endocrinology</u> 113:1913-1915.
42. Leung, F.C., J.E. Taylor, and A. Van Iderstine (1984) Effects of dietary thyroid hormones on growth, serum, T_3, T_4 and growth hormone in sex-linked dwarf chickens. <u>Proc. Soc. Exp. Bio. Med.</u> 177:77-81.
43. Leung, F.C., J.E. Taylor, S.L. Steelman, C.D. Bennett, J.A. Rodkey, R.A. Long, R. Serio, R.M. Weppelman, and G. Olson (1984) Purification and properties of chicken growth hormone and the development of a homologous radioimmunoassay. <u>Gen. Comp. Endocrinol.</u> 56:389-400.
44. Leung, F.C., J.E. Taylor, and A. Van Iderstine (1985) Effects of dietary thyroid hormones on growth, plasma T_3 and T_4, and growth hormone in normal and hypothyroid chickens. <u>Gen. Comp. Endocrinol.</u> 59:91-99.
45. Li, C.H., H.M. Evans, and M.E. Simpson (1945) Isolation and properties of the anterior hypophyseal growth hormone. <u>J. Biol. Chem.</u> 159:353-366.
46. Lilburn, M.S., F.C. Leung, K. Ngiam-Rilling, and J.H. Smith (1986) The relationship between age and genotype and circulating concentrations of T_3, T_4 and growth hormone in commercial meat strain chickens. <u>Proc. Soc. Exp. Bio. Med.</u> (in press).
47. Lilburn, M.S., K. Ngiam, J.H. Smith, and F.C. Leung (1983) Body weight, carcass component characteristics and endocrine comparisons in normal, fast-growing and sex-linked dwarf male and female chickens. <u>Poultry Sci.</u> 62:1458.
48. Lilburn, M.S., K. Ngiam-Rilling, F.C. Leung, and J.H. Smith (1986) The relationship between age and genotype and the growth of commercial meat strain in chickens. <u>Proc. Soc. Exp. Bio. Med.</u> (in press).
49. McMurtry, J.P., R.W. Rosebrough, and N.C. Steele (1983) A homologous radioimmunoassay for chicken insulin. <u>Poultry Sci.</u> 62:697-701.
50. Nagra, C.L., and R.K. Meyer (1963) Influence of corticosterone on the metabolism of palmitate and glucose in cockerels. <u>Gen. Comp. Endocrinol.</u> 3:131-138.
51. Palmiter, R.D., and R.L. Brinster (1985) Transgenic mice. <u>Cell</u> 41:343-345.
52. Palmiter, R.D., G. Norstedt, R.E. Gelinas, R.E. Hammer, and R.L. Brinster (1983) Metallothionein-human GH fusion genes stimulate growth of mice. <u>Science</u> 222:809-814.
53. Proudman, J.A., and B.C. Wentworth (1980) Ontogenesis of plasma growth hormone in large and midget white strains of turkeys. <u>Poultry Sci.</u> 59:906-913.
54. Ringer, R.K. (1976) Adrenals. In <u>Avian Physiology</u>, P.D. Sturkie, ed. Springer-Verlag, New York, pp. 372-382.

55. Roth, J., and C. Grunfeld (1981) Endocrine systems: Mechanisms of disease, target cells and receptors. In Textbook of Endocrinology, R.H. Williams, ed. W.B. Saunders, Philadelphia, pp. 15-72.

56. Russell, S.M., and E.M. Spencer (1985) Local injections of human or rat growth hormone or of purified human somatomedin-C stimulate unilateral tibial epiphyseal growth in hypophysectomized rats. Endocrinology 116:2563-2567.

57. Samuels, H.H., F. Stanley, and L.E. Shapiro (1976) Dose dependent depletion of nuclear receptors by L-triiodothyronine: Evidence for a role in induction of growth hormone synthesis in cultured GH cells. Proc. Natl. Acad. Sci., USA 73:3877-3881.

58. Scanes, C.G., J. Marsh, E. Decuypere, and P. Rudas (1983) Abnormalities in theplasma concentrations of thyroxine, tri-iodothyronine and growth hormone in sex-linked dwarf and autosomal dwarf white leghorn domestic fowl (Gallus domesticus). J. Endocrinol. 97:127-135.

59. Solomon, J., and R.D. Greep (1959) The effects of alterations in thyroid function on the pituitary growth hormone content and acidophil cytology. Endocrinology 65:158-164.

60. Simpson, M.E., C.N. Asling, and H.M. Evans (1950) Some endocrine influences on skeletal growth and differentiation. Yale J. Biol. Med. 23:1-27.

61. Souza, L.M., T.C. Boone, D. Murdock, K. Langley, J. Wypych, D. Fenton, S. Johnson, P.H. Lai, R. Everett, R.Y. Hsu, and R. Rosselman (1984) The application of recombinant DNA technologies to study chicken growth hormone. J. Exp. Zool. 232:465-473.

62. Stewart, P.A., K.W. Washburn, and H.L. Marks (1984) Effect of the dw gene on growth, plasma hormone concentrations and hepatic enzyme activity in a random bred population of chickens. Growth 48:59-73.

63. Underwood, L.E., and J.J. Van Wyk (1981) Hormones in normal and aberrant growth. In Textbook of Endocrinology, R.H. Williams, ed. W.B. Saunders, Philadelphia, pp. 1149-1191.

64. Wagner, T.E., F.A. Murray, B. Minhas, and D.C. Kraemer (1984) The possibility of transgenic livestock. Theriogenology 21:29-44.

65. Wilhelmi, A.E. (1975) Chemistry of growth hormone. In Handbook of Physiology, Sect. 7, Vol. 4, R.O. Greep and E.B. Astwood, eds. American Physiol. Society, Washington, D.C., pp. 59-78.

HISTORY OF GENETIC ENGINEERING OF LABORATORY AND FARM ANIMALS

Caird E. Rexroad, Jr.

U.S. Department of Agriculture
ARS, Reproduction Laboratory
Beltsville, Maryland   20705

INTRODUCTION

In the last few hundred years, man has learned that the tools of genetic selection are a powerful means of improving the utility of livestock. Genetic engineering of animals appears to be a logical extension of the man-animal relationship; albeit one that places more responsibility on man in his efforts to develop an ecosystem in which both man and animals persist. The prospects for genetic engineering of farm animals have improved dramatically because of recent discoveries in the field of gene regulation. Two fundamental tools have made possible the recently reported successful insertion of a cloned human growth hormone gene into pigs and sheep (21). These tools or biotechnologies are embryo micromanipulation and recombinant DNA technology. Judicious application of embryo manipulation and recombinant DNA will provide an opportunity to increase the utilization of animals for food, fiber, and biomedical products. Combining genetic manipulation with the well-established biotechnologies of semen and embryo preservation has increased man's options for preserving genetic resources that at present seem to be dwindling through the loss of animal species in a technology-dominated world. In this chapter, I will discuss methods of genetic manipulation developed since 1950 that appear to have made the most impact on efforts to genetically engineer farm animals. I will also discuss the potential and problems that have arisen with different approaches to engineering animals. The development of recombinant DNA techniques is outside the scope of this chapter, but is obviously an essential part of genetic engineering.

The fundamental questions for developmental biologists are: how does the single-cell, fertilized ovum give rise to the complex mature adult and what roles do changes in the nucleus and cytoplasm play in differentiation? To answer these questions researchers in the field of developmental biology have devised techniques for the experimental transfer of genetic material into animals.

NUCLEAR TRANSFER

Transfer by Microinjection

Nuclear transfer in frogs. In 1952, Briggs and King (3) aspirated nuclei from frog blastula cells into a pipette along with surrounding membrane and a small amount of cytoplasm. The nucleus was then transferred into an activated egg from which the nucleus had been removed. Living embryos were obtained that showed that nuclei could be subjected to manipulation and that nuclei of blastulae were not irreversibly differentiated. King and Briggs also showed in 1955 (28) that the ability of donor nuclei to give rise to developing frogs decreased as the donor frog developed. In another study with frogs (Xenopus laevis) (12), Fischberg et al. produced feeding tadpoles by this technique at the low frequency of 6.7%. Gurdon, in 1962 (19), reported that at least 4% of gut cell nuclei of hatched tadpoles could give rise to normally functioning adults. However, embryonic growth was slow and many abnormalities occurred during development. Primordial germ cell nuclei from Rana pipiens, when placed into enucleated eggs, also gave rise to normally developing tadpoles (50). The growth of these frogs was nearly normal. Nuclei from cultured epithelial cells of tadpoles have also provided nuclei that, in a few instances (<1%), gave rise to adults after transfer into enucleated eggs (20). In experiments with frogs the success rate for development was small, usually less than 6%, and many developmental abnormalities were observed. Although adult frogs have been obtained by transplantation of nuclei from early embryos and cultured epithelial cells of tadpoles, adults have not been obtained from the transplanted nuclei of adult frog cells.

Nuclear transfer in mice. Successful removal of nuclei by micropipette and transfer into enucleated eggs has been reported in mice (23). Inner cell mass and trophectoderm cells were separated from mouse blastocysts and their nuclei were transferred into enucleated fertilized mouse eggs. Illmensee and Hoppe reported that 3 mice were born as a result of transfer of inner cell mass nuclei but no mice were born after transfer of nuclei from trophoblast cells.

Nuclear Transfer by Karyoplast Fusion

McGrath and Solter (29) devised a technique for fusing karyoplasts with enucleated mouse embryos to avoid damage that arises from the penetration of the embryo's membrane with a micropipette. To remove a pronucleus they positioned a pipette at the surface of the vitelline membrane and, by aspiration, drew out a pronucleus. They then withdrew the pipette until the vitelline membrane pinched off. The pipette containing the membrane-bound pronucleus (karyoplast) was filled with a solution of inactivated sendai virus and then its contents were injected into the perivitelline space of an enucleated egg. After karyoplast and egg vitelline membranes fused, the embryos were cultured until they were blastocysts. Sixty-four embryos were transferred to recipient mice. Seven of these engineered offspring developed to adulthood and 5 were fertile. McGrath and Solter (30), using the technique of karyoplast fusion to transfer nuclei of blastomeres into enucleated mouse eggs, found that donor nuclei from embryos having more than 2 blastomeres could not support full development. The results of karyoplast transfer then do not agree with the results obtained by nuclear transplantation.

The application of nuclear transfer techniques to farm animal embryos has not been reported. It is reasonable to assume that, when appropriately modified, the karyoplast fusion technique could be used to successfully transfer nuclei into farm animal embryos. However, for nuclear transfer to be a practical method of cloning, an abundant source of totipotent nuclei must be found or methods devised to reprogram differentiated nuclei into undifferentiated nuclei.

CHIMERA FORMATION

Blastocyst Fusion

In mice. Another method developed to answer fundamental questions in developmental biology is fusion of blastocysts to produce chimeras composed of cells of 2 or more lineages. Tarkowski (54) produced the first fusion chimeras in mammals by the fusion of eggs. He removed zonae pellucidae from 8-cell mouse eggs by pipetting and placed them in close proximity in a small paraffin-covered drop for 5-30 min. They were then cultured for at least 24 hr. By 16 hr, they had begun to fuse, and by 24 hr nearly all were blastocysts. Thirty-six of 153 transferred blastocysts gave rise to fetuses or young. Some of the fetuses were mosaics and 2 young of non-defined origin were born. Proof of success was determined by the formation of intersexes and/or mixed pigmentation.

Mintz (33) digested the zona pellucida of 8-cell eggs with pronase and improved the conditions for fusion. Mintz, in thorough and extensive studies, refined mouse chimera technology and produced adult mice that could be demonstrated to be chimeric by their isozyme contents (for review see Ref. 34).

In farm animals. The technique of aggregating blastomeres to produce chimeras was attempted in sheep (42,55). Success was limited until Fehilly et al. (10) developed a procedure for placing aggregated blastomeres within zonae pellucidae and embedded the zonae pellucidae in agar blocks. Blocks were transferred to sheep oviducts and collected 4-5 days later. Blastocysts of normal appearance were then transferred to recipient ewes. Thirty-six chimeras were obtained from an initial transfer of 110 aggregates of 2- to 8-cell embryos. The same researchers have used their fusion process to produce sheep-goat interspecific chimeras (11). Bovine chimeras have been achieved by fusing blastomeres in surrogate zona pellucida (2).

Blastomere Injection

In mice. A technique closely related blastomere aggregation was developed by Gardner (14) in order to produce mouse chimeras. Blastocysts were treated with pronase to remove the zonae pellucidae and donor cells were disaggregated by gentle pipetting in a calcium- and magnesium-free medium or in the presence of versene. The donor cells were pipetted into the blastocoel of recipient embryos. This technique resulted in 6 chimeras of 47 young and fetuses. Rossant and Frels (43) were able to produce interspecific mouse-mouse hybrids by injecting inner cell mass cells that had been removed immunosurgically from blastocysts into recipient blastocysts.

In farm animals. Immunosurgical removal of the trophectoderm from donor inner cell masses followed by injection of inner cell mass cells into recipient blastocysts was used by Butler et al. (6) to produce ovine chimeras. Fifteen lambs were born by this procedure and 6 were chimeric.

PLURIPOTENT CELLS

Teratoma Cells

Production of chimeras has been a valuable tool for understanding the development of mammalian embryos, especially mice. Although chimeras may not possess the attributes desired in production animals, this technology may provide a convenient means for introducing new genetic material into farm animal embryos. If cells could be engineered to contain desirable genetic traits, they might then be injected into blastocysts in the hope of obtaining chimeras. In cases in which the introduced cells become part of the germline the host animal would provide valuable breeding stock. This scenario requires that the introduced cells be totipotent and susceptible to genetic engineering.

Stevens (52) developed a mouse cell line that provides a model for this protocol. He grafted 3- and 6-day mouse embryos under the testicular capsules of mice. The embryos became disorganized and gave rise to growths composed of many types of tissues (teratomas). When cells from these teratomas were injected intraperitoneally into mice they gave rise to a layer of ectoderm enveloped by endoderm. These embryoid bodies could be maintained for many generations. Brinster (4) used the core cells from these embryoid bodies as a source of cells for injection into mouse blastocysts. He demonstrated the presence of coat color mosaicism in a mouse derived from the experiment. Mintz and Illmensee (35) tested the totipotency of cells from the embryoid bodies developed by Stevens by injecting them into mouse blastocysts. Three (of 48) living mice were shown to be cellular mosaics. One of the male mice subsequently was bred and shown to harbor germ cells derived from the teratoma. Papaioannou et al. (40) injected cultured embryonic tumor cells into blastocysts. They derived 3 lines of chimeras of which at least one line formed postnatal tumors. Teratoma cells have been useful in studying development, but these cells may not be applicable to all animals that one might want to engineer. Low rates of colonization of the germline as well as the potential for tumor formation restrict the usefulness of teratoma cells as carriers for specific genes. Also, the technique of teratoma formation might not be applicable to all species.

EK (Evans and Kaufman) Cells

Evans and Kaufman (9) developed an in vitro culture technique for formation of "pluripotent" cells from mouse embryos. Blastocysts from progestin-treated mice were collected, between days 4 and 6 after estrus, and cultured. After attachment of the trophoblast to the culture dish, the inner cell mass developed into an "egg-cylinder"-like structure that could be removed and separately cultured. These were termed EK cells [as opposed to embryonic carcinoma (EC) cells]. They can give rise to teratomas and thus are pluripotent. The cells had stable and normal chromosome counts. Bradley et al. (1) demonstrated that EK cells could give rise to chimeras when injected into blastocysts. About 50% of the resulting offspring (of 330) were chimeric. Mating demonstrated that about 20% (7 of 35) were chimeric in their germline. Thus, the establishment, then, of EK-type cell cultures might be valuable in providing the cellular substrate for genetic engineering of animals.

INSERTION OF GENES INTO CELLS

Gene Vectors

Genetic engineering of animals depends on the ability to insert genes of interest into some cell. I shall briefly discuss techniques for insertion of genes into cultured cells that could be used in tandem with the injection of EK or teratoma cells into blastocysts. Two techniques that seem to offer the best opportunity for inserting cloned genes into cultured cells are DNA-mediated gene transfer and viral transfection.

DNA-mediated gene transfer. Graham and van der Eb (18) found that the infectivity of adenovirus 5 DNA in a human monolayer KB cell culture could be greatly enhanced by precipitating the viral DNA with calcium phosphate. The reaction was shown to be specific for DNA rather than for viruses by loss of infectivity after DNAse digestion of the viral extract. The technique was also effective for other DNA including simian virus 40 (SV40). Employing DNA-mediated gene transfer, Wigler et al. (60) were able to transfer DNA containing a thymidine kinase (TK) gene cleaved from herpes simplex virus-1 (HSV-1) into cultured mouse L cells deficient in thymidine kinase. The resulting clones contained the viral thymidine kinase and the cell lines were stable for hundreds of generations. Because the efficiency of DNA-calcium phosphate co-precipitation is low, Wigler et al. (60) used the HAT (hypoxanthine, aminopterin, and thymidine) technique to select the clones of cells that had been transformed with TK. The cells to be transformed were deficient in both TK and hypoxanthine-guanosine phosphoribosyl transferase and after transformation were cultured in media containing HAT. Only clones with functional TK survived. Resistance to specific substances such as neomycin can also be used for selection. Many of the cell lines of interest require co-transfer of genes for resistance if the resistance gene is not the gene of interest. The gene of interest must still be identified by techniques other than clone survival. An excellent review of the application of DNA-mediated gene transfer to cells in culture has been presented by Scangos and Ruddle (46).

Viral infection. Viral infection is a "natural" way of introducing genetic information into cells (animals) and viruses have been used to transform cells with genes of interest. Jaenisch and Mintz (26) initiated much of the interest in this area by injecting SV40 viral DNA into the blastocoel of mouse embryos and transferring the embryos into uteri of pseudopregnant mice. DNA was recovered from 25 of 29 mice (of 80 injected embryos) one year later. DNA hybridization analysis identified 10 mice that carried the SV40 DNA. The viral DNA was not uniformly distributed among tissues, perhaps due to unequal infection (uptake) by inner cell mass cells or to elimination of some embryonic cell lines containing virus during development.

Jaenisch (24) found that mice infected with Moloney leukemia virus at the 4- to 8-cell stage stably incorporated a single copy of the virus into their genome. The virus was transmitted to offspring as a Mendelian dominant trait. The mice of the second generation were viremic and developed leukemia. The expression of the virus during adult life is variable (27): some strains of mice developed early leukemia (Mov-3) while another only developed viremia later in life (Mov-2). The difference in expression was not related to the state of the incorporated DNA which appeared to be intact and unrearranged in both strains. First-generation mice blastocysts were chimeric. Harbers et al. (22) isolated the Mov-3 gene and cloned it after excising it from the pBR322 cloning vector. They injected the excised gene into the cytoplasm of embryos 12-16 hr after fertilization. Thirty mice developed from 75 injected embryos, 3 of which were viremic at 4 weeks of age. In one mouse, all organs contained a single copy of the viral DNA but pBR322 sequences that had been attached to the Mov-3 DNA at the time of injection were missing. This was interpreted to mean that the injected DNA was transcribed before integration and that the virus was integrated into the genome very early in development, probably at the 1- or 2-cell stage.

One potential effect of gene insertion by any technique could be insertional mutagenesis: that is, altering the function of an endogenous gene caused by the insertion of an introduced gene within or near to it. Jaenisch et al. (25) and Schnieke et al. (47) found that one strain of mice transformed with the Moloney leukemia virus could not be bred to homozygosity. Early embryonic death of homozygous embryos was attributable to the nonproduction of a major gene product during development. Jaenisch's group found that the virus had inserted near the 5' end of the α-1(I) collagen gene thus blocking its expression. Although insertional mutagenesis may not be desirable in an engineered animal, it is a valuable tool for genetic studies.

Using a retrovirus to transform a mammalian germline with a functional nonviral gene has not been reported; however, Miller et al. (32) have described the synthesis of a selectable retrovirus containing a minigene for rat growth hormone (GH) that can infect cultured cells and produce a high yield of appropriately regulated growth hormone. The viruses inserted into the cellular genomes in a single location without rearrangement. In one cell line, the production of growth hormone was constitutive rather than inducible and analysis of the DNA recovered from the cell culture suggested that the original site of incorporation of the gene rather than some modification of the gene resulted in changing its mode of expression.

Failure to report the successful incorporation of a gene into the germline via a retroviral vector may be more related to the germline than to the vector. Stewart et al. (53) reported that when murine Moloney viruses were used to infect embryonal carcinoma cells, which are pluripotent as are mouse embryonic endodermal cells, the virus which integrated up to a 100 copies per cell was not expressed due to methylation. However, when the DNA was inserted into differentiated cells, it was expressed.

Transposons. Nature has provided elements other than viruses that might act as vectors for gene transfer: the transposable elements. In 1983, O'Hare and Rubin (36) isolated and characterized the DNA from the Drosophila melanogaster transposable P elements. These elements sometimes insert at specific base sequences in the Drosophila genome. They can also be excised with precision allowing the expression of an insertionally mutated gene. Spralding and Rubin (51) injected putative P element DNA into the posterior end of eggs of M type flies just prior to pole cell formation. Injection resulted in a high degree of mutability at the singed locus which controls the morphology of the hairs and bristles on the cuticle of adult flies. The mutated lines could be shown by in situ hybridization to have integrated P DNA. The integrated elements did not contain the flanking Drosophila DNA of the original construct or its pBR322 vector sequences, thus demonstrating that the P element had been transposed into the genome. Rubin and Spralding (44) then introduced the rosy gene (coding for xanthine dehydrogenase) onto the P element and injected this construct into embryos. Adult flies passed the trait for rosy on to their offspring, thus showing that the gene was heritable. The researchers suggested 2 modifications in the P element DNA to improve its utility as a vector: inclusion of more restriction sites in the element and construction of a defective P element that codes for transposase but does not insert into the genome. The latter would help gene-carrying elements to insert their gene into the genome without transforming the mutated flies to the P type. P elements were used to insert 2 genes, dopa decarboxylase (48) and alcohol dehydrogenase (15) into Drosophila. Both genes were expressed at appropriate developmental stages and expression was restricted to tissues that normally express the gene. Functional transposable elements have not been discovered in mammals though transposon-like DNA has been identified (41).

Embryo Microinjection

Injection into frog embryos. Currently, the technique par excellence for the study of gene expression is microinjection of recombinant DNA into 1-cell embryos. Oocytes of Xenopus laveis were first injected with well-characterized DNAs. In 1975, Colman (7) injected Xenopus eggs and oocytes with purified DNAs and studied their expression to determine if the machinery for transcription was present in both oocytes and eggs. He found that, contrary to expectations, both transcribed synthetic DNAs. Mertz and Gurdon (31) showed that the Xenopus oocyte could transcribe several purified DNAs efficiently if the DNA was injected into the germinal vesicle. In 1981, Rusconi and Schaffner (45) injected the rabbit β-globin gene into fertilized Xenopus eggs. The injected DNA was replicated extrachromosomally by the egg through the gastrula stage. By 6 months of age, the foreign gene was only present in the chromosomal DNA; all nongenomic foreign globin DNA had been degraded during development. The gene was present in the genome as long tandem arrays which were correctly transcribed only when the arrays were plasmid-free.

<u>Injection into mouse embryos.</u> Mice have become the mammal of choice for the study of gene expression by microinjection because recombinant DNA can be introduced into the 1-cell eggs, as demonstrated by many researchers. Tissue-specific expression has been demonstrated with a variety of cloned genes. The mouse has a relatively well-mapped genome that permits evaluation of the expression of inserted genes relative to genetic background. In 1980, Gordon et al. (17) injected a plasmid, pBR322, containing an SV40 insert and the HSV-TK gene into a pronucleus of mouse eggs. In one set of experiments, 3 of 78 mice had DNA that hybridized strongly to probes for the injected sequences. These results suggested that chromosomal integration had occurred but that the DNA was rearranged. Wagner, Stewart, and Mintz (56) microinjected the human β-globin gene/HSV-TK construct into pronuclear mouse eggs. Genomic DNA from 5 of 33 fetuses collected on day 16 or 17 hybridized with probes for both the human β-globin gene and HSV-TK. Several copies per cell of each DNA were found and they existed in association with high molecular weight DNA. One of the HSV-TK-positive fetuses appeared to express the thymidine kinase gene. Wagner, Hoppe, Jollick, Scholl, Hodinka, and Gault (58) reported a microinjected mouse egg with the rabbit β-globin gene; the globin gene was expressed in 5 mice and in the offspring of 2 of those mice as determined by immunofluorescence and immunoprecipitation. Southern hybridization confirmed that injected DNA was indeed present in one mouse though the location in the genome was not defined. Brinster et al. (5) microinjected mouse eggs with a construct that had the mouse metallothionein-I (MT) promoter linked to the HSV-TK gene. They found that 7 of 41 mice integrated the gene but only 4 expressed HSV-TK. Expression was most pronounced in the liver and kidney of affected animals, thus reflecting the specificity of the promoter sequence. The inserted construct was tandemly duplicated several times as direct repeats. Constantini and Lacy (8) microinjected a rabbit β-globin gene into mouse ova and obtained 8 positive mice of 18 born. The mice contained, in their liver, DNA that hybridized with the injected DNA. The number of copies of the gene per cell varied from 2 to more than 20, but there was no correlation between the amount of DNA injected and the number of copies integrated. They, too, concluded that the injected DNA was incorporated as tandem arrays. When these mice were bred, 5 transmitted the injected DNA to their offspring. Some progeny contained a significantly greater copy number of the β-globin gene than their parents which could be explained if the parents were mosaic.

Also, in 1981, Gordon and Ruddle (16) microinjected mouse eggs with 2 types of genetic material. One was a fusion of SV40 and HSV-TK while the other contained cDNA for human leukocyte interferon. They coined the term "transgenic" to describe the resulting offspring which had incorporated the injected DNA into their genomes. They identified 3 mice that had met their requirements to be defined as transgenic, which included: (a) the acquisition of restriction sites by the injected DNA when recovered from the host genome (flanking sequences), (b) the characteristic that injected DNA recovered from the host genome should not have the same mobility as plasmid sequences in agarose gels when undigested DNA was applied to the gel, and (c) the ability to transmit the sequences through the germline. Nine of 15 progeny inherited the new restriction sequences present in the parents and the sequences were unaltered. The potential of microinjection to create transgenic mice with new phenotypes was dramatically demonstrated by Palmiter et al. (38). They injected a fusion gene construct containing mouse MT promoter linked to a rat GH structural gene. Twenty-one mice were born and 7 of those integrated the injected sequences. Interestingly, the genes altered the phenotype of 6 mice which grew more rapidly than controls. Gene expression was greatest in the liver, again demonstrating the tissue specificity of the mouse MT promoter in an in vivo system. A number of other constructs have been injected into mouse eggs, further demonstrating the power of the technique. Fusion of the rat elastase gene promoter to the structural gene for human GH results in production of human GH in pancreatic acinar cells (37). Similarly, rat myosin gene injected into eggs results in muscle-specific expression (49), and injection of a pig major histocompatibility complex gene into mice results in a transplantation response to the antigen coded by the injected gene (13).

Problems with gene injection. Clearly, transgenic mice will be enormously useful for the study of gene expression. On the other hand, certain types of problems can occur. Palmiter et al. (39) described a strain of transgenic mice containing a MT-TK construct that did not transmit the gene through males in 5 generations of breeding. The gene may have been integrated in such a way as to interfere with the function of an important gene for spermatogenesis. This important finding has not yet resulted in describing that gene. They also found that the expression of the fusion gene was variable over several generations of mice, perhaps reflecting variation in chromatin structure from generation to generation. Wagner, Covarrubias, Stewart, and Mintz (57) found another type of problem with a strain of mice that arose from the microinjection of embryos. They found that, when they tried to breed 6 strains of transgenic mice to homozygosity for the inserted gene, 2 of the strains did not produce homozygotes, apparently because of prenatal death. The precise nature of the insertional effect is not known.

Microinjection of farm animal eggs. The mouse has been and will continue to be an excellent model system for study of gene function; however, there is much interest in manipulation of the genome of agricultural species, both to engineer farm animals and to study the genome of these species. The fertilized ova of farm species have a dense cytoplasm which makes the visualization of pronuclei more difficult than for mice. Wall et al. (59) overcame this problem for pigs by centrifuging 1-cell eggs. Centrifugation made the pronuclei visible and did not decrease the viability of the eggs. Pronuclei of cow eggs could be visualized after centrifugation. Using centrifugation for pig eggs and interference contrast optics with a high-intensity light source, Hammer et al. (21), in 1985, produced transgenic livestock for the first time. The efficiency of the process was low, requiring 1,032 injected ova to obtain a single transgenic sheep, and 2,035 injected ova to obtain 20 transgenic pigs. The gene studied was a fusion product in which an MT promoter was linked to the structural gene for human GH. Eleven of the 20 pigs expressed the structural gene as determined by radioimmunoassay for plasma human GH. The number of inserted sequences varied from one to about 490. The transgenic pigs are currently being studied to determine the effects of the gene insertion and expression on their physiology.

Recent progress in understanding gene function has been rapid because of the development of recombinant DNA and biological technologies. Now these technologies have been applied to species of agricultural interest resulting in the formation of chimeras and transgenic livestock. On the other hand, predictions about the development of useful livestock are difficult to make. To use gene insertion technology in farm animals, we should understand how the inserted genes work. We must know how promoters and enhancers alter gene expression. We must understand and achieve tissue-specific expression of a gene of interest. We must isolate and determine the functions of genes that control physiological processes important for production. We must have more knowledge of the genome of the animals into which we wish to insert genes. Each transgenic animal potentially is a new strain because the gene may be located in a different position in the genome. We need to be able to relate the location of inserted genes in order to understand position effects that might occur. We need to know if gene expression will be stable over several generations. In addition, scientists must also accept the responsibility of educating the consuming public about the nature of genetic engineering so that the public and regulatory agencies can make informed decisions on the usefulness of genetically engineered livestock.

ACKNOWLEDGEMENTS

The author extends his thanks to William Frels, Harold Hawk, and Robert Wall for their excellent reviews of this manuscript and helpful suggestions.

REFERENCES

1. Bradley, A., M. Evans, M.H. Kaufman, and E. Robertson (1984) Formation of germ line chimaeras from embryo-derived teratocarcinoma cell lines. Nature 309:255-256.
2. Brem, G., H. Tenhumberg, and H. Kraublich (1984) Chimerism in cattle through microsurgical aggregation of morulae. Theriogenology 22:609.
3. Briggs, R., and T.J. King (1952) Transplantation of living nuclei from blastula cells into enucleated frogs' eggs. Proc. Natl. Acad. Sci., USA 38:455-457.
4. Brinster, R.L. (1974) The effect of cells transferred into the mouse blastocyst on subsequent development. J. Exptl. Med. 140:1049-1056.
5. Brinster, R.L., H.Y. Chen, M.E. Trumbauer, M.K. Yagle, A.W. Senear, R. Warren, and R.D. Palmiter (1981) Somatic expression of herpes thymidine kinase in mice following injection of a fusion gene into eggs. Cell 27:223-231.
6. Butler, J.E., G.B. Anderson, R.H. BonDurant, and R.L. Pashen (1985) Production of ovine chimeras. Theriogenology 23:183.
7. Colman, A. (1975) Transcription of DNA's of known sequence after injection into the eggs and oocytes of Xenopus laevis. Eur. J. Biochem. 57:85-96.
8. Constantini, F., and E. Lacy (1981) Introduction of a rabbit β-globin gene in the mouse germ line. Nature 294:92-94.
9. Evans, M.J., and M.H. Kaufman (1981) Establishment in culture of pluripotential cells from mouse embryos. Nature 292:154-156.
10. Fehilly, C.B., S.M. Willadsen, and E.M. Tucker (1984) Experimental chimerism in sheep. J. Reprod. Fert. 70:347-351.
11. Fehilly, C.B., S.M. Willadsen, and E.M. Tucker (1984) Interspecific chimerism between sheep and goats. Nature 307:634-636.
12. Fischberg, M., J.B. Gurdon, and J.R. Elsdale (1958) Nuclear transplantation in Xenopus laevis. Nature 181:424.
13. Frels, W.I., J.A. Bluestone, R.J. Hodes, M.R. Capecchi, and D.S. Singer (1985) Expression of a microinjected porcine class I major histocompatibility complex gene in transgenic mice. Science 228:577-580.
14. Gardner, R.L. (1968) Mouse chimeras obtained by injection of cells into the blastocyst. Nature 220:596-597.
15. Goldberg, D.A., J.W. Posakony, and T. Maniatis (1983) Correct developmental expression of a cloned alcohol dehydrogenase gene transduced into the Drosophila germ line. Cell 34:59-73.

16. Gordon, J.W., and F.H. Ruddle (1981) Integration and stable germ line transmission of genes injected into mouse pronuclei. <u>Science</u> 214: 1244-1246.

17. Gordon, J.W., G.A. Scangos, D.J. Plotkin, J.A. Barbosa, and F.H. Ruddle (1980) Genetic transformation of mouse embryos by microinjection of purified DNA. <u>Proc. Natl. Acad. Sci., USA</u> 77:7380-7384.

18. Graham, F.L., and A.J. Van Der Eb (1973) A new technique for the assay of infectivity of human adenovirus 5 DNA. <u>Virology</u> 52:456-467.

19. Gurdon, J.B. (1962) Adult frogs derived from the nuclei of single somatic cells. <u>Biology</u> 4:256-273.

20. Gurdon, J.B., and R.A. Laskey (1970) The transplantation of single nuclei from single cultured cells into enucleate frogs' eggs. <u>J. Embr. and Exptl. Morph.</u> 24:227-248.

21. Hammer, R.E., V.G. Pursel, C.E. Rexroad, R.J. Wall, D.J. Bolt, K.M. Ebert, R.D. Palmiter, and R.L. Brinster (1985) Production of transgenic rabbits, sheep and pigs by microinjection. <u>Nature</u> 315:680-683.

22. Harbers, K., D. Jahner, and R. Jaenish (1981) Microinjection of cloned retroviral genomes into mouse zygotes: Integration and expression in the animal. <u>Nature</u> 293:540-542.

23. Illmensee, K., and P.C. Hoppe (1981) Nuclear transplantation in <u>Mus musculus</u>: Developmental potential of nuclei from preimplantation embryos. <u>Cell</u> 23:9-18.

24. Jaenisch, R. (1976) Germ line integration and Mendelian transmission of exogenous Moloney leukemia virus. <u>Proc. Natl. Acad. Sci., USA</u> 73:1260-1264.

25. Jaenisch, R., K. Harbers, A. Schnieke, J. Lohler, I. Chumakov, D. Jahner, D. Grotkopp, and E. Hoffman (1983) Germline integration of Moloney Murine Leukemia virus at the Mov-13 locus leads to recessive lethal mutation and early embryonic death. <u>Cell</u> 32:209-216.

26. Jaenisch, R., and B. Mintz (1974) Simian Virus 40 DNA sequences in DNA of healthy adult mice derived from preimplantation blastocysts injected with viral DNA. <u>Proc. Natl. Acad. Sci., USA</u> 71:1250-1254.

27. Jahner, D., and R. Jaenisch (1980) Integration of Moloney leukemia virus into the germ line of mice: Correlation between site of integration and virus activation. <u>Nature</u> 287:456-458.

28. King, T.J., and R. Briggs (1955) Changes in the nuclei of differentiating gastrula cells as demonstrated by nuclear transplantation. <u>Proc. Natl. Acad. Sci., USA</u> 41:321-325.

29. McGrath, J., and D. Solter (1983) Nuclear transplantation in the mouse embryo by microsurgery and cell fusion. <u>Science</u> 220:1300-1302.

30. McGrath, J., and D. Solter (1984) Inability of mouse blastomere nuclei transferred to enucleated zygotes to support development in vitro. <u>Science</u> 226:1317-1319.

31. Mertz, J.E., and J.B. Gurdon (1977) Purified DNAs are transcribed after microinjection into Xenopus oocytes. <u>Proc. Natl. Acad. Sci., USA</u> 74:1502-1506.

32. Miller, A.D., E.S. Ong, M.G. Rosenfeld, I.M. Verma, and R.M. Evans (1984) Infectious and selectable retrovirus containing an inducible rat growth hormone minigene. <u>Science</u> 225:993-998.

33. Mintz, B. (1962) Formation of genetically mosaic mouse embryos. <u>Am. Zool.</u> 2:432.

34. Mintz, B. (1971) Allophenic mice of multi-embryo origin. In <u>Methods in Mammalian Embryology</u>, J. Daniel, ed. W.H. Freeman, San Francisco, pp. 186-214.

35. Mintz, B., and K. Illmensee (1975) Normal genetically mosaic mice produced from malignant teratocarcinoma cells. <u>Proc. Natl. Acad. Sci., USA</u> 72:3585-3589.

36. O'Hare, K., and G.M. Rubin (1983) Structures of P transposable elements and their sites of insertion and excision in the <u>Drosophila melanogaster</u> genome. <u>Cell</u> 34:25-35.

37. Ornitz, D.M., R.D. Palmiter, R.E. Hammer, R.L. Brinster, G.H. Swift, and R.J. MacDonald (1985) Specific expression of an elastase-human growth hormone fusion gene in pancreatic acinar cells of transgenic mice. Nature 313:600–602.

38. Palmiter, R.D., R.L. Brinster, R.E. Hammer, M.E. Trumbauer, M.G. Rosenfeld, N.C. Birnberg, and R.M. Evans (1982) Dramatic growth of mice that develop from eggs microinjected with metallothionein-growth hormone fusion genes. Nature 300:611–615.

39. Palmiter, R.D., T.M. Wilkie, H.Y. Chen, and R.L. Brinster (1984) Transmission distortion and mosaicism in an unusual transgenic mouse pedigree. Cell 36:869–877.

40. Papaioannou, V.E., M.W. McBurney, R.L. Gardener, and M.J. Evans (1975) Fate of teratocarcinoma cells injected into early mouse embryos. Nature 258:70–73.

41. Paulson, K.E., N. Deka, C.W. Schmid, R. Misra, C.W. Schindler, M.G. Rush, L. Kadyk, and L. Leinwald (1985) A transposon-like element in human DNA. Nature 316:359–361.

42. Pighills, E., J.L. Hancock, and J.G. Hall (1968) Attempted induction of chimerism in sheep. J. Reprod. Fert. 17:543–547.

43. Rossant, J., and W.I. Frels (1980) Interspecific chimeras in mammals: Successful production of live chimeras between Mus musculus and Mus carol. Science 208:419–421.

44. Rubin, G., and A.C. Spralding (1982) Genetic transformation of Drosophila with transposable element vectors. Science 218:348–353.

45. Rusconi, S., and W. Schaffner (1981) Transformation of frog embryos with a rabbit β-globin gene. Proc. Natl. Acad. Sci., USA 78:5051–5055.

46. Scangos, G., and F.H. Ruddle (1981) Mechanisms and applications of DNA-mediated gene transfer in mammalian cells--A review. Gene 14:1–10.

47. Schnieke, A., K. Harbers, and R. Jaenisch (1983) Embryonic lethal mutation in mice induced by retrovirus insertion in the α-1(I) collagen gene. Nature 304:315–320.

48. Scholnick, S.B., B.A. Morgan, and J. Hirsh (1983) The cloned Dopa decarboxylase gene is developmentally regulated when reintegrated into the Drosophila genome. Cell 34:37–45.

49. Shani, M. (1985) Tissue-specific expression of rat myosin light-chain 2 gene in transgenic mice. Nature 314:283–286.

50. Smith, L.D. (1965) Transplantation of nuclei of primordial germ cells into enucleated eggs of Rana pipiens. Proc. Natl. Acad. Sci., USA 54:101–107.

51. Spralding, A.C., and G. Rubin (1982) Transposition of cloned P elements into Drosophila germ line chromosomes. Science 218:341–347.

52. Stevens, L.C. (1970) The development of transplantable teratocarcinomas from intratesticular grafts of pre- and postimplantation embryos. Dev. Biol. 21:364–382.

53. Stewart, C.L., H. Stuhlmann, D. Jahner, and R. Jaenisch (1982) De novo methylation, expression, and infectivity of retroviral genomes introduced into embryonal carcinoma cells. Proc. Natl. Acad. Sci., USA 79:4098–4102.

54. Tarkowski, A.K. (1961) Mouse chimaeras developed from fused eggs. Nature 190:857–860.

55. Tucker, E.M., R.M. Moor, and L.E.A. Rowson (1974) Tetraparental sheep chimeras induced by blastomere transplantation: Changes in blood type with age. Immunology 26:613–621.

56. Wagner, E.F., T.A. Stewart, and B. Mintz (1981) The human β-globin gene and a functional viral thymidine kinase gene in developing mice. Proc. Natl. Acad. Sci., USA 78:5016–5020.

57. Wagner, E.F., L. Covarrubias, T.A. Stewart, and B. Mintz (1984) Prenatal lethalities in mice homozygous for human growth hormone gene sequences integrated into the germ line. Cell 35:647–655.

58. Wagner, T.E., P.C. Hoppe, J.D. Jollick, D.R. Scholl, R.L. Hodinka, and J.B. Gault (1981) Microinjection of a rabbit β-globin gene into zygotes and its subsequent expression in adult mice and their offspring. Proc. Natl. Acad. Sci., USA 78:6376-6380.
59. Wall, R.J., V.G. Pursel, R.E. Hammer, and R.L. Brinster (1985) Development of porcine ova that were centrifuged to permit visualization of pronuclei and nuclei. Biol. Reprod. 32:645-651.
60. Wigler, M., S. Silverstein, L.S. Lee, A. Pellicer, Y.C. Cheng, and R. Axel (1977) Transfer of purified herpes simplex virus thymidine kinase genes to cultured mouse cells. Cell 11:223-232.

GENETIC ANALYSIS IN MAMMALS: PAST, PRESENT, AND FUTURE

Stephen J. O'Brien

Laboratory of Viral Carcinogenesis
National Cancer Institute
Frederick, Maryland 21701

In the mid-1960s, when I first became interested in animal genetics as an undergraduate student, the revolution in biological and genetic technology which is realized today was a mere twinkle in the eye of practicing geneticists. The human gene map contained only a few loci, mostly on the X-chromosome, and Drosophila was the undisputed leader in genetic analysis of metazoans. The mouse genome was the prototype of mammalian gene studies because of the terrific advantages offered by the development of inbred mice. Mendel's laws were (and still are) the cornerstone of eukaryote genetics and intuitive genetic analysis was a talent practiced by such pioneers as Bridges, Morgan, McClintock, Dobzhansky, and Green. Since that period, the field of animal genetics has grown appreciably due in no small way to the terrific explosion of molecular biology, parasexual genetic analysis, and DNA methodology. Today, we have in our scientific repertoire the ability to examine gene action virtually by direct observation and to put to the test the scores of intriguing hypotheses that have emerged during the deductive period of genetics.

I would like to begin this chapter with a quotation appropriately enough from Theodosius Dobzhansky, a former professor at the University of California, Davis, who influenced my thinking tremendously in my youth and shall continue to do so as I grow older:

"Nothing in biology makes sense except in light of evolution."

This concept will be apparent throughout this chapter for several reasons. First, the biological systems we all study are indeed products of evolution and reconstructing their origins can only help in resolving developmental mechanisms. Second, a primary interest of my own involves genetic events in neoplastic transformation, and I am often struck by the similarities of tumor development and evolutionary processes. One might even go so far as to say that cancer progression is merely an evolutionary process speeded up! Finally, the internal genetic variation, the etiologic agents which afflict animals, and the interacting genetic systems (e.g., immune system, the major histocompatibility complex, disease resistances) evolve and change according to evolutionary principles.

Today, the best known mammal in a genetic sense is man. The human map has grown to be virtually unreadable with over 1,000 different loci identified and chromosomally mapped (9,21). This number, though impressive, probably represents only about 1% of human structural genes with 99,000 loci yet to be identified. In a recent review, Shows et al. (21) listed over 40 different classes of genes that comprise the human gene map and Frank Ruddle discusses many of the technological advances that contributed to the human map elsewhere in this Volume (see Ruddle and Fries). The major technologies have included pedigree analysis, especially with the recently employed restriction fragment length polymorphisms (RFLPs), heterologous somatic cell hybrids, isozyme procedures, and in situ hybridization of cloned DNA fragments to metaphase chromosomes. The progress of the human mapping has been encouraging, since in several cases independent mapping of seemingly different functions has led to confirmation of their identity. For example, the receptor for epidermal growth factor and the proto-onc gene erb-B were both shown to reside on the short arm of chromosome 7 (19,22) before their DNA sequence alignments proved they were coded by the same gene (2).

Techniques similar to those used to construct the human map have also been applied to a number of other mammals as well. A recent workshop on comparative gene mapping (7) reviewed the genetic maps of 29 different mammals including 11 primates and 18 nonprimate species. Certainly, we have our greatest database in the mouse, but several other species have produced sizable genetic maps as well (Tab. 1). In most of the genetic maps referred to in Tab. 1, the majority of genes mapped are homologous to genes previously assigned to specific chromosomal positions in man and, to a lesser extent, in mouse. This means that the linkage relationships between the same genes in other species can be compared directly to the

Tab. 1. Mammalian gene maps (7,12).

Species	Haploid number	Number of linkage groups	Number of mapped genes
Man	23	23	824
Mouse	20	20	750
Rat	14	14	65
Cat	19	16	53'
Rabbit	17	16	53
Chinese hamster	11	10	48
Gorilla	24	20	37
Orangutan	24	14	36
Cattle	30	24	36
Dog	39	28	36
Pig	19	17	36
Vervet	30	17	36
Mouse lemur	33	13	34
Rhesus monkey	21	14	33
Owl monkey	25	14	33
Baboon	21	10	30
Gibbon	26	16	25
Capuchin monkey	27	14	25
Sheep	27	12	24
Horse	32	5	18

organization in man. These kinds of comparisons form the basis for the small but fascinating field of comparative gene mapping (15,18,23).

A dramatic finding of this new field was the demonstration of a striking conservation in linkage/syntenic associations in the great apes and an even more striking conservation in the Old World monkeys, New World monkeys, and Lemuridae. An abridged phylogeny of the primates is presented in Fig. 1. By and large, and with only a few notable exceptions, the linkage maps of all primates were concordant (15,18). A cytogenetic reinforcement of this conclusion was provided by the extensive and elegant cytogenetic analysis of various primate species described by Bernard Dutrillaux, Jean de Grouchy, and in the great apes, Jorge Yunis (1,3,4,27,28). Dutrillaux and his colleagues performed high-resolution R banding on over 70 different primate species and found (again with a few exceptions) a striking cytological homology between simian karyotypes. In fact, they were able to construct what they considered to be a primitive simian karyotype (3,4). From this karyotype, a very few rearrangements led to the extant chromosomal rearrangements that are seen in modern primates. Two major exceptions to the conservation were seen in gibbons and owl monkeys. Both of these species have an extremely rearranged karyotype and a disruption of most of the syntenies observed in other primates. The chromosomal shuffle in these two primate lineages violates the "rule" of chromosomal conservation and is a good example of the discontinuous mode of chromosomal evolution in mammals (15). As shall be seen below, it appears that cytogenetic evolution is not continuous or steady at all; rather, it is conservative

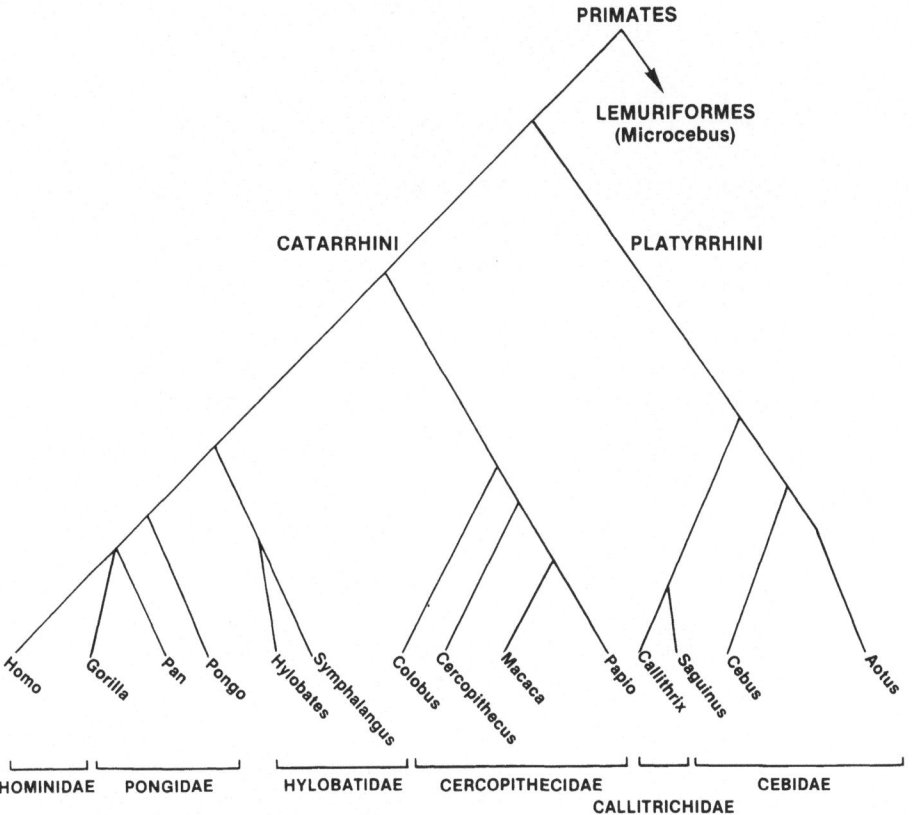

Fig. 1. A consensus phylogeny of the primates based upon morphological and molecular considerations.

for long periods and "punctuated" by extensive chromosomal shuffles in selected lineages such as the gibbon and the owl monkey. Apparently, karyo-
logical divergence, like morphological evolution (5,6), can be static for
entire family radiations with only occasional breaks that are cytologically
extreme.

The next link in the story involves a favorite of my own laboratory:
the domestic cat. The cat has 19 chromosome pairs. A few years back we
began the construction of a gene map using panels of somatic cell hybrids
that lost feline chromosomes in different combinations (13). Isozyme markers and DNA segments detected with specific probes were mapped, and today
we have over 50 loci on the cat map. When we started, the cat map had one
locus mapped: orange on the X-chromosome. The cat is in a different mammalian order, Carnivora, from primates so we were rather surprised to dis-
cover that the extent of chromosome conservation between cat and man was
extensive (11,13). In fact, of the 33 loci that were mapped in both man
and cat, only 3 exceptions to synteny were obtained (Fig. 2).

In light of similarities in linkage association observed between the
cat and man, the extent of cytological homology between the syntenically
homologous chromosomes in the two species was examined (11). G-banded
chromosomes of each human type were compared with feline chromosomes
thought to be homologous based on the linkage homologies shown in Fig. 2.
The results of these analyses have been described in detail elsewhere (11)
and will be summarized here. Essentially, only those band-for-band alignments that were apparent upon high-resolution banding and that include re-

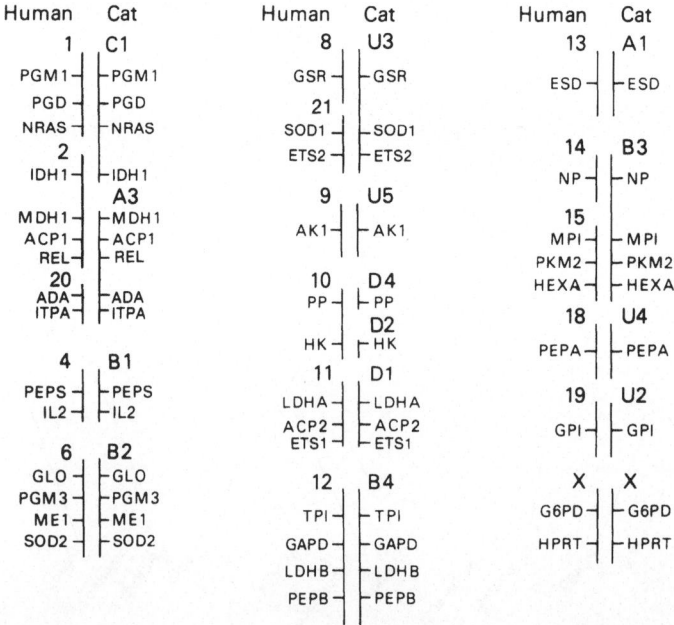

Fig. 2.   Linkage group associations comparing the cat and man. The gene
order given is that determined in man. Feline linkage groups U2
and U4 have not been assigned to feline chromosomes. Gene names
used are those of homologous human enzyme loci as recommended by
the international system for human gene nomenclature (20).

gionally mapped loci (in man and/or cat) were included. We have also taken advantage of the abundant cytological data that has been collected on primate cytogenetics (1,3,4) and feline cytogenetics (25,26). We have already discussed the conservation of large portions of the human genome within the primate order. Then, one could identify those portions of the modern human genome that are "ancestral," and the parts that were recently rearranged. This was possible largely because a sizable portion of the human genome has apparently been preserved intact over the approximately 80 million years of primate evolution.

For the comparison of feline and human chromosomes, another serendipitous situation contributed to the feasibility of such an analysis. It turns out that the karyotype of the cat family is also highly conserved (25,26). The domestic cat has 19 chromosomes, 16 of which are invariant in all 37 species of the Felidae. Furthermore, of these 16 chromosomes, 15 are present in several other carnivore orders, either intact or only slightly modified. The domestic cat is characterized by a highly primitive karyotype, which, like the human karyotype, can be segmented into ancestral and rearranged portions. For the comparative analysis of human and feline syntenies, only the ancestral portions of chromosomes from each species were compared (11). High-resolution, G-banded preparations of homologous chromosomes were carefully examined, and several regions of band-for-band homology were identified. In fact, we could align between 20 and 25% of the human genome band for band with the feline karyotype despite the passage of 100 million years since these species shared a common ancestor (11).

The mammal with the most genes mapped, other than human, is the house mouse, with over 800 loci mapped over 20 acrocentric chromosomes (16). The field of comparative gene mapping really began with extensive comparisons of mouse and man. Today's maps allow the comparison of over 150 homologous loci between man and mouse. Figure 3 is an abridged human map with the regional positions of those genes whose homologous counterparts have also been mapped in the mouse. The murine chromosomal positions are indicated by the numeral next to the gene abbreviation. Several points become evident upon close examination of this comparative map. First, a considerable number of rearrangements have occurred between human and mouse. Nadeau and Taylor (10) estimated that a minimum of 178 rearrangements have occurred during the divergence between man and mouse. We can count over 40 linkage segments in the human map that are homologous to murine syntenic groups. It is clear that the segments are rather small: usually less than 10 map units. A second striking conclusion is that despite the degree of chromosome shuffle that has occurred, it is not beyond resolution, and the conserved segments are by and large contiguous (e.g., human chromosomes 1, 2, 7, 10) in both the human and murine genomes.

When one compares the cat-human alignment vs the mouse-human syntenies, it is worth noting that the extent of shuffle seems to be 2 to 3 times greater in the rodents than in the carnivores (15). Thus, we have concluded that the primitive carnivore-primate linkage arrangements have been rearranged to a lesser extent than in the evolution of rodents.

There are several other examples of chromosomal stasis punctuated by dramatic chromosomal rearrangement (reviewed in detail in Ref. 15). One additional illustration of this is seen in the evolution of the 9 families of Carnivora (Fig. 4). Recall that the 37 species of Felidae have 19 pairs of metacentric chromosomes and 16 of these pairs are identical in all cat species. Furthermore, these 16 conserved chromosomes are found intact in several other families including Viverridae, Procyonidae, and Ailuridae. This means that a primitive carnivore karyotype is being maintained in these families.

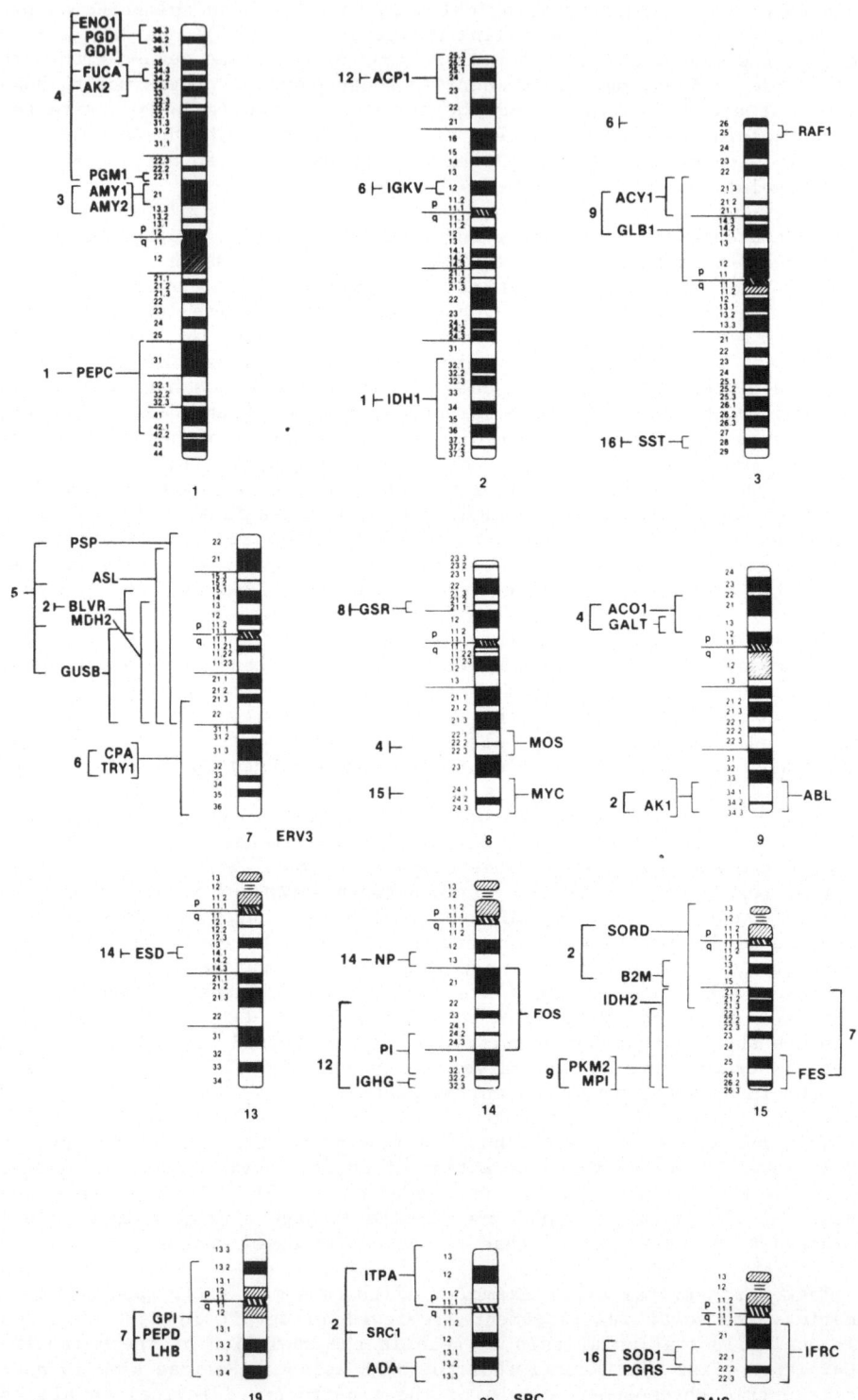

144

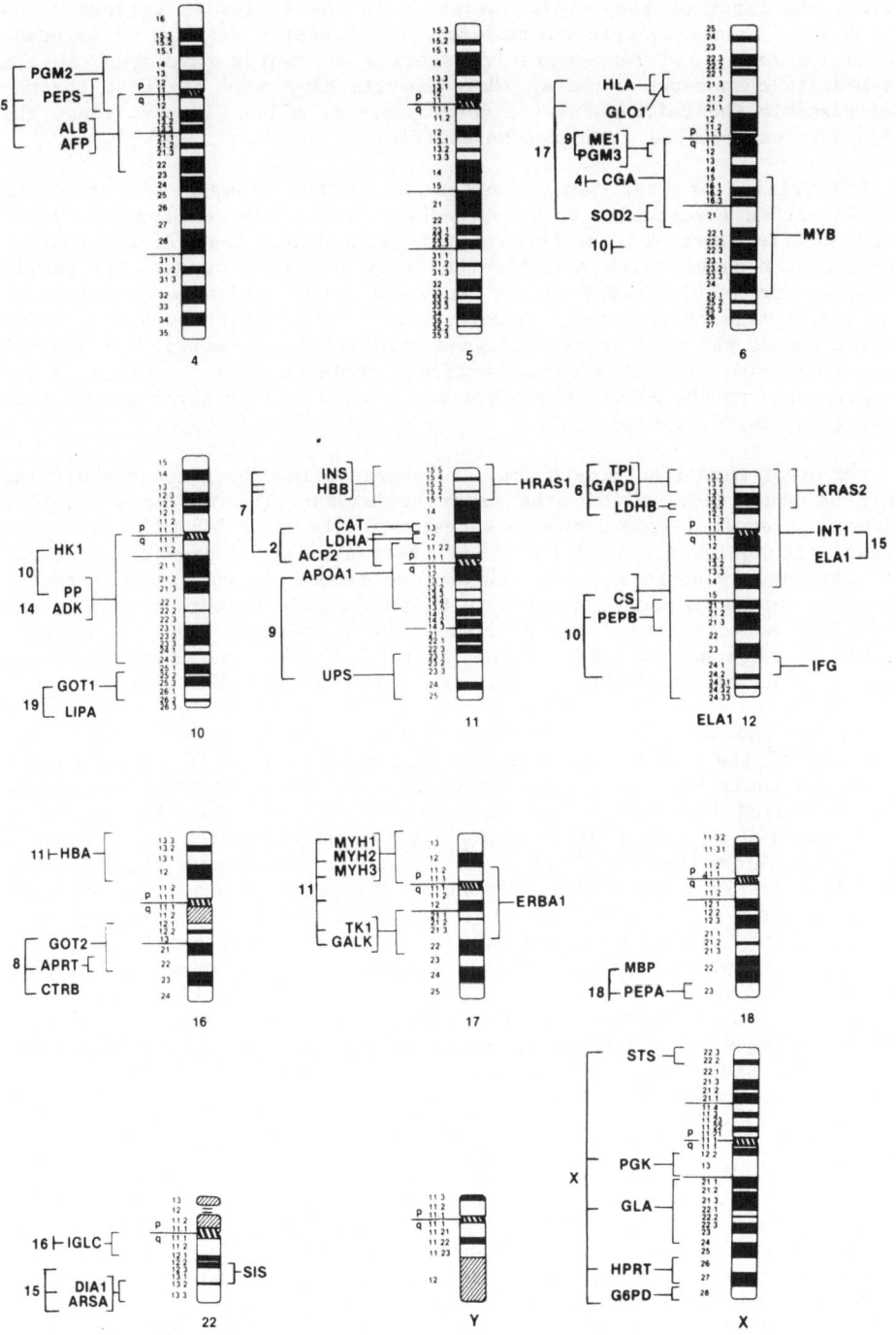

Fig. 3. Regional genetic map of man demonstrating only those loci whose homologous counterparts have been mapped in the mouse. The murine chromosomal position is indicated by the numeral next to the gene name. Conserved linkages between human and mouse are indicated by vertical links connecting syntenic genes.

However, not all carnivores share the conserved karyotype. The Canidae, like the Felidae, has about 40 extant species. Unlike the felids, however, the range of karyotypic variation in the canids is extreme. The red fox has 36 metacentric chromosomes, the domestic dog has 72 acrocentric chromosomes, and representative species of canids have many of the intermediate chromosome numbers. More importantly, none of these chromosomes resemble the felid-carnivore chromosomes at all. Clearly, then, the canids have experienced a chromosome shuffle.

The ursids, or bear family, offer us another example of chromosome shuffles with a fascinating twist. The bear family also underwent a chromosome shuffle since most members of the genus Ursus have 72 acrocentric chromosomes, none of which resemble the carnivore chromosomes. The earliest leg on the ursid lineage led to the giant panda, which has a karyotype of 42 metacentric chromosomes. However, nearly all the chromosome arms of the giant panda can be precisely aligned with ursid acrocentric chromosomes (14). Therefore, the chromosomal shuffle associated with the ursid radiation persisted in the modern bears but was reorganized by chromosome fusion in the giant panda lineage (14).

Two other mammalian orders that are characterized by genomic shuffling should be mentioned. Within the order Perissodactyla, the horse family, Equidae, has evolved from a common ancestor in the last 4-5 million years. In fact, the equids are well known for their ability to produce interspecific hybrids. Nonetheless, the chromosome number in this family ranges from 32 in Hartman's mountain zebra to 66 in the wild precursor to domestic horses, the Mongolian wild horse (17). What is perhaps most striking is the ability to produce hybrids (albeit sterile) between such karyotypically diverse species as the donkey (2n=62) and the mountain zebra (2n=32).

A final and often quoted chromosomal shuffle is apparent in the Cervidae family of the order Artiodactyla. The deer family, Cervidae, has 53 species and their karyotypes are extremely variable. The high extreme is the white-tailed deer with a diploid number of 70. The opposite extreme is seen in the red or Indian muntjac (Muntiacus muntjak) which has a diploid number of 6 in females and 7 in males due to an X-autosome translocation (24). The closely related Chinese muntjac (M. reevesi) has a karyotype of 46 chromosomes. The observation that hybrid (although sterile) offspring were produced by crossing the two muntjac species illustrates the possible rapidity in evolutionary terms of the chromosomal reassortment that led to these modern groups. A comparative analysis of the two muntjac chromosomal complements present in the hybrid revealed that the huge Indian muntjac chromosomes are derived from a series of tandem and Robertsonian translocations of the chromosomes seen in the Chinese muntjac (8). Eight of the latter species' chromosomes combined to produce chromosome 1 in the Indian muntjac.

The cumulative data on comparative cytology, gene mapping, and gene family dispersal in mammals suggest several major generalizations about the character of genome organization. First, although there are certain disagreements between cytological and syntenic homologies, the most common experience has been the independent confirmation of chromosomal homologies. In several mammalian orders, most notably primates and carnivores, a precise cytological alignment of large numbers of the chromomeres is possible, suggesting a preservation of the integrity of chromosomal morphology since the emergence of these taxa.

A second important observation is that chromosomal evolution in mammals is punctuated in a manner reminiscent of morphological variation. Thus, several phylogenetic lineages are rather conserved with a high degree of chromosomal homology, while other lines are characterized by

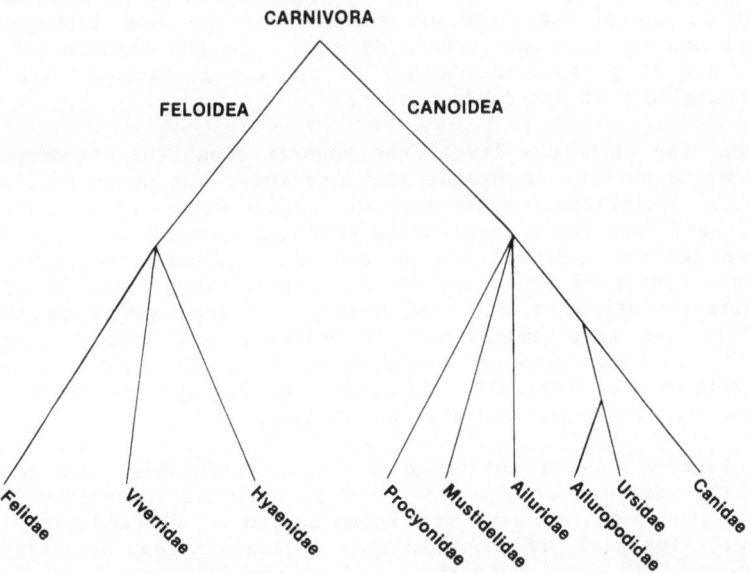

Fig. 4. A consensus phylogeny of carnivores based upon morphological considerations.

abundant chromosomal rearrangements. The primates show levels of extreme conservation, with multiple chromosomal homologs being identified between such distant groups as man, New World monkeys, and mouse lemurs. In fact, the conservation of primate cytological integrity extends beyond primates to the carnivores, in which the genome of one species, the domestic cat, shows striking syntenic and cytological homologies with that of man. When several stringent criteria for cytogenetic homology were employed, over 20% of the human genome was aligned band for band with that of the cat. We have interpreted these concordant cytological and syntenic homologies as reflecting ancestral chromosomal arrangements that have been preserved intact within the modern groups that share them (15).

Conversely, several mammalian families are not at all conserved chromosomally and have chromosomal organizations that are dramatically rearranged or "shuffled," both between and within their respective groups. Within the primates, the gibbon family, Hylobatidae, and the New World monkey, _Aotus_, are particularly shuffled, with many diverse karyotypes between and within genera when compared to the human map. Several carnivore families (e.g., Canidae and Ursidae) are also karyotypically diverse when compared to conserved Felidae. The rodents are particularly diverse chromosomally and vary greatly between and even within certain species. It is important to note, however, that the "shuffled" rodent genome is not rearranged beyond resolution in a comparative sense. Thus, although cytological homologies between rodents and primates are not apparent, the comparison of the human and mouse syntenic maps does reveal the linkage of short regions of homologous loci between the two species. Thus, the human and mouse maps can be segmented into nearly 200 homologous chromosome segments, which vary in length from 1 to 25 centimorgans. These segments contain groups of homologous loci that are contiguous in both species and, as such, illustrate the limits of cytogenetic divergence that occurred between the conserved human prototype arrangement and the shuffled rodent prototype.

Although a cohesive picture of the patterns and processes of chromosomal evolution is beginning to take shape, many new questions emerge from

147

our conclusions. Among the more intriguing uncertainties are the following. What is the evolutionary explanation for why some lineages exhibit chromosomal conservation and others do not? Are the chromosomal shuffles adaptive? Are they rapid mechanisms of species isolation? Are they the cause or the effect of speciation events?

On the fine structure level, one wonders about the chromosomal break junctions which persist in evolution. Are these hot spots of chromosomal breakage, the evolutionary equivalent of fragile sites, often seen in human somatic tissue? Are there interesting genes at these junctions? A goal of comparative genetic analysis has become the ultimate reconstruction, at least within conserved lineages, of the parsimonious sequence of chromosomal rearrangements that occurred during the development of the radiations. This has been approached in primates and felids, and may be possible in conserved carnivore families as well. The more shuffled groups will, of course, be difficult, although the largely contiguous, syntenic segments between mouse and man are encouraging.

What happens beyond eutherian mammals, marsupials, and monotremes? What does the gene map of a wombat or a platypus or an echidna look like? Is there a steady-state level for introduction of chromosomal rearrangements in all lineages? If so, does this indicate strong selection against rearrangements? If not, could this explain the "jerkiness" of chromosomal evolution? We simply do not know the answers to these questions.

Finally, is there any relationship between the ancestral breakpoints in human/primate chromosomes and the specific chromosomal breakpoints observed in human tumors? Do the genes at the junctions of human tumors (in some cases proto-oncogenes) bear any resemblance functionally or selectively to the genes at the chromosomal breakpoints? The field of animal genetics is ripe for addressing all these questions at last, and it is not beyond expectation to envisage a modern synthesis of evolution, development, and neoplastic processes by students of these 3 parallel disciplines.

ACKNOWLEDGEMENTS

I am grateful to James Womack, David Wildt, Hector Seuanez, Glen Collier, William Nash, William Modi, Janice Martenson, and Mary Eichelberger for several stimulating discussions around the concepts here discussed.

REFERENCES

1. Couturier, J., B. Dutrillaux, C. Turleau, and J. de Grouchy (1982) Comparaisons chromosomiques chez quartre especes on sous-especes de gibbons. Ann. Genet. 25:5-10.
2. Downward, J., Y. Yarden, E. Mayes, G. Scrace, N. Totty, P. Stockwell, A. Ullrich, J. Schlessinger, and M.D. Waterfield (1984) Close similarity of epidermal growth factor receptor and v-erb-B oncogene protein sequences. Nature 307:521-527.
3. Dutrillaux, B. (1979) Chromosomal evolution in primates: Tentative phylogeny from Microcebus murinus (Prosimian) to man. Hum. Genet. 48:251-314.
4. Dutrillaux, B., and J. Couturier (1981) The ancestral karyotype of platyrrhine monkeys. Cytogenet. Cell Genet. 30:232-242.
5. Gould, S.J. (1980) Is a new and general theory of evolution emerging? Paleobiology 6:119-130.
6. Gould, S.J., and N. Eldridge (1977) Punctuated equilibria: The tempo and mode of evolution reconsidered. Paleobiology 3:115-151.

7. Lalley, P., and V. McKusick (1985) Report of International Committee on comparative gene mapping. Cytogenet. Cell Genet. (in press).

8. Liming, S., Y. Yingying, and D. Xingsheng (1980) Comparative cytogenetic studies on the red muntjac, Chinese muntjac and their F_1 hybrids. Cytogenet. Cell Genet. 26:22-27.

9. McKusick, V.A. (1984) The human gene map. In Genetic Maps, Vol. 3, S.J. O'Brien, ed. Cold Spring Harbor Press, New York, pp. 417-441.

10. Nadeau, J.H., and B.A. Taylor (1984) Lengths of chromosomal segments conserved since the divergence of man and mouse. Proc. Natl. Acad. Sci., USA 81:814-818.

11. Nash, W.G., and S.J. O'Brien (1982) Conserved regions of homologous G-banded chromosomes between orders in mammalian evolution: Carnivores and primates. Proc. Natl. Acad. Sci., USA 79:6631-6635.

12. O'Brien, S.J., ed. (1984) Genetic Maps, Vol. 3, Cold Spring Harbor Press, New York, 584 pp.

13. O'Brien, S.J., and W.G. Nash (1982) Genetic mapping in mammals: Chromosome map of domestic cat. Science 216:257-265.

14. O'Brien, S.J., W.G. Nash, D.E. Wildt, M.E. Bush, and R.E. Benveniste (1985) A molecular solution to the riddle of the giant panda's phylogeny. Nature 317:140-144.

15. O'Brien, S.J., H.N. Seuanez, and J.E. Womack (1985) On the evolution of genome organization in mammals. In Evolutionary Biology, R.J. MacIntyre, ed. Plenum Press, New York, pp. 518-589.

16. Roderick, T., and M. Davisson (1984) The linkage map of the mouse (Mus musculus). In Genetic Maps, Vol. 3, S.J. O'Brien, ed. Cold Spring Harbor Press, New York, pp. 343-355.

17. Ryder, O.A., N.C. Epel, and K. Benirschke (1978) Chromosome banding studies of the Equidae. Cytogenet. Cell Genet. 20:332-350.

18. Seuanez, H.N. (1984) Evolutionary aspects of human chromosomes. Subcell. Biochem. 10:455-537.

19. Shimizu, N., M.A. Behzadian, and Y. Shimuzu (1980) Genetics of cell surface receptors for bioactive polypeptides: Binding of epidermal growth factor is associated with the presence of human chromosome 7 in human-mouse cell hybrids. Proc. Natl. Acad. Sci., USA 77:3600-3604.

20. Shows, T.B., A.Y. Sakaguchi, and S.L. Naylor (1982) Mapping the human genome, cloned genes, DNA polymorphisms and inherited disease. Adv. Hum. Genet. 12:341-452.

21. Shows, T.B., P.J. McAlpine, and R.L. Miller (1984) The 1983 catalogue of mapped human genetic markers and report of the Nomenclature Committee. Cytogenet. Cell Genet. 37:340-393.

22. Spurr, N.K., E. Solomon, M. Jansson, D. Sheer, P.N. Goodfellow, W.F. Bodmer, and B. Vennstrom (1984) Chromosomal localization of the human homologues to the oncogenes erbA and B. EMBO J. 3:159-163.

23. Womack, J.E. (1982) Linkage of mammalian isozyme loci: A comparative approach. In Isozymes: Current Topics in Biological and Medical Research, Vol. 6, M.C. Rattazzi, J.G. Scandalios, and G.S. Whitt, eds. Alan R. Liss, New York, pp. 207-246.

24. Wurster, D.H., and K. Benirschke (1970) Indian muntjac, Muntiacus muntjak: A deer with a low diploid chromosome number. Science 168: 1364-1366.

25. Wurster-Hill, D.H., and W.R. Centerwall (1982) The interrelationships of chromosome banding patterns in canids, mustelids, hyena and felids. Cytogenet. Cell Genet. 34:178-192.

26. Wurster-Hill, D.H., and C.W. Gray (1973) Giemsa banding patterns in the chromosomes of twelve species of cats (Felidae). Cytogenet. Cell Genet. 12:388-397.

27. Yunis, J.J., and O. Prakash (1982) The origin of man: A chromosomal pictorial legacy. Science 215:1525-1530.

28. Yunis, J.J., J.R. Sawyer, and K. Dunham (1980) The striking resemblance of high-resolution G-banded chromosomes of man and chimpanzee. Science 208:1145-1148.

INTRODUCTION AND REGULATION OF CLONED GENES

FOR AGRICULTURAL LIVESTOCK IMPROVEMENT

Thomas E. Wagner

Edison Animal Biotechnology Center
Ohio University
Athens, Ohio 45701

INTRODUCTION

The advent of gene transfer technology, which allows the introduction of well-characterized, cloned genes into the permanent genetic make-up of mammalian species, including laboratory mice (17,56) and domestic farm animals (20), holds the promise of providing a new methodology for the genetic improvement of livestock. Indeed, using these recombinant genetic procedures, greater genetic improvement may soon be achieved in a single generation of an animal than has previously been possible using classical genetic selection over a period of decades. But, in order to realize this promise, it will be necessary not only to introduce specific genes into the genetic composition of domestic animals but also to regulate the expression of these transgenes in concert with the existing physiological requirements of the animal.

Although cloned genes have been introduced into transgenic animals and shown to function to produce their protein products, demonstration of external, or regulated internal, control of genetic expression of these genes has not, as yet, been fully achieved. The design, construction, and testing of genetic control sequences in transgenic animals is the largest and most significant challenge facing those who wish to use gene transfer technology to improve livestock animals. Although several economically important animals could result from the addition of continuously expressing native or modified genes to transgenic livestock, the most far reaching animal agricultural applications of gene transfer technology will require specific control of the time and site of expression of added genes.

Already, several of the protein products of cloned livestock genes have been shown to dramatically affect the performance of farm animals. Probably the best example of such a genetic system is that for growth hormone. A wealth of data shows clearly that an increase in the growth hormone level in meat-producing animals increases growth rate and feed efficiency and, in dairy animals, increases milk production. The effects of exogenously added growth hormone on growth rate have been known for more than 50 years since the original studies of Evans and Simpson (14) and Lee and Schaffer (27) who demonstrated that extracts from pituitary glands caused dramatic increases in the growth rates of experimental animals. More recently, Machlin (28) and Etherton and Kensinger (13) have clearly demonstrated that purified porcine growth hormone increases the growth rates and feed efficiencies of young growing swine. The data presented by these researchers clearly show that elevated levels of growth hormone in the pig are advantageous in terms of rate of growth, feed efficiency, and body composition, all important economic traits. Also, growth hormone has been shown, first by Machlin (29) and later by Peel et al. (38-40) and Fronk et al. (15), to increase milk production from dairy cattle from 15%, in early lactation, to as much as 40% in late lactation when administered to adult cows. Interestingly, administration of growth hormone to these adult dairy animals, over an entire lactation, has shown no adverse effects on bone growth or on any other aspect of the physiology of the animal (15). Therefore, it is clear that transgenic delivery of growth hormone to grow-ing swine or lactating cattle would result in a superior producing animal.

Studies with transgenic mice, harboring growth hormone fusion genes with simple, strong constitutive promoters and expressing high levels of growth hormone in a wide range of tissues, show continuously through the lifespan of these animals that the uncontrolled presence of growth hormone, in addition to substantially enhancing growth rate, interferes with the re-productive capacity of some female animals (20,55). While this effect would be of little concern in swine and other animals destined for slaugh-ter, it would be a serious impediment to the propagation of improved genet-ic lines of any livestock. Therefore, development of genetic constructions that offer the regulated expression of growth hormone genes in transgenic animals, allowing expression in young animals being grown for slaughter while remaining "dormant" in breeding stock, would greatly aid the practi-cal use of growth hormone gene transfer in swine and other farm animals. The requirement of regulated genetic delivery of growth hormone would be even more stringent in dairy livestock, where the negative implications of enhanced growth effects in the young animal and potential reproductive problems resulting from the continuous delivery of growth hormone in the animal would far outweigh the advantages of increased lactation.

Advancements in molecular biology and recombinant genetic technology which have occurred over the past several years suggest that exquisite con-trol of the expression of added genes in transgenic animals may be a real possibility in the near future. Whereas, until recently, most of the studies aimed toward understanding the molecular mechanisms of genetic reg-ulation have been carried out using specifically transformed, cultured mam-malian cells, studies with transgenic animals within the last year have advanced this understanding dramatically. The principles of genetic regu-lation in mammals elucidated by studies with experimental transgenic ani-mals now offer several means to regulate the expression of transgenes in livestock animals for enhanced agricultural production. Below, a review of the development of mammalian gene transfer and genetic regulation in trans-genic animals is presented prior to a discussion of its potential applica-tion in the animal sciences.

GENE TRANSFER AND THE PRODUCTION OF TRANSGENIC MAMMALS

Transgenic mammals are produced by microinjecting recombinant DNA di-rectly into the pronuclei of zygotes shortly after fertilization, using a concept and methodology developed at the beginning of this decade (17,56). This method allows virtually any DNA molecule to be introduced into the chromosomal complement of the zygote and, therefore, into each of the cells of the resulting animal. Since the development of this seminal technology, mouse lines have been established containing cloned transgenes coding for β-globin for several species (7,10,26,47,53,56), herpes simplex virus thy-midine kinase (3,57), growth hormone from several species (20,35,36,45,58), chicken transferrin (30), simian virus 40 (SV40) viral T-antigen (4,21), immunoglobins μ and κ (18,42,49), rat elastase (34,50), the cellular oncogene myc product (48), rat myosin light chain (46), and mouse alphafetoprotein (25). These studies carried out over the last several years have not only demonstrated the generality of the gene transfer methodology, but have also shown that when appropriate regulatory elements are included in the sequences introduced, transgenes show extremely specific native regulation and tissue-specific expression.

An important aspect of appropriate native regulation of a genetic system is tissue-specific expression (i.e., the production of the protein gene product exclusively, or highly preferentially, within the cells of one or more specific tissues). A number of cloned intact genes are expressed in a tissue-specific manner in transgenic animals. A list of these genetic sequences, the tissue of specific expression, the degree of specificity, and reference is presented in Tab. 1.

Interestingly, the first transferred gene expressed in a transgenic animal showed a high degree of tissue-specific expression (56). In this early experiment the rabbit β-globin gene was shown to be expressed in a highly preferential manner in the erythroid cells of transgenic mice containing this gene in their chromosomal complement. More recently, several other laboratories have confirmed this remarkable globin gene tissue specificity (7,53) and have demonstrated for human β-globin, through 5' deletions to -48 which continue to show expression in appropriate cells, that the sequences regulating tissue-specific expression reside either within the structural gene or 3' to it (53). The highest degree of tissue-specific expression has been demonstrated for the rat elastase gene, which shows absolute specificity of expression to the acinar cells of the pancreas, expressing in excess of 10,000-fold more message in these cells than any other in transgenic mice harboring this gene (34,50). Also, genetic constructions with the elastase promoter/enhancer element fused to either the SV40 T-antigen gene or the growth hormone gene express specifically in the pancreas (34a). In addition to the globin and elastase gene systems, considerable work has established that the immunoglobin μ and κ genes are expressed with a high degree of specificity in lymphoid cells, immunoglobin μ genes equally well in B and T cells, and κ light chain genes only in B cells, while neither are expressed to any observable extent in nonlymphoid cells.

Although the ability to designate the tissue of expression of added genes in transgenic animals is beginning to be developed, little or no progress has been made in regulating the time during the life span of the transgenic animal when an introduced gene is operative. Temporal regu-

Tab. 1. Tissue-specific gene expression.

Gene	Cell type	Expression (max. %)	Name/Reference
Rabbit β-globin	Erythroid	>50	Wagner et al. (56)
Human β-globin	Erythroid	100	Townes et al. (53)
Mouse/human β-globin	Erythroid	2	Chada et al. (7)
Rat elastase I	Pancreatic	1,200	Swift et al. (50)
Rat myosin light chain	Skeletal muscle	500	Shani (46)
Mouse alphafetoprotein	Yolk sac and liver	25	Krumlauf et al. (25)
Mouse kappa light chain (Ig)	B cells	50	Storb et al. (49)
Mouse μ heavy chain (Ig)	B and T cells	85	Grosschedl et al. (18)

lation of gene expression, like tissue specificity, involves complex molecular mechanisms which must be understood in order to effect temporal regulation of transgenes in experimental or commercial animals.

MECHANISMS OF GENETIC REGULATION

Although some of the molecular mechanisms controlling genetic expression in prokaryotic organisms are essentially understood, little is known about the discrete molecular mechanisms and interactions that regulate eukaryotic and, specifically, mammalian gene expression. One of the arguments put forth in support of the detailed study of prokaryotic gene regulatory mechanisms has been that such a study would shed light upon these mechanisms in higher organisms, including mammals. Unfortunately, this expectation has not been realized since genetic regulatory mechanisms appear to differ greatly between lower life forms and higher eukaryotic organisms. In large measure, regulation of prokaryotic gene expression is by negative control. The ability of RNA polymerases to enter the DNA and "read" the genetic message is blocked by the presence of specific repressors in order to inhibit gene expression. Activation of gene expression is accomplished by the removal of repression. A similar basic mechanism does not seem to explain any major portion of the control mechanisms operative in the mammalian genome. Modulation of the basic negative control mechanism in prokaryotes, allowing increases in transcriptional activity in genetic systems where repression has been removed, may more closely resemble the mechanisms in higher eukaryotes, since the promoters, or sites of entry for RNA polymerase enzymes, in higher eukaryotes appear to, alone, allow only very inefficient transcription of structural genes. Without additional influences these genes may express at a very low, continuous, constitutive level. Although the constitutive level of protein production may vary for different higher eukaryotic genetic systems, most appear to be very low, producing very low effective protein levels. A group of genes constantly required for the maintenance of all cells and often referred to as "housekeeping genes" may have substantially higher constitutive levels.

A hypothesis for the mechanism of mammalian genetic control is that, unlike prokaryotes, activation of gene expression in higher organisms utilizes a positive mode of control. In this model, activation of gene expression is the result of an "opening" of the gene promoter or RNA polymerase II entry site sequences adjacent to the structural gene, allowing enhanced access of the polymerase to the initiation site for transcription. Although the actual molecular mechanisms for the "opening" of the polymerase entry sites in this positive gene regulation are not fully understood, they appear to affect the local helical structure of the entry site, making helical unwinding of the DNA by RNA polymerase II a less thermodynamically unfavorable process.

Two elements which may comprise much of such a mammalian genetic regulational mechanism have been identified and studied extensively. For more than a decade it has been known that all mammalian genes contain remarkably similar sequences immediately 5' to the structural gene known as the Goldberg-Hogness [thymidine-adenine-thymidine-adenine (TATA)] box, believed to be the RNA polymerase II binding site from which transcription is initiated. Because of the near identity of these elements in genes with widely divergent tissue specificities, temporal expression programs, and transcriptional levels, it seems unlikely that these elements are in any way associated with regulation of gene expression. But, the Goldberg-Hogness box is a prime candidate for the constitutive promoter which may yield continuous, very low basal levels of transcription from most mammalian genes. Clearly, other elements in the genome must act to regulate or "open" these transcriptional initiation sites, or to allow facilitated entry of RNA polymerase II at other sites from which this transcribing enzyme may readily migrate to the Goldberg-Hogness box. Over the last several years the identity of several of these specific regulational elements may have been established. DNA sequences associated with the specific enhancement of transcription of viral and mammalian genes, termed enhancer elements, have been identified (1,9,12,16,24,41).

Analysis of genes transcribed by polymerase II has revealed, among other promoter/regulatory elements, elements that appear to increase transcriptional efficiency in a manner relatively independent of their position and orientation with respect to a nearby gene (24). The first of these enhancer elements were discovered in animal viral genes, initially as sequences 5' to the SV40 viral structure, and later in many other viral systems including retroviruses (52). The SV40-enhancer, a 72-base pair tandem repeat of SV40 DNA, was first identified as a cis-acting element located more than 100 bp 5' of the Cap site of the early viral genes in SV40 (2,19). Deletion of this element dramatically reduced T-antigen gene expression in SV40 virus mutants, and linkage of the SV40-enhancer to heterologous genes, including the herpes simplex thymidine kinase mouse metallothionein gene and the p21 transforming gene from Harvey murine sarcoma virus, resulted in increased transcriptional activity of, at least, several orders of magnitude (24). Also, enhancer elements have been identified in cellular genes including the immunoglobin heavy chain genes (1,16,41). The immunoglobin enhancer element is naturally found in the intron joining the constant region of the gene to the variable region, but movement of this element either 5' or 3' of the structural gene allows retention of enhancer activity (16). The immunoglobin enhancers, as is the case with viral enhancers as well, may also be inverted in orientation and function equally as well (1,24). The enhancers in the immunoglobin gene system provide tissue-specific activation of somatically rearranged immunoglobin genes and possibly cause abnormal expression of other genes such as c-myc that become translocated to their domain of influence (1).

Recently, Church et al. (9) have demonstrated that the tissue-specific character of immunoglobin enhancers is due to the interaction of lymphocyte— specific cellular proteins with enhancer sequences, rendering the enhancers operative only on lymphoid tissue where appropriate, active, enhancer/cell protein complexes can form.

Although enhancers can influence the transcriptional activity of higher eukaryotic genes, presumably by either decreasing the free energy necessary to "open" the polymerase II entry site at the Goldberg-Hogness box or by providing other more readily useful entry sites for the polymerase, from as far away as several kilobases (perhaps >10 kb) in either direction from the transcriptional initiation site, the ability of enhancer sequences to effect gene transcription appears to be dependent upon the presence of other intervening sequences (perhaps strong promoters themselves) which can inhibit enhancer activity (24). Therefore, enhancers, in their native chromosomal environment, may be very specific, based upon a hierarchy dependent upon relative position and distance.

Although other genetic regulatory systems (some perhaps operating by totally different mechanisms) most probably exist to control the complex expression of the mammalian genome, exquisite regulation of gene expression is achieved solely through the use of the primary Goldberg-Hogness box promoter and associated enhancer sequences; recombinant genetic systems using these elements may be constructed which could offer elegant and precise regulation of transgene expression in transgenic laboratory and livestock animals. It is presently believed that the tissue specificity of those transgenes that show tissue specificity when transferred into experimental transgenic animals is the result of enhancer elements that are active, like the immunoglobin enhancers, only in a specific tissue(s) where cell-specific proteins bind to and activate the enhancer. Identification of these enhancer sequences and their incorporation in recombinant DNA constructions will allow designated tissue-specific expression for any desired structural gene.

An aspect of mammalian genetic regulation that has been thoroughly studied for more than 2 decades is steroid hormone regulation of genetic expression. A recent review of this field, which provides an up-to-date account of research results and outlines of established mechanisms of steroid hormone regulation of gene expression, has been published (33). The general mechanism of steroid hormone action is by activation of transcription at the genomic level. Steroid hormones bind to protein receptor molecules in the cytoplasm of specific, hormone-responsive target cells. The resulting steroid hormone/receptor complex migrates to the nucleus where it binds to specific DNA binding sites within steroid-responsive genes, substantially increasing the transcriptional activity of these genes. Because of the established base of knowledge of the molecular biology of steroid hormone action, one of the best opportunities to effect the regulation of transgene expression in transgenic animals is through the use of steroid hormones and their receptorbinding DNA sequences.

Our laboratory has, for the past several years, been studying the regulation of recombinant growth hormone genes in transgenic mice by the glucocorticoid steroid hormones. The results of these studies (45) demonstrate, for the first time, the regulation of transgene expression by an exogenous agent in a genetic system in which the interacting regulatory molecules and the DNA sequences with which they interact are well-characterized (55). Using a partially purified rat liver glucocorticoid receptor, Moore et al. (32) have identified and characterized the specific DNA sequences to which the glucocorticoid/receptor complex binds in the human growth hormone gene. These sequences have been identified by these workers to reside within the first intron of the human growth hormone structural gene. This DNA binding site shows a high degree of sequence homology with a subset of glucocorticoid receptor binding sites reported for the mouse mammary tumor virus gene (37,44) and for the human metallothionein IIA gene (22). Not only do these binding sites impart glucocorticoid inducibility to their homologous structural genes, but also minimal sequences containing only the binding sites from these genes are able to confer glucocorticoid inducibility to heterologous constitutive promoters (8,31).

In our studies, a fusion gene consisting of the structural gene for human growth hormone, containing the glucocorticoid receptor binding site in its first intron, ligated to the mouse metallothionein promoter sequences, shown to be devoid of glucocorticoid receptor binding sequences (11,35), was constructed to provide an observable constitutive level of human growth hormone expression in transgenic mice harboring this genetic construction. Glucocorticoid inducibility of human growth hormone gene expression was shown by several lines of these transgenic mice in response to either injected dexamethasone or orally administered triamcinolone. The serum growth hormone levels following stimulation with exogenous dexamethasone or triamcinolone increased 4- to 6-fold in mice containing the mouse metallothionein-I/human growth hormone fusion gene. Also, a 4-fold increase in growth hormone messenger RNA in the livers of these mice, as determined by primer extension and Northern hybridization analysis, was displayed in response to the glucocorticoid treatment (45). These results suggest the potential for transgene regulation in transgenic animals with natural steroids or synthetic analogs.

A discussion of the means of application of steroid regulation of transgenes to the practical regulation of genes in domestic transgenic animals has been presented elsewhere (55), and a brief outline of this discussion is appropriate to the present review.

Although many steroid hormones such as the glucocorticoid hormones enhance transcription of responsive genes only 10-fold or less, usually acting to regulate the level of gene expression and not serving as a strict on/off signal, some steroids do act to activate gene expression from essentially a zero level to very high levels of transcription (33). Therefore, the basic mechanisms of steroid hormone regulation could be used for the absolute regulation of transgene expression in transgenic animals. In the case of the glucocorticoids, although these agents have other important actions on other natural genes and act to provide low levels of regulation, it may be possible to use modified DNA binding sequences and synthetic glucocorticoid analogs as a means of external transgene regulation. Since the natural glucocorticoid/DNA binding site pairs have evolved to provide subtle regulation of gene expression in order to provide their physiological function, it may be that synthetic modification of both the DNA binding sequences and the steroids themselves could yield pairs that may provide a more absolute regulatory mechanism.

It is well-established that chemically altered steroid analogs, including glucocorticoid analogs, bind to receptor molecules to produce steroid/receptor structures that interact in a different manner than do the native steroid/receptors to the DNA binding sequences, eliciting dramatically different effects on the transcriptional expression of target genes (23,43,51,59). Also, synthetic modification of DNA sequences to alter the nature of the binding of DNA binding proteins is well-established (5). Caruthers and coworkers have altered, in prokaryotic systems, the RNA polymerase binding to lambda-phage cro promoter sequences several orders of magnitude by synthetic alteration of these promoter sequences (6). The small, 15-bp, glucocorticoid regulation sequence in the human growth hormone gene could serve as an ideal target for such a synthetic modification to potentially enhance its sensitivity to glucocorticoids or their analogs. Using this approach, it may be possible to identify a unique glucocorticoid analog (43,51) paired with an altered glucocorticoid DNA binding sequence that results in maximal transcriptional enhancement of transgenes. Such an approach, involving concomitant modification of transcriptional regulatory molecules and their DNA binding sequences (55), while requiring extensive research and development efforts, provides a unique opportunity and challenge for agricultural and pharmaceutical research which was not apparent prior to the development of mammalian gene transfer technology.

POTENTIAL APPLICATIONS OF REGULATED TRANSGENES IN LIVESTOCK

In addition to the use of regulated growth hormone transgenes in young meat-producing animals and in lactating dairy animals (see Ref. 54 and 55), a broad range of livestock applications of specific regulated transgenes is apparent. These applications would necessarily involve systems affecting the principal economic traits of livestock animals including growth, efficiency of feed utilization, reproductive efficiency, and disease resistance.

Genes coding for proteins such as growth hormone, growth hormone-releasing hormone, somatomedins, somatostatin, and others influencing the growth process could have a direct influence on growth and feed efficiency, while those coding for gonadotropins or influencing the gonadotropins, such as the booroola gene in sheep, may be used to influence the reproductive performance of livestock, either in an enhancing or repressing (sterilization) mode. Possibly the most fertile ground for transgenetic improvement in livestock may be in the area of immune-regulation where the controlled expression of transgenes for immune-modulating factors could upgrade the natural immunological defenses in livestock. In addition to these more obvious examples of potentially useful transgenic systems are numerous other genetic systems coding for protein factors, shown by domestic animal physiologists to influence the performance of farm animals.

CONCLUSIONS

It seems inevitable that biotechnology, in general, and gene transfer in domestic species, in specific, will have a dramatic impact upon the future of the animal sciences and livestock agriculture. As described herein, one of the keys to bringing this technology from the research laboratory to agricultural field testing and on to the farmer will be the development of systems for the appropriate regulation of transgene expression in transgenic animals. Although much progress has taken place in identification of enhancer sequences imparting tissue-specific regulation of transgenes and a good beginning has also been established in the temporal regulation of added genes, development of useful and foolproof genetic systems for the regulation of transgenes in animals may be the most time-consuming and difficult step in bringing to fruition the promise of recombinant genetics for livestock agriculture. It is important that the rapid progress in this field in the recent past does not raise the expectations of the agricultural research establishment or the farming public beyond reasonable levels, because although much has been accomplished, much more remains to be done before useful transgenic animals are available.

Accomplishing these goals will require the integrated efforts of agricultural scientists in the areas of animal physiology, biochemistry, endocrinology, and pharmacology with molecular biologists who have developed the basic technology. The greatest challenge facing animal agricultural research now is to develop programs that will bring this diverse group of scientists together in focused efforts to meet the goal of animal improvement through gene transfer and other biotechnological methods.

REFERENCES

1. Banerji, J., L. Olson, and W. Schaffner (1983) A lymphocyte-specific cellular enhancer is located downstream of the joining region in immunoglobulin heavy chain genes. Cell 33:729-740.
2. Benoist, C., and P. Chambon (1981) In vivo sequence requirements of the SV40 early promotor region. Nature 290(5804):304-310.
3. Brinster, R., H. Chen, M. Trumbauer, A. Senear, R. Warren, and R. Palmiter (1981) Somatic expression of herpes thymidine kinase in mice following injection of a fusion gene into eggs. Cell 27:223-231.
4. Brinster, R.L., H.Y. Chen, A. Messing, T. Van Dyke, A.J. Levine, and R.D. Palmiter (1984) Transgenic mice harboring SV40 T-antigen genes develop characteristic brain tumors. Cell 37:367-379.
5. Caruthers, M.H., S.L. Beaucage, J.W. Efcavitch, E.F. Fisher, M.D. Matteucci, and Y. Stabinsky (1980) New chemical methods for synthesizing polynucleotides. Nucl. Acids Symp. Ser. 7:215-223.
6. Caruthers, M.H., S.L. Beaucage, J.W. Efcavitch, E.F. Fisher, R.A. Goldman, P.L. deHaseth, W. Mandecki, M.D. Matteucci, M.S. Rosendahl, and Y. Stabinsky (1982) Chemical synthesis and biological studies on mutated gene-control regions. Cold Spring Harbor Symp. Quant. Biol. 47(part 1):411-418.
7. Chada, K., J. Magram, K. Raphael, G. Radice, E. Lacy, and F. Constantini (1985) Specific expression of a foreign beta-globin gene in erythroid cells in transgenic mice. Nature 314(6009):377-380.
8. Chandler, V.L., B.A. Maler, K.Y. Yamamoto (1983) DNA sequences bound

specifically by glucocorticoid receptor in vitro render a heterologous promoter hormone responsive in vivo. Cell 33:489-499.

9. Church, G.M., A. Ephrussi, W. Gilbert, and S. Tonegawa (1985) Cell-type-specific contacts to Ig enhancers in nuclei. Nature 313:798.

10. Constantini, F., and E. Lacey (1981) Introduction of a rabbit β-globin gene into the mouse germ line. Nature 294:92-94.

11. Durnam, D.M., J.S. Hoffman, C.J. Quaiff, E.P. Benditt, H.Y. Chen, R.L. Brinster, and R.D. Palmiter (1984) Induction of mouse metallothionein-I mRNA by bacterial endotoxin is independent of metals and glucocorticoid hormones. Proc. Natl. Acad. Sci., USA 81:1053-1056.

12. Ephrussi, A., G.M. Church, S. Tonegawa, and W. Gilbert (1985) B line-age--Specific interactions of an immunoglobulin enhancer with cellular factors in vivo. Science 227:134-140.

13. Etherton, T.D., and R.S. Kensinger (1984) Endocrine regulation of fetal and postnatal meat animal growth. J. Anim. Sci. 59:511-517.

14. Evans, H.M., and M.E. Simpson (1931) Hormones of the anterior hypophysis. Am. J. Physiol. 98:511-546.

15. Fronk, T.M., T.J. Peel, D.E. Bauman, and R.C. Gorewit, (1983) Comparison of different patterns of exogenous growth hormone administration on milk production in Holstein cows. J. Anim. Sci. 57(3):699-705.

16. Gilles, S.D., S.L. Morrison, V.T. Oi, and S. Tonegawa (1983) A tissue-specific transcription enhancer element is located in the major intron of a rearranged immunoglobin heavy chain gene. Cell 33:717.

17. Gordon, J.W., G.A. Scangos, D.J. Plotkin, J.A. Barbosa, and F.H. Ruddle (1980) Genetic transformation of mouse embryos by microinjection of purified DNA. Proc. Natl. Acad. Sci., USA 77(12):7380-7384.

18. Grosschedl, R., D. Weaver, D. Baltimore, and F. Constantini (1984) Introduction of a mu immunoglobulin gene into the mouse germ line: Specific expression in lymphoid cells and synthesis of functional antibody. Cell 38:647-658.

19. Gruss, P., R. Dhar, and G. Khoury (1981) Simian virus 40 tandem repeated sequences as an element of the early promoter. Proc. Natl. Acad. Sci., USA 78(2):943-947.

20. Hammer, R.E., R.D. Palmiter, and R.L. Brinster (1984) Partial correction of murine hereditary growth disorder by germ-line incorporation of a new gene. Nature 311(5981):65-67.

21. Hanahan, D. (1985) Heritable formation of pancreatic beta-cell tumors in transgenic mice expressing recombinant insulin/simian virus 40 oncogenes. Nature 315(6015):115-122.

22. Karin, M., et al. (1984) Characterization of DNA sequences through which cadmium and glucocorticoid hormones induce human metallothionein-IIA gene. Nature 308:513-519.

23. Katzenellenbogen, B.S., M.A. Miller, R.L. Eckert, and K. Sudo (1983) Antiestrogen pharmacology and mechanism of action. J. Steroid Biochem. 19(1A):59-68.

24. Khoury, G., and P. Gruss (1983) Enhancer elements. Cell 33:313-314.

25. Krumlauf, J., R.E. Hammer, S.M. Tilghman, and R.L. Brinster (1985) Mol. Cell. Biol. (in press).

26. Lacy, E., S. Roberts, E.P. Evans, M.D. Burtenshaw, and F.D. Constantini (1983) A foreign beta-globin gene in transgenic mice: Integration at abnormal chromosomal positions and expression in inappropriate tissues. Cell 34:343-358.

27. Lee, M.O., and N.K. Schaffer (1934) Anterior pituitary growth hormone and the composition of growth. J. Nutr. 7:337-363.

28. Machlin, L.J. (1972) Effect of porcine growth hormone on growth and carcass composition of the pig. J. Anim. Sci. 35:794-800.

29. Machlin, L.J. (1973) Effect of growth hormone on milk production and feed utilization in dairy cows. J. Dairy Sci. 56(5):575-580.

30. McKnight, G.S., R.E. Hammer, E.A. Kuenzel, and R.L. Brinster (1983) Expression of the chicken transferrin gene in transgenic mice. Cell 34:335-341.

31. Majors, J., and H. Varmus (1983) A small region of the mouse mammary tumor virus long terminal repeat confers glucocorticoid hormone regulation on a linked heterologous gene. Proc. Natl. Acad. Sci., USA 80:5866-5870.

32. Moore, D.D., A.R. Marks, D.I. Buckley, G. Kapler, F. Payvar, and H.M. Goodman (1985) The first intron of the human growth hormone gene contains a binding site for glucocorticoid receptor. Proc. Natl. Acad. Sci., USA 82(3):699-702.

33. O'Malley, B.W. (1984) Steroid hormone action in eukaryotic cells. J. Clin. Inv. 74:307-312.

34. Ornitz, D.M., R.D. Palmiter, R.E. Hammer, R.L. Brinster, G.H. Swift, and R.J. MacDonald (1985) Specific expression of an elastase-human growth hormone fusion gene in pancreatic acinar cells of transgenic mice. Nature 313(6003):600-602.

34a. Palmiter, R.D., and R.L. Brinster (1985) Transgenic mice. Cell 41: 343-345.

35. Palmiter, R.D., R.L. Brinster, R.E. Hammer, M.E. Trumbauer, M.G. Rosenfeld, N.C. Birnberg, and R.M. Evans (1982) Dramatic growth of mice that develop from eggs microinjected with metallothionein-growth hormone fusion genes. Nature 300:611-615.

36. Palmiter, R.D., G. Norstedt, R.E. Gelinas, R.E. Hammer, and R.L. Brinster (1983) Metallothionein-human GH fusion genes stimulate growth of mice. Science 222:809-814.

37. Payvar, F., D. DeFranco, G.L. Firestone, B. Edgar, O. Wrange, S. Okert, J. Gustafsson, and K.R. Yamamoto (1983) Sequence-specific binding of glucocorticoid receptor to MTV DNA at sites within and upstream of the transcribed region. Cell 35:381-392.

38. Peel, C.J., D.E. Bauman, R.C. Gorewit, and C.J. Sniffen (1981) Effect of exogenous growth hormone on lactational performance in high yielding dairy cows. J. Nutr. 111(9):1662-1667.

39. Peel, C.J., T.J. Fronk, D.E. Bauman, and R.C. Gorewit (1982) Lactational response to exogenous growth hormone and abomasal infusion of a glucose-sodium caseinate mixture in high-yielding dairy cows. J. Nutr. 112(9):1770-1778.

40. Peel, C.J., T.J. Fronk, D.E. Bauman, and R.C. Gorewit (1983) Effect of exogenous growth hormone in early and late lactation on lactational performance of dairy cows. J. Dairy Sci. 66(4):776-782.

41. Queen, C., and D. Baltimore (1983) Immunoglobulin gene transcription is activated by downstream sequence elements. Cell 33:741.

42. Ritchie, K.A., R.L. Brinster, and U. Storb (1984) Allelic exclusion and control of endogenous immunoglobulin gene rearrangement in kappa transgenic mice. Nature 312:517-520.

43. Rochefort, H., J.L. Borgna, and F. Evans (1983) Cellular and molecular mechanism of action of antiestrogens. J. Steroid Biochem. 19:69-74.

44. Scheidereit, C., S. Geisse, H.M. Westphal, and M. Beato (1983) The glucocorticoid receptor binds to defined nucleotide sequences near the promoter of mouse mammary tumor virus. Nature 304:749-752.

45. Seldon, R.F., T.E. Wagner, J.S. Yun, D.D. Moore, and H. Goodman (1985) Glucocorticoid regulation of human growth hormone expression in transgenic mice (submitted for publ.).

46. Shani, M. (1985) Tissue-specific expression of rat myosin light-chain 2 gene in transgenic mice. Nature 314(6008):283-286.

47. Stewart, T.A., E.F. Wagner, and B. Mintz (1982) Human β-globin gene sequences injected into mouse eggs, retained in adults, and transmitted to progeny. Science 217:1046-1048.

48. Stewart, T.A, P.K. Pattengale, and P. Leder (1984) Spontaneous mammary adenocarcinomas in transgenic mice that carry and express MTV/myc fusion genes. Cell 38:627-637.

49. Storb, U., R.L. O'Brien, M.D. McMullen, K.A. Gollahon, and R.L. Brinster (1984) High expression of cloned immunoglobulin kappa gene in

transgenic mice is restricted to B lymphocytes. <u>Nature</u> 310(5974): 238-241.

50. Swift, G.H., R.E. Hammer, R.J. MacDonald, and R.L. Brinster (1984) Tissue-specific expression of the rat pancreatic elastase I gene in transgenic mice. <u>Cell</u> 38:639-646.

51. Tate, A.C., G.L. Greene, E.R. DeSoombre, E.V. Jensen, and V.C. Jordan (1984) Differences between estrogen- and antiestrogen-estrogen receptor complexes from human breast tumors identified with an antibody raised against the estrogen receptor. <u>Cancer Res.</u> 44:1012-1018.

52. Temin, H.M. (1982) Function of the retrovirus long terminal repeat. <u>Cell</u> 28(1):3-5.

53. Townes, T., L. Lingrel, H. Chen, R. Brinster, and R. Palmiter (1985) Erythroid specific expression of human β-globin genes in transgenic mice. <u>EMBO J.</u> 4:1715-1724..

54. Wagner, T.E. (1985) The role of gene transfer in animal agriculture and technology. <u>Can. J. Anim. Sci.</u> (in press).

55. Wagner, T.E., and W. Jochle (1985) Recombinant gene transfer in animals: The potential for improving growth in livestock. In <u>Control and Manipulation of Animal Growth</u>, P.J. Buttery, ed. Butterworths, London (in press).

56. Wagner, E., T. Stewart, and B. Mintz (1981) The human β-globin gene and a functional viral thymidine kinase gene in developing mice. <u>Proc. Natl. Acad. Sci., USA</u> 78:5016-5020.

57. Wagner, T., P. Hoppe, J. Jollick, D. Scholl, R. Hodinka, and J. Gault (1981) Microinjection of a rabbit β-globin gene into zygotes and its subsequent expression in adult mice and their offspring. <u>Proc. Natl. Acad. Sci., USA</u> 78:6376-6380.

58. Wagner, T.E., F.A. Murray, B. Minhas, and D.C. Kraemer (1984) The possibility of transgenic livestock. <u>Theriogenology</u> 21:29-44.

59. Wakeling, A.E., B. Valcaccia, E. Newbolilt, and L.R. Green (1984) Nonsteroidal antiestrogens--Receptor binding and biological response in rat uterus, rat mammary carcinoma and human breast cancer cells. <u>J. Steroid Biochem.</u> 20:111-120.

A TECHNIQUE FOR BISECTION OF EMBRYOS TO PRODUCE IDENTICAL TWINS

Susan E. Donahue

Department of Animal Science
University of California
Davis, California 95616

INTRODUCTION

A successful technique for bisection of postcompaction-stage embryos has been devised that combines methods of micromanipulation developed by other workers. Several papers describe generally and in detail various equipment, microtools, and methods of negotiating the bisection of bovine and ovine embryos (6). Willadsen (4,5) used a fine glass needle to penetrate the zona pellucida and to sever the embryo against a holding pipette using a Wild M5 Stereozoom microscope. Ozil (3), using 6 micromanipulators, opened the zona pellucida with needles, ejected the embryo with medium from a suction pipette, then cut the embryo with a specially machined microscalpel. Needles were also used by Lambeth et al. (2) and Gatica et al. (1) to access the embryo and cut it against the bottom of a petri dish. Williams and Seidel (7) used a fine microsurgical blade for opening the zona pellucida and bisecting the embryo in a single step. These methods were attempted with various difficulties in our laboratory. Spatial visualization was difficult when attempting needle cutting against the holding pipette with a Leitz Diavert inverted microscope. The alternative microsurgical blade made a cut in the zona pellucida suitable for introducing a suction pipette, but the embryo tended to stick when the blade was made to cut completely through it. The blade approach was generally more favorable but only for the initial stage of cutting. A method was needed to finish cutting the embryo without the problem of sticking. The fine glass needle seemed suitable for this task, which could be accomplished against the floor of the petri dish after the embryo was gently removed from the zona by blowing in medium from the suction pipette. The resultant technique utilizes a holding pipette, a microsurgical blade for splitting the zona and embryo, and a suction pipette for withdrawing half a severed embryo and inserting it into an empty zona pellucida or for ejecting the semidivided embryo to the floor of the petri dish for completion of the separation with a fine glass needle. Since the blade and needle tools do not require a pressure line to a microinjection syringe, they are easily interchangeable, and only one holder is required for them. Only 2 micromanipulators are required with one single and one double holder for all 4 microtools.

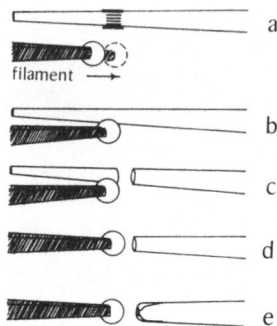

Fig. 1. Manufacturing a holding pipette: (a) Pipette diameter is measured with a microscope eyepiece reticle; (b) glass bead on filament fused to pipette at desired diameter; (c) pipette breaks as
filament cools and contracts; (d) pipette faces glass bead for
firepolishing; and (e) completion (repolishing) of holding pipette.

MATERIALS

Equipment

A Leitz Diavert microscope, with 10X oculars and a 4X objective, and 2
micromanipulators are mounted on a base plate. The left manipulator has an
attached assembly for one microinstrument holder, while the right manipulator has an assembly for 2 microinstrument holders. One microinstrument
collar specially modified to accommodate Fisher glass microtubing is mounted in one of the holders on each manipulator. To each of these is attached
a pressure line and syringe apparatus that is placed on the opposite side
such that one hand controls motion while the other hand simultaneously controls pressure when a suction or holding pipette is in use. This injec-
tion/suction apparatus for each manipulator is made up of an instrument
collar, a Hamilton gas tight syringe with a Luertip and a blunt 18-gauge
needle attached, an inclined mount anchoring the syringe to the base plate,
a screw-in plunger with control knob, a 2-ft length of tygon microbore tubing (formulation S-54-HL; 0.04-in ID, 0.07-in OD) inserted through the collar, and a 3/8-in length of Dow Corning silastic medical grade tubing
(0.04-in ID; 0.085-in OD) over the end of the Tygon tubing to give a pipette a snug fit in the instrument collar. Two extra collars are fitted
with the short length of silastic tubing to hold tools that do not require
pressure lines. The syringe and tubing are filled with paraffin oil. In
order to keep oil from receding in line between uses, a short length of
glass tubing, sealed on each end, can be inserted into the collars. A Kopf
vertical-type pipette puller and DeFonbrune microforge and gas microburner
are used for the manufacture of microtools.

Microtools*

Holding pipette. A piece of tubing is mounted in the pipette puller
and melted at the center into 2 tapered pipettes. One of these is mounted
in the microforge and observed through the attached dissecting microscope.
Using an eyepiece reticle, the point at which the pipette diameter is 125

---

* All 4 microtools are made from Fisher 6-in coagulation capillary glass
tubing.

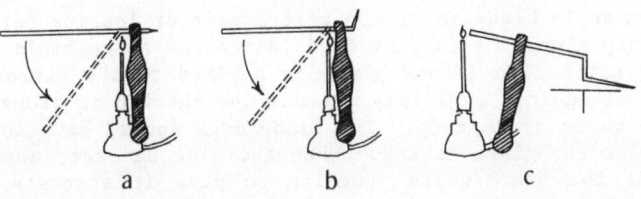

Fig. 2. Bending a microinstrument for use in a petri dish: (a) first
bend; (b) second bend; and (c) firepolishing end of finished microinstrument with desired configuration.

μM is determined and the glass bead-covered filament is brought near that
point (Fig. la). The unit is switched on to the setting at which the filament color is just short of being noticeably orange, and the filament is
allowed to heat up and expand. It is then brought against the glass tubing
at the desired diameter and allowed to fuse (Fig. lb); immediately after
fusion, the filament power is switched off. As the filament cools, it contracts to its original position and, if left untouched, should cause the
glass to break cleanly at the point of contact (Fig. lc). This should result in a pipette with a straight end break of the desired diameter. The
pipette is then repositioned with the end facing the filament evenly (Fig.
ld). The power is turned on and the temperature is increased until the
filament glows orange, at which time the pipette is brought close and evenly toward the filament and firepolished until the edges have melted, leav-
ing only a very narrow lumen (Fig. le).

The pipette is held with forceps and bent over a gas microburner flame
to enable it to clear the side of a 60 x 15 mm petri dish when it is mounted for micromanipulation, and the collar insertion end is firepolished for
ease of entry into the tubing. The first bend in the glass is made by
holding the pipette horizontally over the flame. The forceps are as close
as possible to the finished tip, and the flame is just to the rear of the
forceps, keeping the length of the working end as short as possible. The
barrel of the pipette begins to drop as the glass melts, and the pipette is
pulled away while the angle of the arms of glass is greater than 90° (Fig.
2a). A blunt needle has been mounted on the microburner to raise the
height of the flame to improve the working distance.

The second bend is made about 1 cm from the first, again holding the
barrel horizontally, with the finished tip pointing upward. The resulting
angle is also greater than 90° (Fig. 2b). Finally, the end is firepolished
to facilitate filling and loading into collars (Fig. 2c). The approximate
angles of the finished pipette are shown. The tip must slope enough to approach the bottom of a petri dish in order to pick up embryos, and the pipette should lie in one plane. Holding the center section vertically is a
good reference to check angles.

Suction pipette. The same procedure as above is used to make the suction pipette except the tip diameter is 75 μ ID and is not firepolished.

Fine microsurgical blade. The 6-in glass tubing is hand-pulled using the microburner such that the center is narrow for about 1/4 in (Fig.
3). The center is scored with a diamond pencil and broken. A double-edge

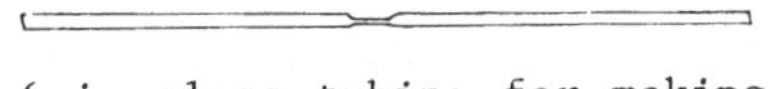

Fig. 3. Hand-pulled, 6-in glass tubing for making microknives is scored
and broken in the center.

surgical razor style blade is broken with a pair of locking forceps to produce a tiny chip along the cutting edge, as narrow as possible (Fig. 4). A cyanoacrylate ester glue (Krazy glue) is applied to the narrow tip of the glass to seal it against capillary action, and the tip is brought into contact with the razor blade chip. The blade edge should be held in position perpendicular to the glass tubing. When the glue has set, another drop is applied across the blade/glass junction to give it strength. The blade shaft is then bent and the insertion end firepolished as with pipettes.

Fine glass needle. The glass tubing is also hand-pulled for the needle, but it is not scored and broken. Instead, it is mounted in the pipette puller with the prepulled area centered in the Nichrome coil. This tubing is pulled with the usual settings, although some adjustment may be necessary to get the desired point taper. The resulting tip should be very narrow, of uniform taper, and should have a sealed point to avoid capillary action. The needle is bent over the microburner but the angle of the tip should be slightly greater than for the other tools (Fig. 5).

The blade and needle can be reused after rinsing with 95% ethanol and washing with sterile purified water. The pipettes can be boiled in water before reuse, or else new ones are manufactured. The microinstruments can all be made in advance, and kept embedded slightly in a strip of modeling clay in a covered box. The pipettes can be distinguished and identified by size with an indelible mark on the barrel.

METHODS

Preparation of Micromanipulators

In order to mount the microinstruments for use, the collars are removed from the holders. To mount pipettes into the collars with the oil lines it is convenient to place each collar atop its micromanipulator with the cap removed. Oil is brought out to the tip of the tubing with pressure from the Hamilton syringe. No air bubbles should be present in the tubing. Inert buffer fluid from a syringe with a short piece of tygon tubing on a blunt needle is forced through the holding pipette, removing all bubbles. The pipette is removed carefully from the filling syringe and held horizontally so that no air replaces leaking fluid. The insertion end is passed through the cap and inserted in the collar tubing, buffer fluid to oil. The cap is screwed into place securely, and the collar with the holding pipette is mounted in its holder and tightened. For mounting the suction pipette on the opposite micromanipulator, the same procedure is followed.

Both blade and needle microtools are inserted into collars with only a short piece of silastic tubing under the cap. The blade-bearing collar is mounted first in the remaining holder of the double set on the right, while the needle-bearing collar is set aside. Each holder assembly is adjusted on the micromanipulator with the collar(s) horizontal, about 5-mm above the top surface (Fig 6). The guide levers and adjustment knobs are centered

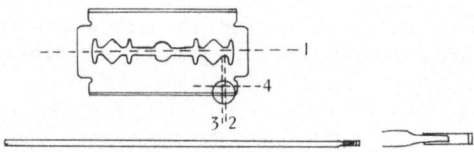

Fig. 4. A thin chip of razor blade is glued to the narrowed end of the glass tubing to make the microknife.

such that there is ample range of movement in their appropriate directions. This is especially important for the vertical control which provides cutting action for the needle. The collars are all roughly positioned such that the tips of the microinstruments approach the center of the microscope field.

The knife and suction pipette are mounted such that one can be raised and lowered relative to the other. There is a screw mechanism on the rear holder that lowers it at an angle. If, for instance, the blade is set in the rear holder, it must reach a working position below the level of the suction tip and be clear of the suction tip when raised up out of the way with the screw mechanism. If this is not possible, due to variations in the hand-bent angles of their respective barrels, switching the blade and suction pipette in the 2 holders will usually facilitate this spatial relationship. The micromanipulation equipment is now ready for use. The large black tilt adjustment knobs are used to lower the microinstruments into the dish and into a drop of medium. Any motion of the microinstruments during their approach to a focused embryo should be watched directly until they have entered the microscopic field. The medium drop or drops must be central in a 60 x 15 mm petri dish to allow ample room for the pipettes and some degree of motion. A larger petri dish with a row of drops down the center will also suffice. The pipettes are at an angle when the microinstruments are positioned in the medium drop, but the bending puts the working ends of each instrument relatively horizontal. This approach allows the microinstruments to be raised out of the petri dish with enough clearance to easily place and remove dishes without constantly readjusting the height of the holder.

Preparation of the Embryo

Each embryo to be split is put in a rounded 0.5-ml drop of Dulbecco's phosphate buffered saline (PBS) with 1% Gibco antibiotic antimycotic solution and 1% heat-treated bovine serum (flushing medium) in a Falcon 1007 petri dish without oil, enabling the pipettes to be used for many embryos without clogging. To the drop is added a zona pellucida from degenerated or unfertilized fresh eggs or oocytes collected from ovaries that have had their zonae pellucidae broken and emptied using the splitting procedure, or from a collection of empty zonae pellucidae frozen in flushing medium in a plastic artificial insemination straw. If new tools are used at each session, it is helpful to prepare empty zonae using fresh, unfertilized embryos to check alignment of the microtools. An embryo is focused and centered in the field using the 4X objective. This relatively low power allows visualization of the various tools as they are moved about the petri dish. The holding pipette is lowered into the drop and moved toward the center. Likewise, the blade is lowered into the drop by direct observation until it is visible in the microscope field. A special adjustment is then necessary for the knife blade. It must be loose in the holder and rotated until the

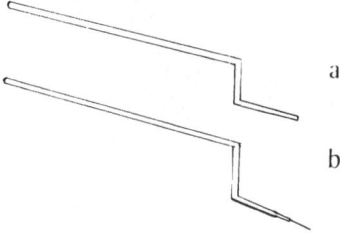

Fig. 5. Finished configuration of microinstruments: (a) microknife and holding and suction pipettes; and (b) fine glass needle.

edge is vertical and can be put in sharp focus. Some medium is drawn into the holding (and suction) pipette initially so that the embryo will not be exposed to inert buffer fluid.

The embryo is examined and pulled onto the holding pipette in the best position for dividing the embryo evenly (Fig. 7). A blastocyst should have the inner cell mass centered over the holding pipette in order to ensure equal division, as invasion by the blade knife will cause the embryo to collapse. The holding pipette with embryo is raised off the bottom of the dish to allow freedom of motion for the blade which, depending on the width of its now vertical cutting surface, may tend to scrape bottom. The narrower the original blade section, the more satisfactory the knife will be in terms of focus and maneuverability. The section of blade is focused that gives the sharpest image. The embryo is brought into this plane of focus and adjusted to the spot where the knife blade makes the most even indentation when pressed gently against the zona. If not centered, the knife will tend to push the embryo off the holding pipette rather than cut through it.

Splitting Technique

With the instruments ready and the embryo positioned on the holding pipette in proper relation to the knife blade, the actual splitting process can begin. Using the control lever, the blade and embryo are brought together in one definite, smooth motion such that the 2 are kept in alignment, should either shift (Fig. 8). It is suggested, at least in initial endeavors, to bring the embryo into the blade, as it is necessary to have the opposite hand free to immediately release the pressure of the holding pipette once the zona is cut through to the opening of the holding pipette. If the pressure across the crack in the zona is not released, the risk of all or part of the embryo being pulled into the narrow lumen of the holding pipette is great. Should that happen, the embryo would be destroyed. A holding pipette with a much larger bore would be safer, but would not provide an end surface suitable for the cutting motion that follows.

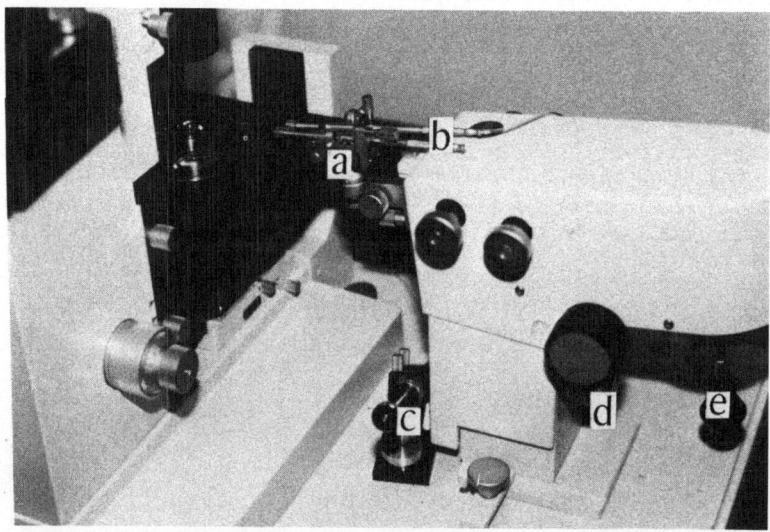

Fig. 6. Micromanipulator: (a) double instrument holder; (b) instrument collars; (c) microinjection syringe with screw plunger; (d) vertical controls; and (e) guide lever.

Once the blade has cut through the central axis of the embryo and rests against the holding pipette, and the pressure has equilibrated, a gentle up and down motion of the blade is used to sever the 2 halves. Using the vertical motion knob, the blade is moved against the embryo in a slicing motion until the embryo appears to be divided, and then the blade is pulled clear. It is safest at this point to achieve a new hold on the embryo to one side of the crack that has just been made. If the attempt is made to slice until the embryo is clearly in 2 halves, the zona may be in 2 halves or gaping open, and the embryonic cells may be stuck to the blade. If the edge of the blade and end surface of the holding pipette are relatively parallel (the approximation should improve as toolmaking by hand becomes consistent), the embryo should be nearly separated with only minimal sticking to the blade. As the blade is withdrawn, the zona can be used to hold back the embryo(s). If it appears that they will come out of the zona attached to the blade, a bit of gentle tapping on the control lever should vibrate them loose (Fig. 9). The blade and suction pipette are then repositioned. The right manipulator can be brought up toward horizontal and the suction pipette lowered or blade raised such that the suction pipette is in the lower position relative to the blade. Medium drawn into the tip will ensure that the embryo does not come in contact with inert buffer fluid to which it will stick.

In the event that the embryo is seen to be completely separated, one of the halves can either be drawn out into the suction pipette or blown out of the zona pellucida. If allowed to come back in contact, the 2 halves will tend to readhere. If it is not clear that separation is complete, the entire embryo can be blown out of the zona pellucida by drawing extra medium into the suction pipette, placing it strategically against the crack in the zona, and gently applying positive pressure until the stream of medium ejects the embryo (Fig. 10). If the crack in the zona is too small the embryo will catch in it and be damaged during evacuation. The embryo should be severed or have a well-defined crease, or it should be in a butterfly configuration, just held together by a strand of cells. The double holder can be brought up out of the dish and the blade replaced with the needle, which is then aligned over the crease in the embryo (the dish can be rotated slightly, or the embryo turned into position with the tip of the needle). The needle is then brought down using the vertical control with guide lever as necessary to keep it centered during descent. It should bisect the embryo along the cut until it reaches the floor of the plate (Fig. 11). The vertical control is used to move the needle up and down until the embryo separates. It is necessary that the needle hit the floor of the dish directly over the embryo. In order to put enough pressure on the embryo to cut it with the fine taper, the needle is at a steep slope with the tip actually curving up after it hits bottom. If the needle will not contact the dish properly, a steeper slope can be achieved by raising the holder on the manipulator. Once the place is found near the tip of the needle which will hit bottom against the embryo, the vertical motion continues until the embryo separates. This may require a number of strokes that have both a vertical and horizontal component due to the angle of the pipette. Finally, when the embryo appears to be severed, the halves can be disengaged from the needle with gentle taps on the guide lever if necessary. The embryo halves may appear ovoid-shaped, but they resume their round shape. It is important when loading a half embryo into the suction pipette to pull it in with its length parallel to the pipette to avoid cell damage by trying to pull it in broadside. Special care must be taken to position the pipette just above the surface of the plate such that a forward motion to gather the embryo will not scrape a spiral of plastic from the bottom. Embryos caught in plastic spirals are generally inextricable.

One half embryo can now be replaced inside its original zona, still attached to the holding pipette. The most secure way of placing the embryo

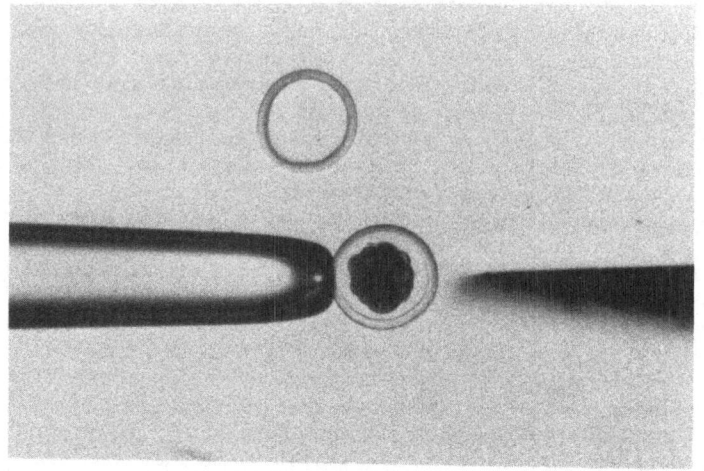

Fig. 7. Bovine morula mounted on holding pipette with microknife and empty zona pellucida nearby.

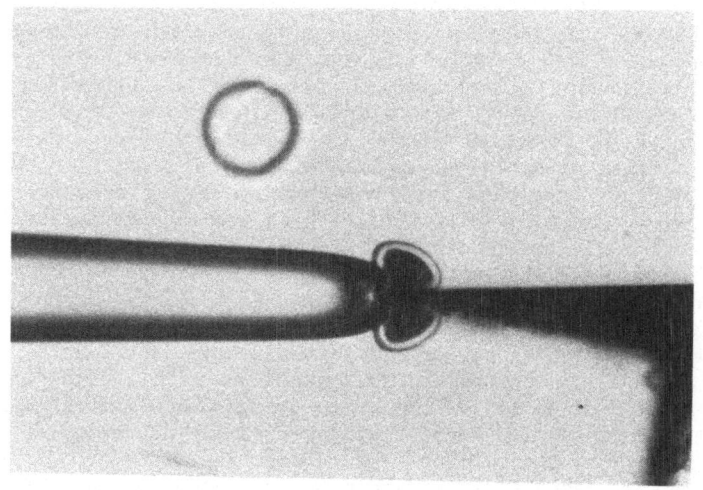

Fig. 8. Embryo is divided with microknife.

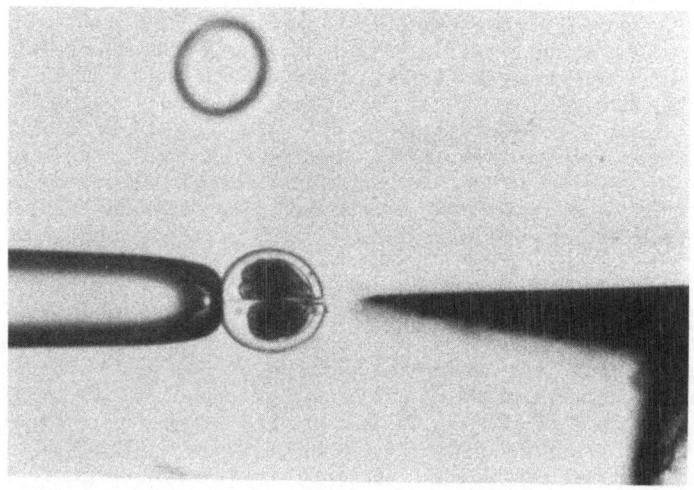

Fig. 9. Embryo has been cut with microknife.

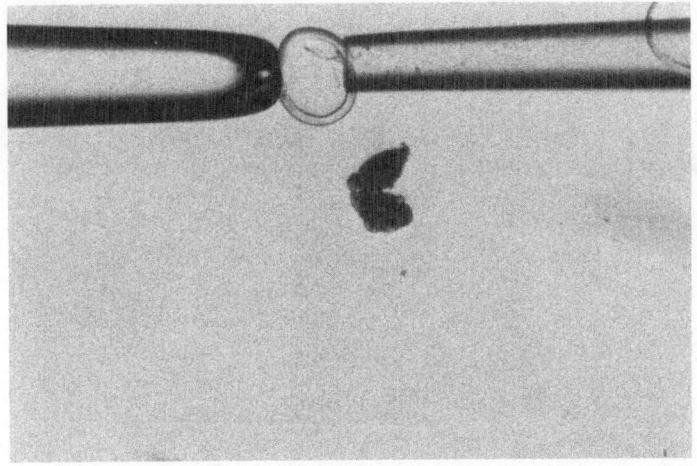

Fig. 10. Nearly severed embryo has been ejected from the zona pellucida.

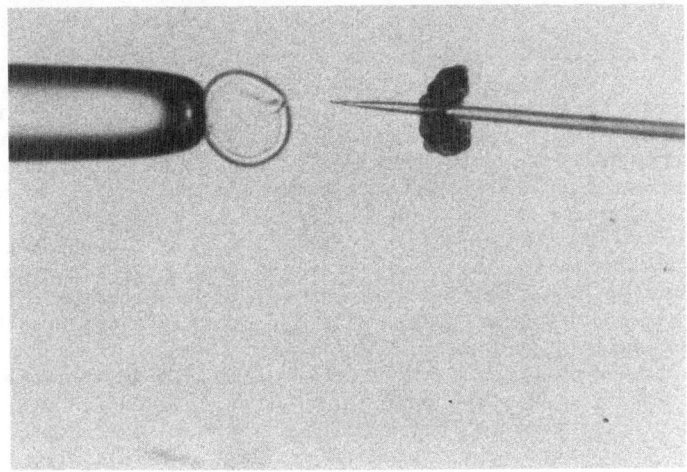

Fig. 11. Glass needle is used to complete separation of the embryo halves.

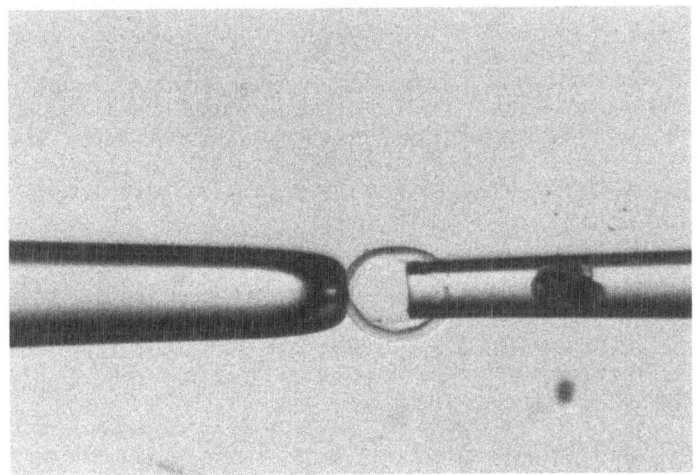

Fig. 12. Half embryo is introduced into the zona pellucida with a suction pipette.

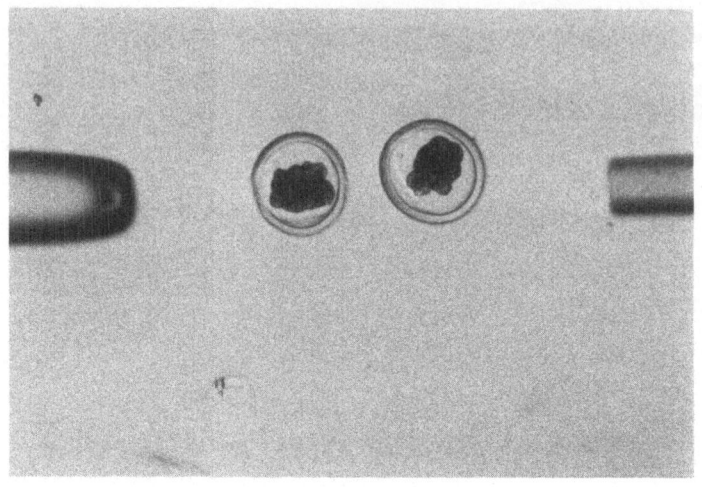

Fig. 13. Identical twin half embryos.

is with the suction pipette fully inside the zona pellucida (Fig. 12). If it is positioned at the edge of the crack, the embryo is just as likely to be placed outside the zona pellucida or become lodged in the crack, in which case it must then be reintroduced to the suction pipette, thus exposing it to further handling and possible damage. Since one hand operates the suction mechanism on the Hamilton syringe and the other the guide lever, the suction pipette can be pulled free as the embryo enters the zona pellucida. The complete half embryo is released from the holding pipette and given a position in the drop away from the center of activity. It is best to pick up the other half embryo and see it safely into the suction pipette, and then mount the new zona pellucida so that the embryo has less chance of being pulled into the holding pipette by accident. The second half embryo is loaded into the remaining zona pellucida in the same manner. This step sometimes requires the most patience as it is often difficult to align the crack in the zona pellucida so that entry by the suction pipette is possible. The zona can be released and realigned repeatedly, and if it still fails to accommodate the suction pipette, the crack can be expanded by tearing it a bit further with the 2 pipettes working against each other. Care again must be taken that the crack in the zona pellucida does not cross the opening in the holding pipette.

The microtools can be lifted out of the petri dish and above the height of the sides with the large black tilt control knob so that the dish can be removed and immediately replaced with another dish containing an embryo for splitting. The 2 half embryos (Fig. 13), each in a zona pellucida, can be loaded directly into an artificial insemination straw for transfer.

REFERENCES

1. Gatica, R., M.P. Boland, T.F. Crosby, and I. Gordon (1984) Micromanipulation of sheep morulae to produce monozygotic twins. Theriogenology 21:555-560.
2. Lambeth, V.A., C.R. Looney, S.A. Voelkel, D.A. Jackson, K.G. Hill, and R.A. Godke (1983) Microsurgery on bovine embryos at the morula stage to produce monozygotic twin calves. Theriogenology 20:85-95.
3. Ozil, J.P. (1983) Production of identical twins by bisection of blastocysts in the cow. J. Reprod. Fert. 69:463-468.

4. Willadsen, S.M. (1982) Micromanipulation of embryos of the large domestic species. In Mammalian Egg Transfer, C.E. Adams, ed. CRC Press, Boca Raton, pp. 185-210.

5. Willadsen, S.M., and R.A. Godke (1984) A simple procedure for the production of identical sheep twins. Veterinary Record 114:240-243.

6. Williams, T.J., and G.E. Seidel, Jr. (1983) Methodology and equipment for microsurgery with mammalian ova. Proc. Owners and Managers Workshop, IX Ann. Conf. Int. Embryo Transfer Soc., International Embryo Transfer Society, pp. 33-52.

7. Williams, T.J., R.P. Elsden, and G.E. Seidel, Jr. (1984) Pregnancy rates with bisected bovine embryos. Theriogenology 22:521-531.

PRODUCTION OF EXPERIMENTAL CHIMERAS IN LIVESTOCK

BY BLASTOCYST INJECTION

James E. Butler

Department of Animal Science
University of California
Davis, California 95616

INTRODUCTION

Mammalian embryos have the interesting property that allows them to be combined to form a normal individual containing cells of 2 distinct genotypes. Although this mixing of embryonic cells or chimerism occurs naturally (see Ref. 3 for review), whole body chimerism is relatively rare. Nicholas and Hall (13) attempted to produce chimeras in the laboratory by aggregating one-cell rat embryos, however, no conclusive evidence could be offered in terms of chimerism. Tarkowski (16) was the first to develop a procedure by which chimeras could be readily produced. Immediately following, Mintz (10,11) published a similar method but with modifications that resulted in a simpler procedure. Briefly, the zonae pellucidae were removed either mechanically or enzymatically from early cleavage-stage embryos, the 8-cell stage being the most common. Two embryos of different strains were then placed in contact with each other and allowed to aggregate into a single embryo. The chimeric embryos were allowed to develop to the blastocyst stage and then transferred to suitable recipients. The ability to readily produce large numbers of whole body chimeras proved to be a valuable tool in the study of embryology and developmental genetics (see Ref. 9 for review).

Although highly successful in mice, the aggregation method yielded only limited success when applied to sheep (14). The primary reason for the low survival rate of sheep aggregation chimeras appears to be that sheep embryos require an intact zona pellucida during early development (17). This problem, however, can be overcome by surrounding the aggregated embryos with an artificial zona pellucida or by using later-stage blastocyst embryos which no longer require an intact zona.

A technique by which ovine chimeras can be produced by placing aggregated embryos in an artificial zona pellucida has recently been published (5). In this method, 8-cell stage embryos were aggregated in the same manner as mouse embryos. The aggregated embryos were then placed in a surrogate zona pellucida which was in turn sealed by embedding it in an agar cylinder. The cylinder was then transferred to the oviduct of an intermediate recipient and allowed to develop to the blastocyst stage. Blastocysts were recovered from the intermediate host, freed from the agar cylinder, and then transferred to the definitive recipient. Although highly successful in terms of numbers of chimeric lambs produced, the technique requires extensive embryo handling and is somewhat expensive in terms of the number of recipients required.

Several techniques have been developed to produce chimeras using later cleavage-stage embryos. Gardner (7,8) devised a method that utilized 3 micromanipulators and 5 microtools to open a triangular hole in the trophectoderm through which cells or tissues could be introduced into the blastocoele. Moustafa and Brinster (12) attempted to simplify this technique by replacing 4 of the microinstruments with a single beveled microinjection pipette. Although successful, the authors found it difficult to consistently produce high-quality beveled pipettes using a microforge. This problem was overcome by Babinet and Bordenave (2) who made use of a mechanical pipette beveler to produce injection pipettes.

To date, most research involving chimera production using later-stage embryos has been done in laboratory species with only limited work being done in domestic livestock. Tucker and co-workers (18) were able to produce 3 chimeric rams following injection of from one to 4 blastomeres into 92 recipient embryos. More recently, Bren and co-workers (4) produced a single chimeric calf by combining 4 half embryos. Interspecific chimeras have also been produced between sheep and goats by injecting mechanically isolated inner cell masses into recipient embryos (6).

This chapter describes a method for production of chimeras in livestock by injection of inner cell masses isolated by immunosurgery into recipient blastocyst-stage embryos. The method is very successful in terms of chimera production, does not require the use of intermediate recipients, utilizes a single injection pipette, and is relatively easy to master.

MICROMANIPULATION SYSTEM

The basic system consists of left and right Leitz micromanipulator units mounted on a baseplate. The Leitz units were chosen as they possess several features necessary for blastocyst injection. Movement can be accomplished either by use of a single joystick or by forward/backward and left/right control knobs. The joystick is used for the routine movement of embryos and the actual injection while the control knobs are used to position the injection pipette relative to blastocyst prior to injection. Second, instruments attached to the manipulator can be raised and lowered without having to tip the manipulator units. By keeping the manipulator units horizontal, movement of the instrument is restricted to one plane. Because only 2 pipettes are required for blastocyst injection, each manipulator can be fitted with a single instrument head. If, however the manipulator is to be used for other purposes, such as embryo splitting, it is recommended that at least one manipulator unit be fitted with a double instrument head. The capillary tubing supplied by Leitz has a smaller diameter than tubing more commonly available. If the common capillary tubing is used, the Leitz instrument holders will have to be enlarged in order to accommodate the larger tubing. Already-modified holders are commercially available (Elden Instrument Co., Lakewood, Colorado).

Blastocyst injection requires the use of 2 microinjectors, one to hold the blastocyst and the other to inject the inner cell mass into the embryo. The injectors must be capable of precisely moving very small volumes of fluid. Several devices for this purpose are available. The author uses 2 modified Hamilton #1750 gastight syringes (Hamilton Syringe Co., Reno, Nevada). The modification consists of replacing the normal syringe plunger with one that is controlled by a turning of a knob that advances the plunger. Modified syringes are commercially available (Eldon Instrument Co.).

Even finer control can be achieved with a DeFonbrune suction and force pump although the Hamilton syringe system has proved adequate for blastocyst injection.

The microtools are connected to the injection syringes by flexible tubing. In choosing the tubing to be used, 2 opposing factors must be considered. The tubing must be flexible enough so that it can be threaded around the various pieces of equipment attached to the baseplate. However, the walls of the tubing must be rigid enough so that the movement of fluid within the tubing does not cause expansion and contraction of the walls which can lead to uncontrolled movement of fluid. Tygon microbore tubing (formulation 5-54-H6) that has an inner diameter of 0.04 in and an outer diameter of 0.07 in satisfies both of these requirements. The tubing is connected to the microinstruments by a collar made from a short piece of silastic medical tubing (0.04-in ID; 0.085-in OD). The tubing is connected so that volume changes in the right-hand syringe are transferred to the left manipulator and vice versa. This allows the operator to control fluid movement within a microtool with one hand while positioning the same microtool with the other hand.

Volume changes in the system are transferred from the syringes by filling the system with paraffin oil. Care should be taken to ensure that no air bubbles are trapped within the system. The presence of air in the syringes or tubing will decrease fine control. The fact that the syringes and Tygon tubing are transparent make it easy to detect the presence of air bubbles. Micropipettes are not filled with paraffin oil but with Fluorinert, an inert buffer fluid (Elden Instrument Co.). The inert buffer acts as a barrier between the saline solution containing the embryos and the paraffin oil that, in some cases, has been found to be toxic.

The last component of the micromanipulator unit is the microscope. For embryo injection, a compound binocular microscope should be used. The Leitz Diavert inverted microscope used by the author possesses several useful features for blastocyst injection. The Diavert attaches to the Leitz manipulator baseplate giving a stable working surface. The stage of the microscope is fixed, so that movement of the focus controls moves the nosepiece and not the stage. The inverted design of the microscope also allows a larger working space on top of the stage.

The overall magnification of the microscope must be great enough so that the injection pipette and blastocyst can be placed in the same plane. However, the magnification cannot be so great that the field is too small for the entire blastocyst and pipette to be viewed at the same time. A good compromise can be achieved using a 32X objective and a pair of 10X wide-range oculars. Finally, if possible, the microscope should be adapted to provide image-erect optics, although injections can be accomplished with image-reversed optics.

MICROTOOLS

Equipment

Several pieces of equipment are necessary for the production of the microtools used in blastocyst injection. Pipettes are made from lengths of coagulation capillary tubing. This tubing has an outer diameter of 1.2 to 1.4 mm and a length of 150 mm (Fischer Scientific). All pipettes are pulled using an automatic pipette puller. Either vertical- or horizontal-type pullers can be used successfully. A DeFonbrune-type microforge is used to break pipettes for use as injection or holding pipettes. The microforge should be fitted with a binocular dissecting microscope with a calibrated ocular micrometer for sizing pipettes. In addition, a micro-burner is needed for bending pipettes.

Several bevelers are available for the production of microelectrodes. With minor modifications, these bevelers can be used to readily produce high-quality beveled injection pipettes. The beveler used by the author is a K.T. Brown-type micropipette beveler (Sutter Instrument Co., San Rafael, California). The standard unit includes a resistance meter used in the manufacture of microelectrodes. This meter is not used in the manufacture of injection pipettes and will be excluded by the supplier upon request. The grinding plates used are coated with 10 micron diamond dust and can be ordered from the supplier. The system includes a 2-stage micromanipulator that permits the pipette to be slowly advanced onto the grinding surface for accurate beveling. In addition, the beveler needs to be equipped with a binocular dissecting microscope capable of a magnification of 500 diam-eters. This microscope, which is not available from the beveler supplier, is used to monitor the progress of the pipette during the beveling proce-dure.

Pulling and Breaking Pipettes

The controls on the pipette puller should be adjusted to produce pi-pettes that have an even taper to the tip. An even taper will tend to keep the ratio of tube diameter to wall thickness more or less constant.

Since it is difficult to accurately determine the inner diameter of a glass pipette when viewed from the side, all measurements given will refer to the outer diameter. In breaking pipettes, care must be taken to assure that the break is even, thus providing a flat working surface. This is especially true of holding pipettes. Pipettes are broken by fusing the capillary tubing to the bead of glass on the microforge filament. By ad-justing the controls, it is possible to obtain a firm union without bending or local thickening of the capillary wall. When the heating filament is turned off, contraction of the filament should break the capillary tubing perpendicular to its long axis. The outer diameters for holding and injec-tion pipettes vary with the size of embryo used. For cattle and sheep, holding and injection pipettes with outer diameters of approximately 150 μm and 50 μm, respectively, are used. In addition, the tips of the holding pipettes must be polished by placing the tips adjacent to the reheated fil-ament. The holding pipette should be polished until the apparent internal diameter is approximately 50 μm. Finally, the shank of the holding pipette needs to be bent so that the tip of the mounted pipette will be parallel to the surface of the petri dish. This is accomplished using a microburner. The bending of injection pipettes is not done until after beveling.

Beveling

In producing an injection pipette, the main consideration is the length of the bevel. The longer the bevel, the easier it will be for the pipette to penetrate the zona pellucida and trophectoderm. However, if the bevel is too long, the opening in the tip of the pipette may extend outside the embryo, making injections very difficult. For blastocyst injection the author uses a 30° bevel. During the beveling process, the diameter of the pipette is increased slightly. Because of this, it is necessary to break injection pipettes at a diameter slightly less than the final diameter re-quired.

Before beveling begins, the grinding wheel should be moistened with enough deionized water so that a thin film of liquid covers the surface of the wheel while it is turning. The pipette to be beveled is placed in the holder on the beveler and the angle set for 30°. Next, tubing is connected to the distal end of the capillary tubing and to a 3 cc syringe filled with deionized water. The syringe is used to maintain a steady flow of water out of the tip of the pipette during the beveling process. The flow of water prevents the tip of the pipette from being clogged with glass shavings. If the tip does become clogged, the pipette should be discarded as attempts at removing the clog will prove fruitless. While viewing through the microscope and using the coarse control on the beveler micromanipulator, the pipette is advanced until the tip touches the surface of the water on the grinding wheel. At this time, water will be pulled into the pipette by capillary action. With one hand, pressure is applied to the 3 cc syringe to maintain a steady flow of water through the pipette while the other hand advances the tip of the pipette on to the grinding surface using the fine control on the beveler. The pipette is slowly advanced using the fine control until the bevel is complete as determined by viewing through the microscope. The pipette is then withdrawn from the grinder using the course control while still maintaining the flow of water.

After beveling, the pipette is ready to be bent. During the injection process, the entire length of the bevel must be visible. This necessitates the bevel be viewed from the side rather than from above or below. The easiest way to achieve this is to mark the top of the pipette after beveling is completed, but before the pipette is removed from the holder. When the pipette is bent, care should be taken to ensure that this mark will be on the side of the pipette when it is mounted on the micromanipulator. The injection pipette is then bent in the same manner as the holding pipette.

Storage

The final step in the manufacture of micropipettes is to siliconize them prior to storage. Pipettes are siliconized by pulling a 1:100 dilution of Prosil (Specialty Chemical, Gainesville, Florida) from the tip to the distal end using a syringe. Fluid should always be pulled from the tip to the shank end so that any debris in the pipette is pulled away from the tip to prevent clogging. Next, 3X glass-distilled, or the equivalent, water is pulled through the pipette as a rinse. Finally, air is pulled through to aid in drying. Pipettes should then be left to dry overnight or can be placed in a 100°C oven for 10 min prior to use. Pipettes are stored in a clean box with a closed lid to prevent contamination by dust. The shanks of the pipettes are embedded in a substance such as modeling clay so that the tips are not broken by contact with the surface of the box. Using the above procedures, one should be able to produce enough microtools in an afternoon to last through several injection procedures.

ANTISERUM DEVELOPMENT

An inner cell mass (ICM) used for blastocyst injection can be obtained either by mechanically dissecting out the ICM or by immunosurgery. Immunosurgery is technically less demanding in terms of embryo manipulation and yields an ICM with little or no contamination by trophoblast cells. The antiserum used does not have to be specific for trophectoderm as the structure of the embryo protects the ICM from damage. In most embryos, the goat being a possible exception (1), the ICM is completely surrounded by trophectoderm. Tight junctions between the cells of the trophectoderm prevent the antibodies from binding to the ICM.

Antisera are made by injecting rabbits with spleen cells from an animal of the same species as that from which ICMs are to be obtained using the method of Solter and Knowles (15). Fresh spleens are finely chopped using a scalpel blade and then homogenized with a small volume of physiological saline in a glass tissue homogenizer. The resulting homogenate is then either gently centrifuged for approximately 30 sec or filtered through several layers of cheesecloth to remove any large pieces of tissue. The number of cells per ml of the spleen cell suspension is then determined using a hemocytometer. Once the cell concentration has been determined the suspension is diluted with saline to give a final concentration of 4 x 10^8 cells per ml. Each rabbit is given 1 ml of the spleen cell suspension intravenously through the marginal ear vein. Three weekly injections are given with a fresh spleen cell preparation used each time. Ten days following the last injection, the rabbits are bled either from the ear or by heart puncture. After clotting and centrifugation, the serum is harvested and heat-treated at 56°C for 30 min in order to inactivate complement. The heat-treated antisera are then divided into small aliquots and stored at −70°C. Aliquoting of the antisera is necessary as repeated freezing and thawing will significantly decrease the activity of the antisera.

One bleeding should provide sufficient antisera to perform a large number of ICM isolations. If additional antisera are required, the animal can be boosted with an additional injection of 4 x 10^8 spleen cells and bled 7 to 10 days later.

Since embryos are often a limiting factor, they are usually not used to titrate the antisera. However, a test to determine if the rabbit has reacted to the injected cells can be performed by testing the ability of the antisera to lysis red blood cells of the same species as the spleen cells. Briefly, one drop of a 2.0% solution of washed red blood cells is placed in a small test tube along with 2 drops of a 1:2 dilution of the antiserum to be tested. This is allowed to incubate at room temperature for 30 min. One drop of normal guinea-pig serum is then added to the tube and the tubes held at room temperature for 1 hr. If an antibody to the species tested is present, the tubes should show hemolysis of the red blood cells.

INNER CELL MASS ISOLATION

The procedure used is essentially that developed by Solter and Knowles (15). The zonae pellucidae are removed from the embryos to be treated. Zona removal can be accomplished mechanically with a micromanipulator but this can become tedious with a large number of embryos. A more rapid method is to incubate the embryos in a 0.5% solution of pronase (11). Usually within 2 min of being placed in the solution, the zonae will have visibly thinned and expanded away from the embryo. At this point, the embryos should be removed from the pronase and vigorously washed through 3 drops of culture medium containing a protein source (i.e., bovine serum albumin or serum). The washing frees the embryo from the partially digested zona while the protein acts to inactivate the pronase.

The denuded embryos are then individually placed in drops of heat-treated antiserum diluted with culture medium. If the antiserum has been titered, the greatest dilution that will cause complete lysis of the trophoblast cells is used. If the antiserum has not been titered, an antiserum dilution of 1:8 is used to ensure that sufficient antibody is present. Embryos are incubated in the antiserum for 1 hr and then washed to remove any unbound antibody. If the antiserum has been properly heat-treated, the embryos should not show any visible effects.

The washed embryos are then placed in individual drops of normal guinea-pig serum diluted 1:8 with culture medium. The guinea-pig serum serves as a source of complement. In the presence of complement, the trophoblast cells to which the antibody has bound are lysed. Embryos are usually left in the complement solution for 1 hr in order to assure lysis of the trophoblast cells. Inner cell mass cells, which have no antibody attached to their surface, are not affected by this treatment and so timing is not a critical factor. Final removal of lysed trophoblast cells is accomplished by gentle pipetting through a small-bore pipette.

BLASTOCYST INJECTION

Set-up

In setting up the manipulator for blastocyst injection several factors must be considered. Best results are obtained if the manipulators are kept in the horizontal position. To this end, raising and lowering of the pipettes should be accomplished by using the controls that raise and lower the manipulator unit and not those controls that tilt the manipulator. The reason for this is that when the manipulator is tilted away from horizontal, advancing the pipette will cause the tip to move not only forward but downward toward the microscope stage as well. This downward movement can result in the pipette scraping along the bottom of the petri dish and, more importantly, will tend to push the recipient blastocyst off the holding pipette during the injection process. Before attaching the pipettes to the tubing, make certain there is no air in the lines. The presence of air will cause a delay in the suction or injection response while the air in the line compresses or expands. Although it is possible to compensate for this delay, removing air from the lines will prevent much frustration. Next, oil is advanced through the tubing until a drop appears on the end of the pipette collar. The pipette is filled with inert buffer fluid and fitted into the collar, making sure that no air bubbles are trapped in the process. The holding nut is then tightened to hold the pipette steady.

Adjustments of the pipettes are also important for successful injection. Coarse adjustments can be made visually while final adjustments must often be made while looking through the microscope. The pipettes should be adjusted so that the tip of each pipette will touch the bottom of the petri dish while maintaining the pipette as parallel to the microscope stage as possible. It is often necessary to slightly tilt the pipettes to accomplish this. Again, this should not be done by tilting the manipulators but rather by adjusting the instrument holding heads. The pipettes should also be adjusted so that they meet directly head-on. During the injection process, the injection pipette must push the recipient blastocyst squarely against the holding pipette. If not, the injection pipette will simply push the blastocyst off the holding pipette.

Injections

The procedure used is based on that reported by Babinet and Bordenave (2). Injections are performed at room temperature in drops of phosphate buffered saline containing 1% (volume/volume) heat-treated serum and 1% (v/v) antibiotic-antimycotic mixture (100X, Gibco, Grand Island, New York). A number of drops are placed in a 100 x 15 mm petri dish. The drops are covered with a layer of Dow Corning 360 Medical Fluid (20 cs viscosity) to prevent evaporation and to stabilize the drops. One isolated ICM and one recipient blastocyst are placed in each drop. It is important that there be only one blastocyst per drop as it is impossible to distinguish between successfully and unsuccessfully injected blastocysts. Usually, one drop is left empty. This drop is used to make final adjustments of the pipettes.

Before the pipettes are lowered into the drops containing embryos, some fluid should be pushed through the pipettes in order to remove any air bubbles that may be trapped in the tips. The pipettes are lowered into the drop without embryos to check their position. If the 2 pipettes do not meet squarely they should be adjusted at this time.

After the pipettes have been correctly adjusted, they are moved to the first drop containing a blastocyst and an ICM. The blastocyst to be injected is picked up using the holding pipette by the polar trophectoderm overlying the ICM. This allows both the ICM and the blastocoele to be clearly seen. The suction applied to the holding pipette should be sufficient to firmly hold the blastocyst on the end of the pipette, but not so strong as to cause the blastocyst to be completely pulled into the pipette. Next, the isolated ICM is pulled into the injection pipette. This step requires, by far, the most patience as the ICM must be slowly pulled into the pipette in order to prevent extensive cell damage to the ICM. Once the ICM is in the injection pipette, it should be held near the tip so its movements can be observed during the injection process. The above procedures are usually performed at a magnification of 40 diameters.

The magnification is next increased to 200 diameters and the forward movement of the injection pipette adjusted. During the injection process, the injection pipette is rapidly advanced into the blastocoele. To prevent the pipette from being pushed completely through the embryo, the controls on the injection manipulator are adjusted to provide a mechanical stop at the point where the tip of the injection pipette just touches the ICM. This is accomplished using the forward/back and left/right control knobs on the manipulator. The joystick is advanced to its most forward position. The injection pipette is then brought to the middle of the field using the control knobs. Again, using the control knobs, the recipient embryo is brought in just below the injection pipette and its position adjusted so that the tip of the injection pipette will be in contact with the ICM and the bevel on the pipette will be completely within the blastocoele (Fig. 1). The injection pipette is then moved back using the joystick and the recipient blastocyst moved to the center of the field using the control knobs.

If the injections are to be successful it is important that the injection pipette be in the same plane as the widest part of the embryo. To

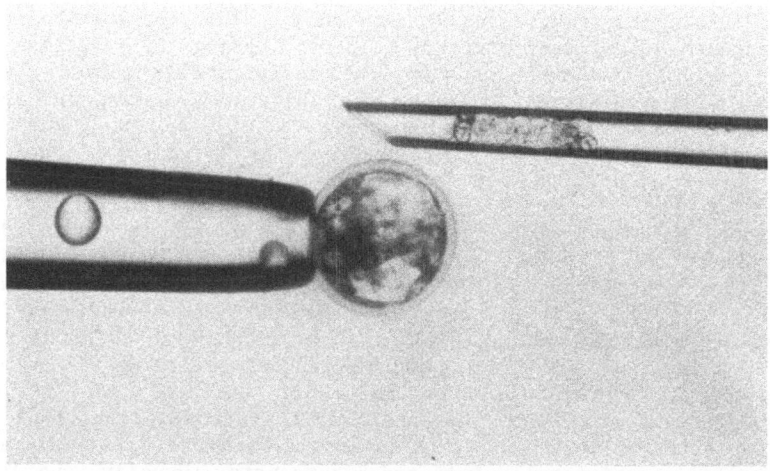

Fig. 1.   The blastocyst is held by the polar trophectoderm with a holding pipette. The injection pipette containing an isolated inner cell mass has been positioned so as to prevent the injection pipette from passing completely through the blastocyst.

accomplish this, the microscope is adjusted to its highest magnification, which reduces the depth of focus. Using the up and down control on the manipulator, the injection pipette is raised or lowered so that it is in the same focal plane as the edge of the embryo. If adjusted properly, the tip of the injection pipette should touch the extreme outer edge of the zona pellucida as it is advanced forward with the joystick (Fig. 2).

The actual injection is carried out at 200 diameters in order to provide a wider field of view. Using the joystick, the injection pipette is moved away from the embryo. In one swift motion the injection pipette is rapidly advanced into the blastocoele, puncturing the zona pellucida and trophectoderm. Using the control knob the pipette is adjusted so that the tip contacts the ICM of the recipient blastocyst (Fig. 3). The ICM is then gently expelled from the injection pipette onto the surface of the ICM. During this time the recipient blastocyst should remain in the expanded state. After the ICM is expelled, the injection pipette is removed. The blastocyst will now begin to collapse upon itself, trapping the injected cells against the ICM. Injected blastocysts are usually cultured overnight and then examined to assess cell incorporation. Blastocysts will usually re-expand within a few hours of injection.

Common Problems

Trophoblast not penetrated. This is perhaps the most common problem encountered, especially if beveled pipettes are not used. The problem is usually caused by either incorrect adjustment of the injection pipette or the pipette not being advanced rapidly enough to penetrate the trophectoderm. In the case of the former, the manipulator must be adjusted so that the tip of the injection pipette will be in contact with the ICM. If not, the pipette will stop before the trophoblast has been completely punctured, making introduction of the isolated ICM difficult if not impossible. If the pipette is properly adjusted, but the trophoblast still is not penetrated, the problem may be due to the injection pipette being advanced too slowly. The trophectoderm, unlike the zona pellucida, is not a rigid structure. While slowly advancing the injection pipette will eventually cause it to "pop" through the zona pellucida, this procedure will only

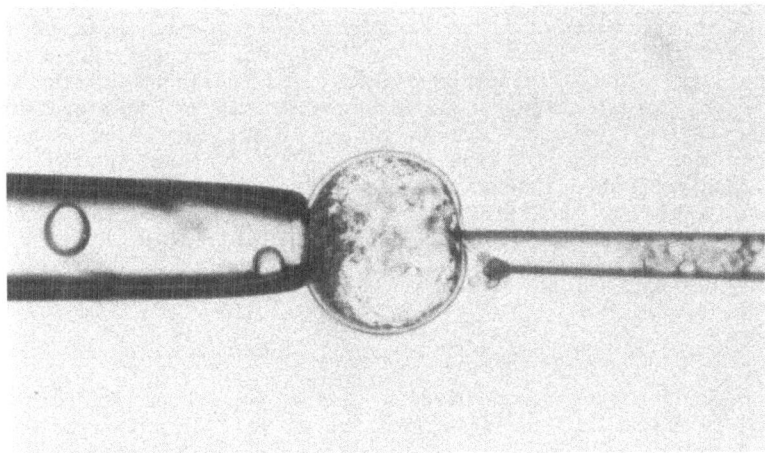

Fig. 2.   Proper positioning of the injection pipette in the plane of the recipient blastocyst. When properly positioned, advancing the injection pipette should cause the tip to contact the recipient blastocyst at its maximum diameter.

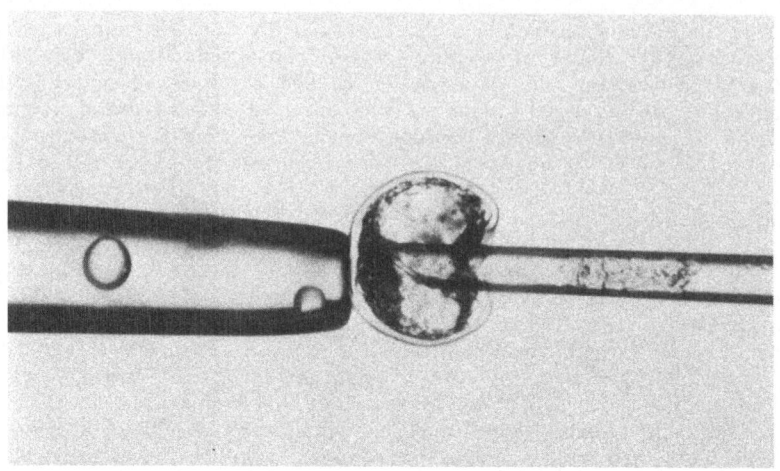

Fig. 3.   Successfully injected blastocyst with the tip of the injection
pipette properly positioned for the introduction of the isolated
inner cell mass.

succeed in pushing the trophectoderm ahead of the pipette and collapsing
the blastocyst. The pipette must be advanced as rapidly as possible to
cleanly penetrate the trophectoderm. This ability can only be developed
with practice.

Pipette completely through blastocyst. This problem is just the opposite of the one discussed above and is more often seen when beveled
pipettes are used. Again, this is a problem of improper pipette adjustment, but unlike the previous problem it often does not result in an unsuccessful injection. If the suction applied by the holding pipette is not
so great as to evacuate the zona through the hole made, the injection pipette can be withdrawn slightly until its tip is in the blastocoele and the
injection completed.

Uncontrolled movement of ICM in injection pipette. This problem can
have a number of causes and can be quite aggravating. Some of the more
common causes are air in the injection lines, an improper seal between the
injection tubing and the injection pipette, and localized heating and cooling of the injection tubing by the microscope lamp or other equipment. If
the problem of cell movement occurs during injection, it can sometimes be
compensated for by holding the isolated ICM further up the injection
pipette. However, when this is done the ICM is often out of view, making
it hard to judge how much pressure should be applied to inject the cells
into the embryo without introducing too much fluid too rapidly so that the
blastocyst explodes.

CONCLUSIONS

The procedures described in this chapter have been used successfully
to produce intraspecific chimeras in sheep as well as interspecific chimeras between sheep and goats. The techniques described do not require an
unusual amount of manual dexterity, but they do require patience and a substantial amount of practice. If done routinely, blastocyst injections can
be performed relatively rapidly and with the loss of few embryos. However,
if a substantial period elapses between experiments, the investigator
should allow a period of practice to "relearn" the technique. The author

has found murine blastocysts to be the best for practice due to the relative ease with which they can be obtained and to the greater resilience of their zonae pellucidae, which makes them slightly more difficult to inject than livestock embryos. Finally, it should be pointed out that while the techniques presented have been used successfully by the author and others, different laboratories may require some modifications and suggestions as to how these techniques may be improved are always welcomed.

ACKNOWLEDGEMENTS

The author would like to thank Dr. Gary Anderson who assisted in the development of this technique as well as in the preparation of this manuscript. The author would also like to thank S. Donahue, M. Dunbar, C. Tores, and D. Anderson for their expert assistance during the course of this research.

REFERENCES

1. Amoroso, E.C., W.F.B. Griffiths, and W.J. Hamilton (1942) The early development of the goat (Capra hircus). J. Anat. 76:377-411.
2. Babinet, C., and G.R. Bordenave (1980) Chimaeric rabbits from immunosurgically-prepared inner-cell-mass transplantation. J. Embryol. Exp. Morph. 60:429-440.
3. Benirschke, K. (1970) Spontaneous chimerism in mammals. A critical review. Curr. Topics Pathol. 51:1-61.
4. Bren, G., H. Tenhumberg, and H. Kraublich (1984) Chimerism in cattle through microsurgical aggregation of morulae. Theriogenology 22:609-613.
5. Fehilly, C.B., S.M. Willadsen, and E.M. Tucker (1984) Experimental chimaerism in sheep. J. Reprod. Fert. 70:347-351.
6. Fehilly, C.B., S.M. Willadsen, and E.M. Tucker (1984) Interspecific chimaerism between sheep and goat. Nature (London) 307:634-636.
7. Gardner, R.L. (1968) Mouse chimaeras obtained by the injection of cells into the blastocyst. Nature (London) 220:595-597.
8. Gardner, R.L. (1978) Production of chimeras by injecting cells or tissue into the blastocyst. In Methods in Mammalian Reproduction, J. C. Daniel, ed. Academic Press, New York, pp. 137-165.
9. McLaren, A. (1976) Mammalian Chimaeras, Cambridge University Press, Cambridge, England.
10. Mintz, B. (1962) Formation of genotypically mosiac mouse embryos. Am. Zool. 2:432.
11. Mintz, B. (1962) Experimental study of the developing mammalian egg: Removal of the zona pellucida. Science 138:594-595.
12. Moustafa, L.A., and R.L. Brinster (1972) The fate of transplanted cells in mouse blastocysts in vitro. J. Exp. Zool. 181:181-192.
13. Nicholas, J.S., and B.V. Hall (1942) Experiments on developing rats. J. Exp. Zool. 90:441-459.
14. Pighills, E., J.L. Hancock, and J.G. Hall (1968) Attempted induction of chimaerism in sheep. J. Reprod. Fert. 17:543-547.
15. Solter, D., and B.B. Knowles (1975) Immunosurgery of mouse blastocyst. Proc. Natl. Acad. Sci., USA 72:5099-5102.
16. Tarkowski, A.K. (1961) Mouse chimaeras developed from fused eggs. Nature (London) 190:857-860.
17. Trounson, A.O., and N.W. Moore (1974) The survival and development of sheep eggs following complete or partial removal of the zona pellucida. J. Reprod. Fert. 41:97-105.
18. Tucker, E.M., R.M. Moor, and L.E.A. Rowson (1974) Tetraparental sheep chimaeras induced by blastomere transplantation. Changes in blood type with age. Immunology 26:613-621.

GENE TRANSFER BY DIRECT PRONUCLEI MICROINJECTION

Albert W. Tam

Department of Molecular Biology
Genentech, Inc.
South San Francisco, California   94080

INTRODUCTION

The technique of micromanipulating and injecting a mouse egg was first described in 1966 by T.P. Lin (8,9). With the advent of recombinant DNA technology, it has become possible to transform cultured cells by modifying this technique to microinject directly into the cell nuclei-specific cloned gene sequences (1,4). This approach is amenable only to established cell lines in culture, and so its general usefulness in the analysis of tissue-specific gene expression has been found to be somewhat limited. Subsequently, transgenic animals have been produced by introduction of foreign genes at the pronuclear stage of fertilized, one-cell zygotes. Most of the successes have been with mouse eggs (3,5,6,11,12), and only recently has the successful production of transgenic animals been extended to the rabbit, pig, and sheep (7). This technique has become a powerful tool for studying gene regulation and physiological functions of gene products in a normal host environment. This chapter summarizes the current methodology used for mouse eggs in gene transfer experiments by direct pronuclear microinjection.

INSTRUMENTS FOR THE MICROINJECTION SET-UP

Micromanipulators

The micromanipulators of choice are Leitz micromanipulators, consisting of a right-handed and a left-handed unit. Each micromanipulator is mounted on the corresponding side of the microscope stage, one controlling the egg-holding pipette and the other controlling the injection pipette. Vertical movement is adjusted by separate coarse and fine drives. Horizontal movements in the X- and Y-axes are actuated by 2 independent coarse controls. The fine adjustments of the microinstruments are controlled with a guide lever, or "joy stick," a most important feature in the Leitz micromanipulator. The use of the guide level allows simultaneous movement of the microinstrument in 2 coordinates. The operator's hand movement of the lever is transmitted to the microinstrument with a gear ratio that can be varied continuously between 1:1/16 and 1:1/800. Micromanipulators made by Narishige (Distributor: Medical Systems Corporation, Greenvale, New York) and Zeiss (Thornwood, New York) may also be acceptable, as they have recently incorporated the "joy stick" feature in their designs.

Microscope

An inverted microscope with a fixed stage, image-erected optics and a long working-distance condenser and objectives is essential for ease of performing microinjection. A long working-distance condenser and objectives are required to provide the space necessary to accommodate the injection chamber (discussed below) and micropipettes. Nomarski Differential Interference Contrast (DIC) optics or Hoffman Modulation Contrast optics (Modulation Optics, Inc., Greenvale, New York) are preferable to the standard brightfield optics and phase contrast, as they enhance greatly the visualization of the outline of the pronucleus, thus allowing much easier injection. With Nomarski optics, a glass injection chamber must be used, as plastic will destroy the Nomarski image. Microinjection of pronuclei is usually carried out at a magnification of 320 to 400X. This can be achieved by using a 32X or 40X objective with 10X eye-pieces. A low-power objective, e.g., 4X, can be used to move eggs generally in and out of and within the injection chamber.

The microscope and micromanipulators should be mounted on a common base plate (e.g., the Leitz base plate made for this purpose). If the Leitz Diavert is the choice of microscope, it and the micromanipulators can be attached simply to this base plate. With other models of microscopes (e.g., Zeiss IM-35, Zeiss ICM-405, or Nikon Diaphot), the micromanipulators may have to be mounted on elevated platforms attached to the base plate such that the micropipettes can be positioned at a height relative to the stage of the microscope.

Another consideration for the setting-up of the apparatus is the amount of vibration at the site of the work area. If extraneous vibration is not a problem, it is possible to set up the instruments on a standard laboratory bench, or on any sturdy table surface. A heavier base plate with some shock-damping materials, such as sheets of foam rubber underneath, may have to be considered if occasional building vibration arises. For severe vibration problems, special tables, such as an air-filled, vibration isolation system MICRO-g table (Technical Manufacturing Corporation, Wobman, Massachusetts), are necessary to damp out all extraneous vibrations. For such tables, fixed arm rests that support the weight of the operator's arms are helpful, as these tables are very susceptible to vibration generated by the operator.

Holding Pipette

The egg-holding micropipette is a suction pipette, with an outside diameter (OD) of 70 to 100 microns (μm) and an inside diameter (ID) of about 10 to 15 μm at the tip. These micropipettes are prepared from glass capillaries of 1-mm OD (e.g., E. Leitz, Rockleigh, New Jersey; Drummond Scientific Company, Broomal, Pennsylvania). The capillaries are heated and hand-pulled to the desired outside diameter. A clean event break is introduced to the drawn-out region of the capillary on a DeFonbrune microforge. The tip is subsequently melted and polished to the desired internal diameter. The holding pipette can be inserted into a length of plastic or polyethylene tubing, housed inside a Leitz instrument collar, and connected to a micrometer with a mechanism for fine movement control. The entire set-up is filled to the tip of the holding pipette with either mineral oil or Fluorinert FC77 (3M Company). No air bubbles should be present in the system. The holding pipette, which is held in place by clamping the instrument collar to the micromanipulator, should be positioned at a slight angle such that it can reach into the bottom of the injection chamber. A slight bend of about 15 to 20° can be introduced 2 to 3 mm from the tip by carefully heating the pipette with the microforge at the point of the intended bend. This facilitates the positioning of the tip horizontally at the bottom of the injection chamber. A number of holding pipettes can be prepared prior to use, dry-heat sterilized, and stored under sterile conditions.

Injection Pipette

Injection pipettes are best drawn from thin-walled glass capillaries by a mechanical pipette puller (e.g., David Kopf Instruments, Tujunga, California). By varying the temperature of the heating filament and tension of the pull, reliable injection pipettes can be prepared. A good injection pipette should have a tip that is fine enough (about 0.5 μm) to minimize damage to the mouse egg, but also rigid enough to allow it to push through the zona pellucida and vitelline or plasma membrane. The necessary settings on the pipette puller must be determined by trial and error, and once determined can be adjusted according to the age and quality of the heating element.

Glass capillaries containing a filament fused to the inner wall (e.g., Kwik-Fil capillaries, W-P Instruments, Inc., New Haven, Connecticut, or Omega Dot capillaries, Frederick Haer and Company, Brunswick, Maine) are often preferred, as they make possible the rapid filling of the injection pipette by capillary action from the blunt end.

Glass capillaries can be routinely used out of the package without prior cleaning or sterilization, especially when the injected eggs are to be transferred immediately after micromanipulation back to foster females. Alternatively, the glass capillaries can be washed, dry-heat sterilized, and stored under sterile conditions until use. Injection pipettes are best pulled and used on the same day, as stored pipettes generally tend to clog more easily.

Two methods have been devised to discharge DNA from the injection pipette. One is a continuous flow system. A DNA solution-filled pipette is hooked up to a pressurized air or gas cylinder by a length of rigid tubing. A gas regulator capable of fine pressure control is placed in the line to regulate the flow. The amount of pressure required to generate a reasonable flow of DNA solution is normally fairly low (a few psi). This should be determined empirically by injecting a number of mouse eggs at various pressures. Often, this may result in the bursting and killing of a few eggs, but the upper limit of exerted pressure can be determined. An advantage of injecting with a continuous flow is the elimination of negative pressure in the system, and thus there is never any sucking back of culture medium into the tip of the injection pipette. This method, however, may be more difficult to master.

The other method used is a positive-pressure, bolus-like injection. Three different variations have been designed to produce this positive pressure. The first involves the use of a large (e.g., 50 ml) ground glass syringe. The injection pipette is connected by plastic tubing to the syringe. The plunger of the syringe is pulled back about half-way, before the pipette filled with DNA is loaded onto the instrument collar. Depressing the plunger exerts pressure on the DNA solution and it flows out of the tip. When the pressure on the plunger is released, it quickly returns to its neutral position, thus preventing negative pressure from building up. The rate of flow is mainly controlled in this case by the size of the orifice of the injection pipette. The second alternative utilizes the same set-up as in the continuous flow system, except a T-tube is used in place of the fine pressure regulator. One branch of the T-tube is normally left open. Then partial blockage (e.g., by the thumb) of this open end during pressurized air-flow through the system creates positive pressure and forces the DNA solution out. The amount and duration of blockage determine the volume of DNA solution injected into the pronucleus. The third set-up uses the same principle as the control mechanism for the holding pipette. A micrometer or a Hamilton 100-μl screw-type syringe is used to provide the pressure. The plastic tubing connecting the injection pipette and the micrometer or syringe in this system is filled with mineral oil or silicone fluid, and DNA solution is generally picked up at the tip of the pipette by suction, rather than filled from the blunt end.

Injection Chamber

Two types of injection chambers are most commonly used. First, one can make use of a regular glass depression microscope slide. A drop of standard mouse egg culture medium (substituting 25 mM Hepes, pH 7.4, for the bicarbonate ions so as to maintain the pH at room temperature independent of CO_2) is placed on the center of the depression of a lightly siliconized slide. Mineral oil is then gently laid over the drop of medium. The medium drop should be made as flat as possible to minimize a lens effect which will distort the optics. With thorough siliconization, the drop of medium will remain too round, and without siliconization, the drop will not hold in place when micromanipulation is attempted. The glass depression slide can be used with both Nomarski and Hoffman optics systems on an inverted microscope.

A second chamber often used is the lid of a disposable plastic petri dish. A large, flat drop of modified culture medium is placed on an inverted lid and covered with mineral oil. This set-up cannot, however, be used with Nomarski DIC optics. The petri dish lid is utilized because it has lower walls which allow insertion of micropipettes at a lower angle. For both types of chambers, the holding and injection pipettes are inserted into the medium drop at an angle of about 10 to 20°, allowing them to reach the eggs at the bottom of the chamber. The bent portion of the holding pipette should be positioned horizontally relative to the bottom. In this manner, eggs can be sucked up and held by the holding pipette without rotation once a pronucleus is visualized and focused upon. The injection pipette should not be inserted at too much of an angle, as viability of the eggs will be affected if they are pierced in a slit-opening fashion while being injected.

OPERATING PROCEDURES

Preparation of Mouse Eggs

Fertilized one-cell mouse eggs can be obtained by either natural matings using female mice in estrus or by superovulating females with gonadotropin injections to induce mating. Eggs are flushed from the oviducts of vaginal plug-positive females about 12 to 14 hr after ovulation, using standard culture medium for preimplantation mouse embryos. Eggs recovered at this stage are usually embedded in cumulus masses and can be treated with hyaluronidase (300 U/ml) to dissociate the cumulus cells. This can be carried out under a dissecting microscope for several minutes until the cumulus cells begin to fall off. The released eggs should be removed promptly from the hyaluronidase treatment, as prolonged exposure may be harmful. Eggs collected are washed free of hyaluronidase in fresh medium and kept in culture at 37°C under 5% CO_2, except during microinjection.

Preparation of DNA Solution

DNA solutions used for microinjection should be free of contaminants that may harm the eggs and particulate matter that may clog the microinjection pipette. Reagents such as phenol, ethanol, chloroform, etc., which are used to purify the DNA, must be removed thoroughly. Supercoil plasmid DNA should be banded twice on ethidium bromide/cesium chloride gradients. Linear fragments generated by endonuclease digestion should be extensively extracted with phenol/chloroform, or gel-purified followed by phenol/chloroform extraction. A final dialysis step to further purify the DNA is suggested.

Purified DNA pellets are dissolved in an injection "buffer." Different buffers, ranging from distilled water to isotonic phosphate buffer or saline, have been used by various researchers. There has been no evidence to suggest the superiority of one buffer over another. Ethylenediaminetetraacetic acid (EDTA), however, has been found to enhance DNA integration frequency in the range of 0.1 to 0.3 mM (2). Two examples of injection buffers used are: (a) 10 mM Tris-HCl/0.25 mM EDTA, pH 7.5, and (b) 5 mM NaCl/5 mM Tris-HCl, pH 7.4/0.1 mM EDTA. All solutions being added to the DNA samples should be filtered sterile, and the final DNA preparation centrifuged for 15 to 30 min in a microfuge to bring down any remaining particulate matter.

Microinjection of the Pronucleus

An injection chamber is set up, as described above, and placed onto the microscope stage. While focusing on the edge of the medium drop, using the low-power objective, the holding pipette is lowered into the drop until the tip just touches the bottom. Slight suction is applied using the micrometer to take up some medium into the holding pipette such that the meniscus between the mineral oil or Fluorinert and culture medium is visible and close to the tip. A group of mouse eggs is transferred by a finely drawn embryo pipette into the drop. Under high magnification, the eggs are examined for the presence of pronuclei. If some or all of the eggs are found to be injectable, the tip of an injection pipette is filled with DNA and inserted slowly into the medium drop at low microscopic power. The tip should come in focus along with the eggs in the medium. Using the injection pipette, an egg is moved next to the tip of the holding pipette. Switching to high magnification, the male and female pronuclei are located. The male pronucleus is the larger of the two and, thus, may be easier to inject. However, a pronucleus can be most easily injected if it is located close to the plasma membrane in the hemisphere of the egg closest to the injection pipette. Ideally, the pronucleus should also be close to the central horizontal axis of the egg. If it is positioned on the top or bottom part of the cytoplasm, an attempt to penetrate the egg with the injection pipette will often rotate it, causing the pronucleus to go out of focus.

Using the injection pipette, the egg is oriented so as to put the male pronucleus in position to be injected. By applying gentle suction on the holding pipette, the egg is sucked onto the opening at the tip. The egg should not be deformed by applying too much suction; only the zona should be pulled into the opening. Next, the outline of the pronucleus and the plasma membrane should be brought back into focus to make sure the egg is still in position to be injected. If not, it should be released, reoriented, and sucked back onto the holding pipette. The tip of the injection pipette is then placed right next to the egg and brought into the same focal plane by using the fine vertical drive of the micromanipulator unit (see Fig. 1). With the "joy stick," the injection pipette is advanced to push through the zona and plasma membrane, and into the pronucleus. The tip of the injection pipette and the pronucleus should remain in sharp focus. The nucleoli that are present within the pronucleus should not be touched, as they are quite sticky and will invariably adhere to the tip of the injection pipette. With the continuous flow method, one should see a fairly rapid expansion of the pronucleus if the injection pipette is successfully

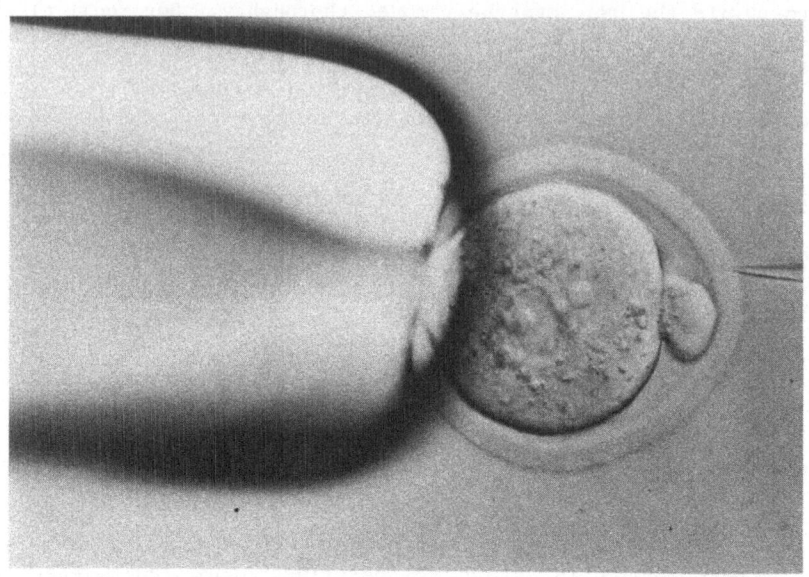

Fig. 1.   A fertilized, one-cell mouse egg is held by suction onto the
holding pipette on the left. Both the male and female pronuclei
are clearly visible, as is the tip of the injection pipette on
the right. (Picture taken with Nomarski DIC optics.)

placed inside the nucleus. When the pronuclear volume has increased by
about 30 to 50%, the injection pipette is quickly withdrawn by pulling back
the "joy stick." With positive-pressure injection, pressure is applied
when the injection pipette has been pushed into the pronucleus. If a micrometer or a Hamilton syringe is used, the mechanism is turned to force
out the DNA solution. A gradual swelling of the pronuclear volume should
be observed. This expansion is much less dramatic and slower than with
continuous flow. If a 50-ml glass syringe is used, the plunger is squeezed
to put positive pressure on the DNA. Pronuclear expansion is, in this
case, rapid and obvious. If compressed air and a T-tube system are utilized, complete or partial blockage of the open end creates a flow, and pronuclear expansion caused by the inflow of DNA is again a fast swelling.

If the pronucleus does not expand upon injection, one of 2 possibilities exists. First, the injection pipette has become clogged and no flow
of DNA is possible. In this case, a new pipette is used, filled with DNA,
and the injection is attempted again. Second, the injection pipette has
not actually pierced through the plasma membrane and, thus, is not inside
the pronucleus. The membrane is highly flexible and can be indented far
into the egg without being pierced. Two signs indicating this are: (a)
the presence of a "blob" of DNA solution around the tip of the injection
pipette when pressure is delivered to express the DNA, and (b) the indentation of the plasma membrane at the point of apparent entry by the injection
pipette rather than the two forming somewhat of a 90° perpendicular. Both
are clear indications that the injection pipette is simply pushing the membrane into the cytoplasm but not piercing it. Often it is possible to get
inside the pronucleus by advancing the injection pipette apparently all the
way through the pronucleus, then pulling back slightly and into the pronucleus. Care must be taken not to hit the injection pipette tip against the
wall of the holding pipette opening. A clear expansion by the pronucleus
upon injection, however, must be the final indication of whether an egg has
been injected or not (see Fig. 2).

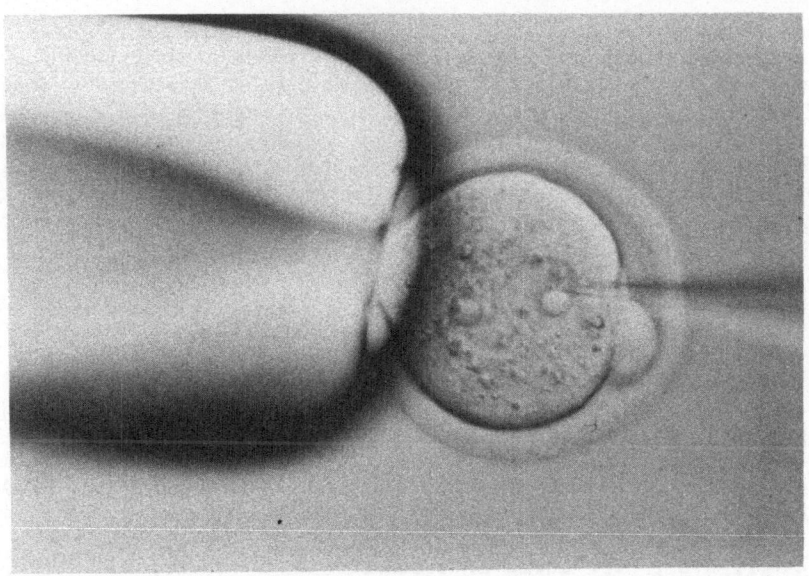

Fig. 2. Expansion of the pronucleus subsequent to microinjection. (Picture taken with Nomarski DIC optics.)

If cytoplasmic granules begin to flow out of the egg as the injection pipette is withdrawn, the egg is lysing and will likely die. The eggs in the chamber should be injected sequentially until all injectable eggs have been injected. Injected eggs should be kept separate, away from the unfertilized or uninjectable eggs (e.g., eggs with very small pronuclei). The injected eggs are then transferred back into culture at 37°C under CO_2 and new eggs are deposited into the chamber. When all available eggs have been injected, they are examined under a dissecting microscope to sort out the lysed eggs from the viable ones. In most experiments, between 60 to 80% of the injected eggs will survive and can be transferred back to the oviducts of synchronous foster females.

CHOICE OF MOUSE STRAINS

F1 hybrid mice are generally used as donor animals. On the average, a higher number of fertilized eggs can be recovered from hybrid females, and the eggs have been found to survive better the physical trauma of the microinjection process, as well as to give an overall higher efficiency of integrating the injected DNA into their genome (2). Several inbred strains (e.g., SJL, CBA, C3H) have been crossed with C57BL/6 to provide the F1 hybrid mice, and each combination has been successfully used to generate transgenic mice. When in some instances a uniform and defined genetic background is desirable, C57BL/6 is the only inbred line that has been used for such experiments.

DETECTION OF TRANSGENIC MICE

Transgenic animals that have integrated the injected foreign gene sequences are detected by performing DNA hybridization experiments. Typically, high molecular weight DNA is isolated from a small segment of the mouse tail. Total tail DNA can be dot-blotted, and hybridized to a specific probe from the introduced gene sequences. Mice that contain these gene sequences can be scored as positives by their hybridization signals. Alternatively, the genomic DNA can be digested with restriction endonuclease and screened for the presence of gene sequences of interest by Southern blot hybridization (10). In most instances, the restriction endonuclease used is selected to give a diagnostic hybridizing DNA fragment of known size. To examine whether a gene has been stably integrated into the genomes of the transgenic animals, progeny are produced by breeding these animals and the young are screened by the same hybridization experiments.

REFERENCES

1. Anderson, W.F., L. Killos, L. Sanders-Haigh, P.J. Kretschmer, and E.G. Diacumakos (1980) Replication and expression of thymidine kinase and human globin genes microinjected into mouse fibroblast. Proc. Natl. Acad. Sci., USA 77:5399-5403.
2. Brinster, R.L., H.Y. Chen, M. Trumbauer, M.K. Yagle, and R.D. Palmiter (1985) Factors affecting the efficiency of introducing foreign DNA into mice by microinjecting eggs. Proc. Natl. Acad. Sci., USA 82:4438-4442.
3. Brinster, R.L., H.Y. Chen, M. Trumbauer, A.W. Senear, R. Warren, and R.D. Palmiter (1981) Somatic expression of herpes thymidine kinase in mice following injection of a fused gene into eggs. Cell 27:223-231.
4. Capecchi, M.R. (1980) High efficiency transformation by direct microinjection of DNA into cultured mammalian cells. Cell 22:479-488.
5. Constantini, F., and E. Lacy (1981) Introduction of a rabbit β-globin gene into the mouse germ line. Nature 294:92-94.
6. Gordon, J.W., G.A. Scangos, D.J. Plotkin, J.A. Barbova, and F.H. Ruddle (1980) Genetic transformation of mouse embryos by microinjection of purified DNA. Proc. Natl. Acad. Sci., USA 77:7380-7384.
7. Hammer, R.E., V.G. Pursel, C.E. Rexroad, Jr., R.J. Wall, D.J. Bolt, K.M. Ebert, R.D. Palmiter, and R.L. Brinster (1985) Production of transgenic rabbits, sheep and pigs by microinjection. Nature 315:680-683.
8. Lin, T.P. (1966) Microinjection of mouse eggs. Science 151:333-337.
9. Lin, T.P. (1971) Egg micromanipulation. In Methods in Mammalian Embryology, J.C. Daniel, Jr., ed. W.H. Freeman and Company, San Francisco, pp. 157-171.
10. Southern, E.M. (1975) Detection of specific sequences among DNA fragments separated by gel electrophoresis. J. Mol. Biol. 98:503-517.
11. Wagner, E.F., T.A. Stewart, and B. Mintz (1981) The human β-globin gene and a functional viral thymidine kinase gene in developing mice. Proc. Natl. Acad. Sci., USA 78:5016-5020.
12. Wagner, T.E., P.C. Hoppe, J.D. Jollick, D.R. Scholl, R.L. Hodinker, and J.B. Gault (1981) Microinjection of a rabbit β-globin gene into zygotes and its subsequent expression in adult mice and their offspring. Proc. Natl. Acad. Sci., USA 78:6376-6380.

APPLICATION OF BIOENGINEERING TO DISEASE DIAGNOSIS

Bennie I. Osburn

School of Veterinary Medicine
University of California
Davis, California 95616

INTRODUCTION

Bioengineering offers great promise for revolutionizing disease diagnosis. Traditional approaches such as clinical, pathological, and etiological methods remain important; however, the newer techniques and reagents can add to the rapidity and specificity of diagnosis. The principal technological advances contributing to these new approaches include methods of recombinant DNA, monoclonal antibody, and immunoassays. The application of these techniques to disease diagnosis in animal diseases is reviewed in this chapter.

CURRENT DIAGNOSTIC TECHNIQUES

Diagnosis of animal diseases is based on clinical signs, morphologic lesions, isolation and characterization of infectious or chemical agents, and evaluation of immunological responses on serum samples for specific antibody activity. Although clinical signs are important indicators of specific diseases, sometimes allowing for definitive diagnosis, there are many diseases in which a number of different causative agents may cause similar clinical signs (3). Under these circumstances it is often necessary to utilize additional diagnostic procedures for definitive diagnosis. Many infectious agents and toxins have primary effects on the morphologic appearance (pathology) of specific organs in the body (11). The distribution and pattern of the lesions characteristic of the causative agent are strong criteria for diagnosing disease. Again, some infectious agents may have similar pathogenetic mechanisms leading to a set of lesions caused by more than one etiological agent. In these cases other diagnostic criteria are needed.

Identification of the etiological agent is the single most important criterion for efficient disease control. Conventional techniques use culturing of bacteria followed by classification of the isolates by biochemical or serological procedures. Viruses are isolated by animal inoculation or culturing in embryonated chicken eggs or cell cultures. Serological procedures such as fluorescent antibody or neutralization assays are applied to make the definitive identification. These methods for diagnosing disease have been in use for at least 20 years.

The application of the principles of biotechnology to diagnostics can greatly increase the rapidity and sensitivity of identifying the causes of disease. A number of improvements have been made in the last few years in marker systems for immunoassays. Radioimmunoassays are one of the most sensitive systems available for identifying antigens through precipitation or competitive binding assays. Detection is either by scintillation counter or through the use of radiographic film. The tagging of fluorescein dyes to antibodies has proven to be a highly specific and useful tool for identifying specific antigens associated with infectious disease agents in tissues or cell cultures. In the past, the antibodies used in these assays have been derived from immunization of animals such as rabbits with the particular virus or bacteria. The immune response of the test animal produces a heterologous antibody response to the many antigens (epitopes) of the infectious agent. These antibodies are then conjugated with fluorescein dye. In order to identify fluorescein conjugated antibody, which attaches to antigens in tissues or cells, an ultraviolet light source is required to excite the fluorescein dye causing visible fluorescence. This requires additional, sometimes expensive equipment, and trained personnel to interpret the results.

Enzyme linked immunoassay (ELISA) systems utilize antibodies as the principal means of identifying antigens. The difference between ELISA and fluorescent antibody systems resides in the marker systems used to identify the location of the antigen-antibody complex (13). ELISA amplification systems have been developed in which linking antibodies to immunoglobins made in sheep or goats increase the sensitivity of the reactions. Once these enzyme-labeled antibodies attach to antigen-antibody complexes in tissues, a substrate is added to the solution. The enzymatic reaction on the dye will cause a color change. These assays can be observed visually or in a spectrophotometer. In the latter case the assays are quantitative.

MOLECULAR BIOCHEMISTRY

Molecular biochemistry provides a useful approach for rapid and sensitive diagnosis of infectious diseases in laboratory and domestic animals. The techniques employed in these analyses detect specific molecular subunits of microorganisms in infected tissues. The usual subunits identified by these methods are specific gene sequences or the products of gene sequences which are specific protein epitopes of the infecting microorganism. With few exceptions, infectious microorganisms contain genetic material in the form of deoxyribonucleic acid (DNA) and/or ribonucleic acid (RNA), as well as proteins that provide the protective membrane coating and the enzymes necessary for replication.

BIOCHEMICAL STRUCTURE OF MICROORGANISMS

Bacteria and fungi consist of cell membranes containing proteins and glycoproteins, and nuclear chromatin consisting of DNA and RNA. The proteins and glycoproteins are antigenic and, when introduced into animals, cause an antibody response. Gram-negative bacteria contain endotoxins which consist of lipopolysaccharides. Endotoxins are antigenic as well. Viruses consist of outer protein coats and either DNA or RNA as the nuclear material.

Peptides associated with microbial cell membranes or outer coat proteins of viruses can be separated by polyacrylamide gel electrophoresis (PAGE) (Fig. 1). Separation of these subunit proteins is based on molecular weight. Identification of subunit proteins can be identified by

protein binding dyes, or radioactively labeled precursors. Specific antibody activity can be detected by radiolabeling proteins and then performing
immune precipitation followed by PAGE. It is also possible to electrophoretically transfer proteins out of gels onto absorptive paper (western
blots). Utilization of these techniques provides a better assessment of
the proteins specifically involved in the immune response and for characterization of diagnostic reagents.

Viral and bacterial genome segments can be separated by PAGE (9,18,
20,25). With many viruses and microorganisms, restriction enzyme digestion
is required before the genome material is placed on PAGE. Migration of
genome segments in a gel is related to the molecular weight and possibly
the secondary structure of the genome segment (10). Two-dimensional electrophoresis of restriction enzyme fragments of DNA or RNA gene segments
have been used to "fingerprint" viruses (4). The restriction enzymes cut
the segments of genes at similar chemical sites giving rise to patterns
specific for individual virus genome segments. Electrophoretic patterns of
genome segments are helpful in classifying some viruses and other microorganisms, however, this system cannot be used on all viruses for diagnostic purposes because of potential co-migrating genes (14).

With viruses, each gene is transcribed and translated into proteins in
the infected cell. These proteins serve as the structural basis of the

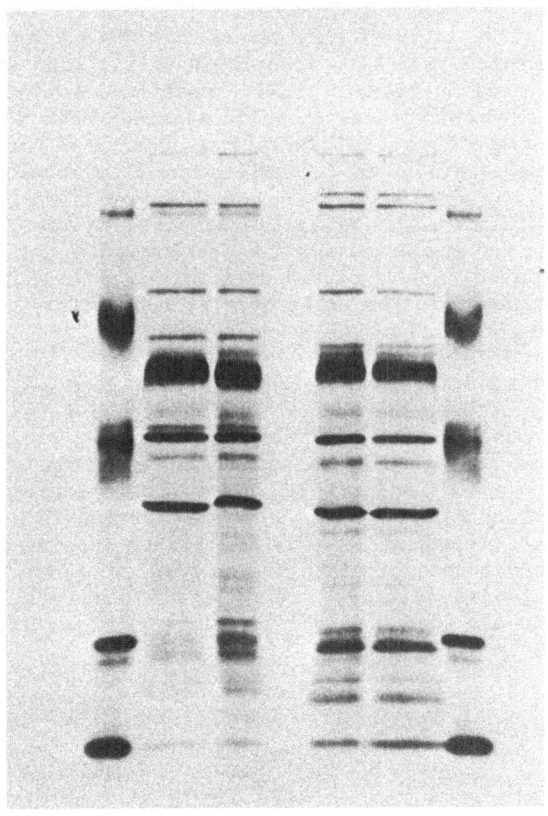

Fig. 1.   Polyacrylamide gel (PAGE) of radiolabeled subunit proteins of
           bluetongue virus. Molecular weight markers are in the two out-
           side lanes with virus in the inner four lanes.

progeny virions or as nonstructural entities. Many of these proteins are responsible for immunological reactions that identify group or serotype reactive epitopes.

MOLECULAR CLONING OF MICROBIAL AND VIRAL GENES

Cloning of gene segments can be carried out by separating segments on PAGE, cutting out and electroeluting the genome segments, and then cloning the segment. Another method is to reverse transcribe and clone all segments simultaneously (19).

Cloning procedures are standard (25). The genomic RNA or DNA is transcribed into complementary strands of DNA (cDNA) and annealed with a complementary copied DNA strand to form double-stranded cDNA. The double-stranded cDNA is inserted into a plasmid which is used to transform bacteria, and bacterial growth allows for the replication and duplication of the insert. Following cloning the genome segment must be identified by hybridization studies.

GENETIC PROBES AS DIAGNOSTIC TOOLS

Cloned cDNA copies of individual genome segments are exact replicas of the original DNA or RNA segment from which they were cloned. These cDNA gene strands consisting of complementary nucleotide sequences can be used as genetic hybridization probes to identify specific gene sequences of the microorganisms or virus in tissues (12). Conditions for DNA/DNA, DNA/RNA, and RNA/RNA genetic hybridizations on solid matrices have been standardized (16). The stringency of the hybridization conditions can be varied by changing reaction temperature and formamide concentrations.

In order to determine that hybridization between the probe and microbial nucleic acids has occurred, one of the reagents has to be labeled. The most commonly used labeling procedure involves the incorporation of radioactive nucleotides into the probe by a standard nick translation reaction on the plasmid containing the cloned insert.

Specific identification of the genome or portions of it requires hybridization testing between the cloned probe and the gene segments of the microorganism. Microbial genome segments are commonly prepared by endonuclease digestion and subsequently denatured and electrophoretically transferred to a solid support medium [i.e., 2-aminophenylthioether (APT), nitrocellulose paper, etc.] (16,17). Naturally segmented genomes may not require endonuclease treatment. The radioactively labeled cloned probe is denatured by heat and then hybridized to the strip of matrix paper containing the microbial genome segments. The cloned probe will hybridize only with the genome segments that have complementary gene sequences. This is a highly specific test.

MOLECULAR GENETIC DIAGNOSTIC TESTS

Southern Blot Hybridization

The Southern blot hybridization technique is used to characterize genome segments of DNA (16,17). The system, originally developed by E.M. Southern, utilizes labeled cDNA probes to hybridize with complementary DNA from microorganisms. The specificity of the system is limited to homologous nucleotide sequences.

Northern Dot Hybridization

Northern blot hybridization assays can be used as diagnostic tests for viral RNAs and messenger RNAs (2,19,22,24). The specific identification of isolates depends on the specificity of the cloned diagnostic probes. The cloned probe, if directed to shared group genes, will identify microorganisms containing these gene sequences (Fig. 2). In other cases, where genome segments code for serotype-specific antigens, the cDNA probe will hybridize only with that particular serotype specific coding gene.

Dot Blot Hybridization

Dot blot hybridization assays work on the same principle as northern and Southern blot hybridization assays (15,18). The difference between the assays is that, in the dot blot system, the genome segments are not electrophoretically separated prior to hybridization with the cDNA probe. The RNA/DNA is extracted from infected tissues, concentrated, and denatured. An aliquot of the denatured extract is dotted onto nitrocellulose paper and the labeled cloned probe is then placed on the extract and hybridization takes place if the complementary gene sequences are in the extracted material. The advantage of this test over other tests is the rapidity by which

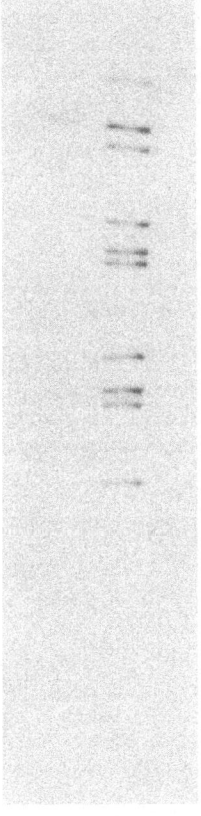

Fig. 2.   Northern blot of bluetongue virus (BTV) which was electrophoresed on a polyacrylamide gel and transferred to APT paper. The left column was transferred unlabeled and the right column was transferred end-labeled with ^{32}P. Both were hybridized with a cDNA clone representing 25% of genome segment 2 of BTV serotype 17 which had been nick-translated.

it can be performed. The label used on the cloned probe can be either radioactively labeled material, such as with ^{32}P or enzymatically labeled with an avidin-biotin complex.

For rapid diagnostic purposes, the dot blot assay applied to tissue extracts can markedly reduce the time that it takes to diagnose the cause of disease (Fig. 3). For example, diagnosing bluetongue virus infection of cattle and sheep is costly and it requires 3 weeks to 3 months before the virus can be isolated and characterized (18). By taking tissue extracts from infected chick embryos, dotting these onto nitrocellulose paper, and then hybridizing with radioactively cloned cDNA probes, a positive identification of infection can be made in less than 3 days. It is now possible to utilize biotinylated labeled nucleotides in the nick translation, or end labeling reaction, and then to detect the biotin with a strepavidin-linked enzyme color reaction. The adaptation of this labeling procedure to genetic hybridization of RNA extracts of bluetongue-infected tissue removes the problems associated with radiolabeled material.

In Situ Hybridization

Another hybridization procedure is the in situ diagnostic test (1,12, 19). The test involves fixing cells on glass slides and exposing the cells to a genetic probe(s) under hybridization conditions. Cells used in this procedure can be from cell cultures or they can be from animal tissues, including peripheral blood. The labeled cDNA probe will hybridize to any complementary genome sequence present in the preparation. Again, either radioactively labeled probes or the biotin-avidin-enzyme color reaction can be used to identify microbial nucleic acids in cells. The cells on the glass slides hybridized with biotin-avidin systems can be viewed under a light microscope (Fig. 4). The principal advantage of this system is that the morphology of the cells and tissues remains intact and the site of viral replication can be identified.

MOLECULAR DIAGNOSIS OF SPECIFIC ANTIGENS

Animal species respond to foreign proteins/antigens by making highly specific antibodies. This basic principle provides the basis for accurate diagnostic tests against microorganisms and viruses. The technology developed for molecular biology has direct application for highly sensitive diagnostic tests.

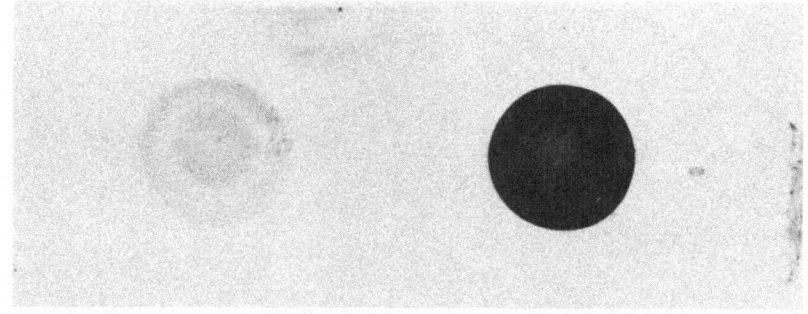

Fig. 3.   Dot blot hybridization of biotinylated cDNA of bluetongue virus gene segment 7 in a negative and positive cell culture extract on nitrocellulose paper.

Western (Immuno) Blotting

Microbial or viral proteins extracted from cultures can be separated by SDS-polyacrylamide gel electrophoresis (Fig. 5) (7,8,10,19,20,23). Separated proteins can be electrophoretically transferred from the gel to nitrocellulose paper. Strips of paper containing proteins can be air-dried and stored desiccated at -20°C. The nitrocellulose strips containing protein can be blocked with gelatin in Tris buffered saline (TBSS) containing Tween 20. The strips can then be treated for 2 hr with sera containing polyclonal or monoclonal antibodies. Following washing, the strips can be incubated with affinity-purified goat anti-IgG specific for the first antibody. The goat antispecies-IgG contains either a radioactive label or a biotinylated marker. Radioactively labeled preparations are then developed on X-ray film. The biotin-treated strips are incubated with avidinperoxidase conjugate, followed by the application of a substrate such as 4-chloro-1-naphthol. Following color development, strips are viewed and then stored in water in a dark place. This system can be used to characterize immune responses and to rapidly identify viruses.

Monoclonal Antibodies

The applications of monoclonal antibody technology to animals have recently been reviewed. Each antibody-producing cell produces antibody to only one epitope. The fact that the antibody-producing cells can be individually cloned and immortalized makes it possible to propagate cloned lines of a cell producing a single antibody. These highly specific antibodies can be used in diagnostic systems to identify serologically similar microorganisms. The advantages of these reagents are that they provide a standardized product with greatly reduced nonspecific reactions. Monoclonal antibodies with markers such as fluorescein, peroxidase-antiperoxidase, biotin-avidin, or radiolabels can be used to identify microbial agents in tissues, cell cultures, and in ELISA (Fig. 6).

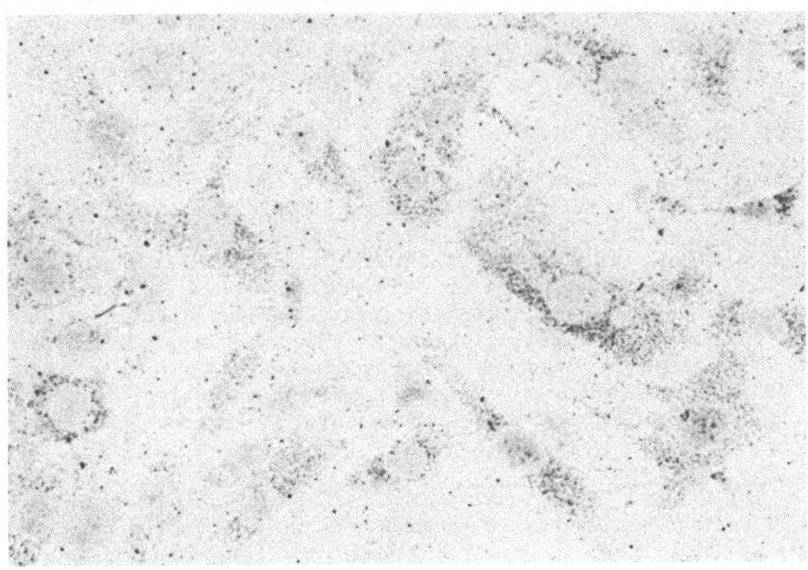

Fig. 4. In situ hybridization of a biotinylated cDNA probe on cell culture infected with bluetongue virus.

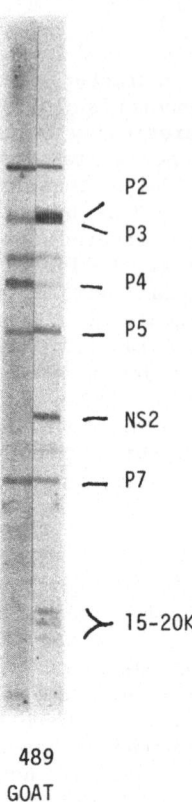

P2

P3

P4

P5

NS2

P7

15-20K

489
GOAT

Fig. 5.   Western blot of bluetongue virus run on PAGE, electrophoretically
transferred to nitrocellulose paper, and treated with goat sera
hyperimmune to bluetongue virus.   The left lane represents post-
infection sera obtained long after (6 months) infection and the
right lane represents sera obtained 4 weeks after reinfection.

Protein Dot Blot Assay

The solid phase medium used in this assay is nitrocellulose strips.
Monoclonal antibodies specific for the microorganism or virus to be tested
are applied to nitrocellulose.   The antibody is permitted to dry for 10 to
15 min and the paper is then blocked with bovine serum albumin (BSA) or
gelatin in TBSS.   Supernatants from infected tissues are then applied to
the paper and incubated for 2 hr.   Following washing, biotinylated antisera, specific for the infectious agent, is added to the paper and incubated for 2 hr followed by washing.   An avidin-enzyme conjugate is added,
followed by addition of substrate, with resultant formation of pigment deposition if the microbial antigens are present (20).

In Situ Immunoperoxidase Procedure

An indirect immunoperoxidase procedure consisting of peroxidase-antiperoxidase as the principal reagent has been used for detecting proteins in
tissues.   Both polyclonal and monoclonal antibodies have been used to identify microbial and viral antigens in formalin-fixed tissues (5,19,20).
Following fixation of tissues, washing, and blocking steps, tissues are incubated with specific antibodies.   Linking antibody consisting of affinity-

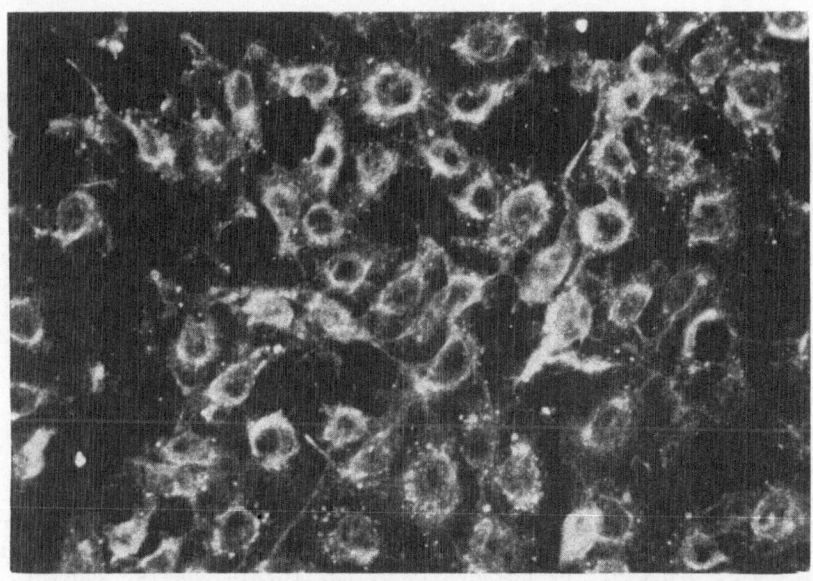

Fig. 6.   Bluetongue virus infected cell culture stained with fluorescein-
conjugated monoclonal antibody that is specific for the nonstruc-
tured bluetongue virus protein NS1.

purified, peroxidase-labeled goat antimouse IgG is followed by the amino-
ethyl-carbazole (AEC) substrate.   The protein antigens of microorganisms in
tissues develop a colored precipitate of the peroxidase substrate.   This
procedure is important for diagnosing the etiology of disease in formalin-
fixed specimens and for studying the pathogenesis of infections.

CONCLUSIONS

    The recent technological advances used in studies on molecular genet-
ics and protein chemistry are having a significant impact on diagnosis of
diseases.   The traditional technology, although reliable, is time-consuming, labor-intensive, and expensive.   The opportunity to develop highly
sensitive probes for nucleic acid sequences or characteristic proteins of
specific organisms provides the basis for amplifying reagents to identify
microorganisms in tissues or cells for direct diagnosis.

    The specific binding of single strands of complementary sequences of
nucleic acids provides a means of labeling a known sequence which may anneal to the suspect viral or microorganismal nucleic acid in tissue.   Large
quantities of labeled nucleic acid probes can now be made by genetically
engineering the desirable nucleic acid into bacteria which in turn make DNA
copies.   These copies can be labeled with radioactive material such as ^{32}P
or with biotin-avidin complexes and substrate.   In many instances, these
systems may replace conventional diagnostic tests.

    Advances in immunology have led to monoclonal antibody technology.
Antibodies derived from this technology are highly specific for defined
epitopes.   These antibodies made in large quantities can be labeled for use
as specific recognition probes.   These antibodies are highly specific, making for standardized diagnostic reagents.

SUMMARY

Traditional approaches to diagnosing disease include clinical observations, pathological changes in tissues, and searches for the etiology, by isolation, or identification of microorganisms, or by serological methods. Development of techniques for studying the molecular biology of microorganisms, manipulation of cellular systems, and improved immunoassays have contributed to better diagnostic technology. Recombinant DNA technology has made it possible to apply highly specific probes consisting of nucleotide sequences that hybridize with complementary sequences of microorganisms. The specific techniques utilized include Southern, northern, and dot blot hybridization and in situ hybridization. Identification of proteins of microorganisms is done by western blot, dot blot, and in situ peroxidase-antiperoxidase techniques. These techniques have the promise of being highly specific and rapid methods for diagnosis of disease.

ACKNOWLEDGEMENTS

The author wishes to thank M. Adkison, J. Cherrington, L. Chuang, R. Chuang, C. Dangler, R. Doi, H. Ghalib, R. Oberst, M. Sawyer, C. Schore, K. Squire, and J. Stott of the University of California, Davis, and W.C. Davis of Washington State University, Pullman, for contributions to this manuscript. This research is supported in part by USDA Special Grants and 1433 Animal Health and Disease Grants, USDA-ARS-Cooperative Agreement and the Livestock Disease Research Laboratory, University of California, Davis.

REFERENCES

1. Ackerman, M., E. Peterhaus, and R. Wyler (1982) DNA of bovine herpesvirus type 1 in the trigeminal ganglia of latently infected calves. Am. J. Vet. Res. 43:36-40.
2. Alewine, J.C., D.J. Kemp, and G.R. Stark (1977) Methods for the detection of specific RNA's in agarose gels by transfer to diazobenzyloxymethyl-paper and hybridization with DNA probes. Proc. Natl. Acad. Sci., USA 74:5350-5354.
3. Blood, D.C., J.A. Henderson, O.M. Radostits (1979) Veterinary Medicine, 5th edition, Lea Febiger, Philadelphia.
4. Brown, F., A.R. Carroll, B.E. Clark, E.J. Ouldridge, and D.D. Rowlands (1985) Chemical basis for antigenic variation in foot-and-mouth disease virus. In Veterinary Viral Diseases of Southeast Asia and the Western Pacific, T. Della-Porta, ed. Academic Press, Sydney, Australia, pp. 265-272.
5. Cherrington, J.M., H.W. Ghalib, W.C. Davis, and B.I. Osburn (1985) Monoclonal antibodies raised against bluetongue virus detect viral antigen in infected tissues using an indirect immunoperoxidase method. In Bluetongue and Related Orbiviruses, T.L. Barber and M.M. Jochim, eds. A.R. Liss, New York, pp. 505-510.
6. Davis, W.C., T. Yilma, L.E. Perryman, and T.C. McGuire (1985) Perspectives on the application of monoclonal antibody and transfection technology in veterinary microbiology. In Progress in Veterinary Microbiology and Immunology, Vol. 1, R. Pandey, ed. S. Karger, Basel, pp. 1-24.
7. Della-Porta, A.J., A.R. Gould, B.T. Eaton, and D.A. McPhee (1985) Biochemical characterization of Australian orbiviruses. In Bluetongue and Related Orbiviruses, T.L. Barber and M.M. Jochim, eds. A.R. Liss, New York, pp. 337-346.
8. Ghalib, H.W., J.M. Cherrington, M.A. Adkison, and B.I. Osburn (1985) Humoral and cellular immune response of sheep to bluetongue virus. In

Bluetongue and Related Orbiviruses, T.L. Barber and M.M. Jochim, eds. A.R. Liss, New York, pp. 489-496.

9. Gorman, B.M., J. Taylor, P.J. Walker, W.L. Davidson, and F. Brown (1981) Comparison of bluetongue type 20 with certain viruses of the bluetongue and eubenangee serological group of orbiviruses. J. Gen. Virol. 57:251-261.

10. Gorman, B.M. (1985) Molecular structure of bluetongue and related orbiviruses. In Bluetongue and Related Orbiviruses, T.L. Barber and M.M. Jochim, eds. A.L. Liss, New York, pp. 329-337.

11. Jubb, K.V., P.C. Kennedy, and N. Palmer (1985) Pathology of Domestic Animals, Vol. 2, 3rd edition, Academic Press, Orlando, Florida, pp. 108-112.

12. Haase, A., M. Brahic, L. Stowing, and H. Blum (1984) Detection of viral nucleic acids by in situ hybridization. Methods in Virol. 8: 1-39.

13. Leary, J.J., D.J. Brigati, and D.C. Ward (1983) Rapid and sensitive colorimetric method for visualizing biotin-labeled DNA probes hybridized to DNA or RNA immunological or nitrocellulose: Bio-blots. Proc. Natl. Acad. Sci., USA 80:4045-4049.

14. Oberst, R.D., S. Dunn, L.F. Chuang, and B.I. Osburn (1986) Identification of segmented contamination in bluetongue virus genome segments individually excised from polyacrylamide gels. Veterinary Microbiology (submitted for publication).

15. Owens, R.A., and T.O. Diener (1984) Hybridization for detection of viroids and viruses. Methods in Virol. 7:173-187.

16. Raub-Traub, N., and J.S. Pagano (1984) Hybridization of viral nucleic acids: Newer methods on solid media and in solution. Methods in Virol. 8:1-39.

17. Southern, E.M. (1975) Detection of specific sequences among DNA fragments separated by gel electrophoresis. J. Mol. Biol. 98:503-517.

18. Squire, K.R.E., R.V. Chang, L.F. Chuang, R.H. Doi, and B.I. Osburn (1985) Detecting bluetongue virus RNA in cell culture by dot hybridization with a cloned genetic probe. J. Virol. Methods 10:59-68.

19. Squire, K.R.E., J.L. Stott, C.A. Dangler, and B.I. Osburn (1986) Application of molecular techniques to the diagnosis of bluetongue virus infection. In Progress in Veterinary Microbiology and Immunology, Vol. 2, R. Pandy, ed. Academic Press, New York (in press).

20. Stott, J.L., K.R.E. Squire, R. Oberst, J.M. Cherrington, and B.I. Osburn (1984) Application of modern biotechnology to the diagnosis of bluetongue virus infection. 27th Proc. American Assn. Veterinary Lab. Diagnosticians, American Association of Veterinary Laboratory Diagnosticians, Madison, Wisconsin, pp. 35-50.

21. Sugiyama, K., D.H.L. Bishop, and P. Roy (1981) Analysis of the genome of bluetongue viruses recovered in the United States. 1. Oligonucleotide fingerprint studies that indicate the existence of naturally occurring reassortant BTV isolates. Virology 114:210-217.

22. Thomas, P.S. (1979) Hybridization of denatured RNA and small DNA fragments transferred to nitrocellulose. Proc. Natl. Acad. Sci., USA 77:5201-5205.

23. Tobin, H., T. Staehelin, and J. Gordon (1979) Electrophoretic transfer of proteins from polyacrylamide gels to nitrocellulose sheets. Procedures and some applications. Proc. Natl. Acad. Sci., USA 76:4350-4354.

24. Wahl, G.M., M. Stern, and G.R. Stark (1979) Efficient transfer of large DNA fragments from agarose gels to diazobenzyloxymethyl-paper and rapid hybridization by using Dextran sulfate. Proc. Natl. Acad. Sci., USA 76:3683-3687.

25. Walker, P.J., J.N. Mansbridge, and B.M. Gorman (1980) Genetic analysis of orbiviruses by using RNA Tl oligonucleotide fingerprints. J. Virol. 34:583-591.

RECOMBINANT DNA APPROACHES TO FELINE

LEUKEMIA VIRUS IMMUNIZATION

Paul Luciw,[1] Debbie Parkes,[1] Gary Van Nest,[1] Dino Dina,[1]
Kathleen Hendrix,[2] and Murray B. Gardner[2]

[1]Chiron Laboratory
4560 Horton Street
Emeryville, California 94608

[2]Department of Medical Pathology
School of Medicine
University of California
Davis, California 95616

INTRODUCTION

Retrovirus vaccines are in the limelight now more than ever before due to the etiologic involvement of retroviruses, human T-lymphotropic virus (HTLV-III/LAV) in the acquired immunodeficiency syndrome (AIDS) of man (1,5) and HTLV-I in certain human lymphomas (14). Experience with retroviruses and other viral vaccines in animal models is thus of great importance in guiding the way toward the most effective procedures for immunizing humans. Conventional vaccine approaches with inactivated or live attenuated retroviruses have been successful in some circumstances (7). However, these vaccines pose a number of economic and safety problems, primarily because of the tissue culture requirement for growing viruses. In view of these considerations, the new recombinant DNA technology offers the advantage of producing synthetic vaccines of greater purity, stability, safety, and cost-effectiveness (13). Researchers in private industry are playing an important role in developing such products.

Feline leukemia virus (FeLV) is a type C retrovirus responsible for considerable morbidity and mortality in domestic house cats (8). Inasmuch as FeLV may also cause immunosuppression in infected cats, it also serves as a model for AIDS. Conventional FeLV vaccines have shown some success in prevention of related feline disease (10,11), and one such commercial vaccine is now available. However, the expense of these vaccines and the uncertainty of their components, safety, and long-term effectiveness have prompted efforts to develop FeLV vaccines using novel recombinant DNA techniques. These tools have only recently become available, and with their application we hope to further fathom the complexities of the virus-host immune response as well as to develop effective vaccines. We have utilized 2 recombinant DNA strategies toward development of an FeLV vaccine:

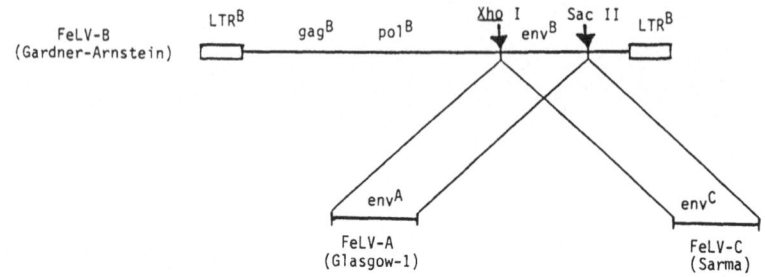

Fig. 1. Molecularly cloned envelope gene recombinants of feline leukemia virus (FeLV) subgroups A, B, and C.

(a) attenuation of the virus by envelope recombination and mutation of transcriptional control elements, and (b) production of subunit envelope protein in yeast.

The rationale for development of an attenuated FeLV vaccine based on envelope recombination is predicated by our knowledge that the envelope gene of retroviruses is the most important determinant for infection and virulence, perhaps by facilitating entry into specific cell types, and is the principal target for immune defense mechanisms (2). The other 2 retroviral genes, gag and pol, are also required for virus replication, and the long terminal repeats (LTRs) at each end of the proviral DNA carry critical promoter and enhancer sequences for regulating viral gene expression. Based on virus interference patterns, host range, and, to a lesser extent, serology (virus neutralization), it has been determined that FeLV consists of 3 distinct envelope subgroups called FeLV-A, FeLV-B, and FeLV-C (14). FeLV-A alone is pathogenic for cats. However, FeLV-B and -C are relatively nonpathogenic, except for experimental FeLV-C infection in newborn kittens (9). Under natural conditions, FeLV-A infection may occur alone, but FeLV-B and -C are always found mixed with FeLV-A. These findings suggest that the subgroup A virus may facilitate the replication of subgroup B or C viruses. The possibility thus arose that an FeLV-B genome, genetically engineered so that its own envelope gene was replaced by the FeLV-A or -C envelope gene, might be somewhat attenuated for replication and pathogenesis, yet capable of presenting the "protective" envelope antigen to the immune system. To further attenuate the potential for replication, the LTR enhancer sequences were deleted by restriction enzyme treatment in the FeLV-B envelope recombinants. In vitro replication was thereby markedly reduced. These FeLV constructs with deleted LTR enhancer sequences, and substituted FeLV-A or -C envelope genes are referred to in this chapter as "AΔ" and "CΔ," respectively.

The rationale for developing a combined vaccine using genetically engineered FeLV envelope protein followed by live altered virus was based on the recent demonstrations involving "priming" of the immune system. A normally subimmunogenic dose of live virus can generate a successful immune response in animals "primed" by a previous immunization with virus-related peptides (4). The development of a human hepatitis B vaccine composed of a viral surface antigen expressed in yeast served as our model for producing the FeLV envelope protein in yeast (16).

In this chapter we summarize the results of our preliminary studies using the attenuated virus and combined subunit-attenuated virus approaches to attempt to protect cats against challenge with virulent FeLV.

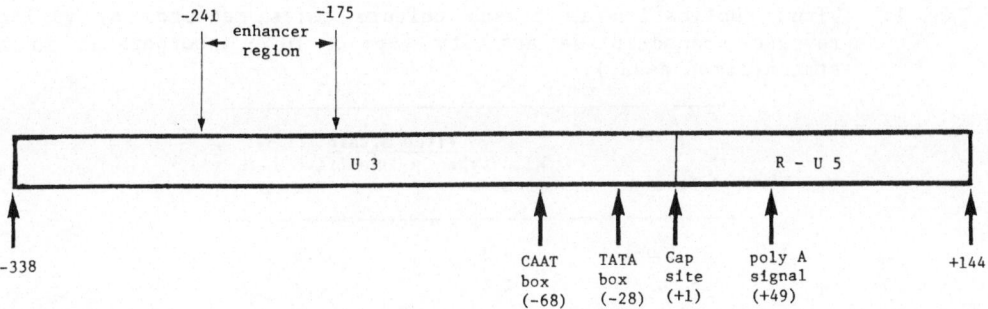

Fig. 2.   Control elements of the FeLV-B LTR are outlined with respect to their positions.

METHODS

Genetic Engineering of Attenuated FeLV Vaccines

Molecularly cloned envelope (env) gene recombinants of FeLV subgroups A, B, and C with enhancer sequences deleted from the LTRs were constructed at Chiron Laboratory. The initial step in these genetic manipulations was the molecular cloning of FeLV subgroups A, B, and C.* FeLV-A was derived from the Jarrett Glascow-1 strain; FeLV-B, from the Gardner-Arnstein strain; and FeLV-C, from the Sarma strain. Each isolate was molecularly cloned in bacteriophage lambda vectors from integrated DNA of infected human RD rhabdomyosarcoma cells. The envelope and LTR regions of FeLV-A, -B, and -C were then sequenced. Published sequence data for the FeLV-B envelope gene was also available (3). Sequence comparison showed ∿97% homology between the env genes of FeLV-A and -C, and ∿90% between FeLV-A and -B, and FeLV-B and -C. The molecularly cloned proviral DNA of FeLV-A and FeLV-B were transfected into cat lung cells and infectious virus was recovered.

Envelope gene recombinants were then constructed by replacing the FeLV-B env sequences with the analogous regions of cloned FeLV-A or -C. This manipulation was facilitated by the presence, in each FeLV subgroup, of identical restriction enzyme sites close to 5' (XhoI) and 3' (SacII) termini of the env gene (Fig. 1). These recombinant viruses contained LTRs and the gag and pol genes from the nonpathogenic FeLV-B, and an env gene (about 1,500 nucleotides) from the pathogenic FeLV-A or -C.

The next step was to make deletions in the enhancer region from both LTRs using Bal 31 nuclease digestion. The deleted region lies upstream of the promoter sequences (CAAT** box and TATA box) in the LTR and removes sequences in excess of the enhancer region (Fig. 2). This manipulation reduced the infectivity of FeLV-B to about 20% of its normal value for cat and dog cells (Tab. 1). These FeLV-B recombinants carrying FeLV-A or -C env genes and deleted LTR enhancer sequences were designated "AΔ" and "CΔ," respectively.

Genetic Engineering of FeLV-A env Protein in Yeast

Construction of plasmids for the expression of FeLV-A envelope protein

---

* FeLV-B clone and tissue culture cells infected with FeLV-A and FeLV-C were kindly provided by Jim Casey, Frederick Cancer Research Center.
** A, adenine; C, cytosine; T, thymine.

Tab. 1.   Virus replication in tissue culture cells measured by virion
          reverse transcriptase activity (cpm x 10^{-4} incorporated in a
          standardized assay).

| | Virus strain | |
Cell line	Wild-type FeLV-B	Modified live Δ FeLV-B
Cat (Lu-1)	5.3	1.6
Dog (D-17)	2.4	0.2

in yeast (<u>Saccharomyces cerevisiae</u>) was also done at Chiron Laboratory, using the same techniques that were successful in expressing the hepatitis B
virus surface antigen in yeast (16).   The FeLV-A env gene fragment was
joined to the yeast pyruvate kinase (PYK) promoter as shown in Fig. 3.   The
plasmid vector contains a replication origin and selectable marker for bacteria derived from pBR322 and a replication origin and selectable marker
from yeast.   The amount of FeLV-A envelope protein made was several mg
per liter of yeast culture, about 5% of the total protein produced.   The
nonglycosylated env protein had a molecular weight of about 53,000 and was
immunoreactive with FeLV-B antiserum.   The FeLV-A envelope preparations
were designated "A-env."

Immunization Groups and Dosage

A total of 62 cats in 6 groups comprised this study (Tab. 2).   Group 1
consisted of 22 controls inoculated with wild-type FeLV-A without prior
vaccination.   Group 2 received AΔ alone.   Group 3 received CΔ alone and
Group 4 received CΔ followed by FeLV-A challenge.   Group 5 was given A-env
followed by AΔ, at a dose one-tenth that given Group 2, and then challenged
with FeLV-A.   Group 6 received the same dose of AΔ given to Group 5, followed by FeLV-A challenge.

The AΔ or CΔ "attenuated" vaccines titered at 1 x 10^5 infectious
units/ml on cat lung cells.   Tissue culture fluids were frozen and thawed
once before use.   The standard dose was 2.0 ml given all at once, one-half
intraperitoneally (i.p.) and one-half subcutaneously (s.q.) (Groups 2, 3,
and 4).

The combined vaccine consisted of 325 μg of yeast-produced A-env per
dose, given half i.p. and half s.q. with complete Freund's adjuvant followed 2 weeks later by AΔ modified live virus, diluted 10-fold (1 x 10^4 infectious units/ml) (Group 5).

Two weeks following vaccination, all vaccinated cats were challenged
with 2.0 ml of molecularly cloned, wild-type FeLV-A grown in cat lung
cells.   The virus titered at 1 x 10^5 infectious units/ml and was given one-
half i.p. and one-half s.q.

Animals

Specific pathogen-free (SPF) cats were obtained from a closed breeding
colony at the University of California, Davis.   All were screened for antibodies to FeLV and for viral core antigen by enzyme-linked immunosorbant
assay (ELISA) as described below.   FeLV vaccines were administered at 8 to
9 weeks of age and FeLV-A challenge was at ∿12 weeks of age.   Cats were
housed in individual cages and fed standard laboratory chow.   All animals
were under the care of a veterinarian.

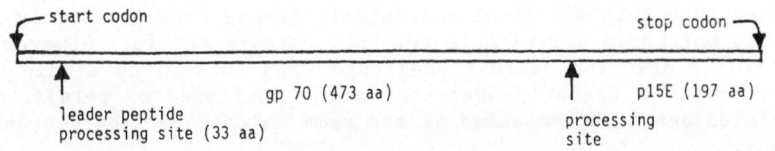

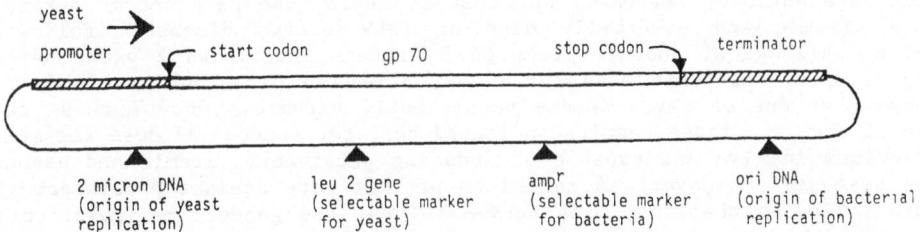

FELV ENVELOPE GENE IN YEAST EXPRESSION PLASMID VECTOR

Fig. 3. Features of the feline leukemia virus (FeLV) envelope gene.

Virus Assays

These tests were done at the University of California, Davis. Viremia (V+) was detected by an ELISA for the major core antigen, p27, using serum diluted 1:7 (11). The validity of this assay was confirmed by the fluorescent antibody test on peripheral blood smears. Antibody to FeLV was measured by an ELISA using a 1:200 serum dilution and whole disrupted FeLV-B (11) and against the yeast A-env protein. Positive ELISA antibody tests were confirmed by immunofluorescence on FeLV-infected cat lung cells or FeLV-infected FL74 cells. FeLV antigen and antibody tests were done at 2- to 3-week intervals for the first 2 to 3 months after challenge, then only periodically. Most cats were thus monitored for virus about 10 to 12 times during the first 12 months following challenge. The term "persistent viremia" will be used to refer to cats that were repeatedly ELISA-positive for FeLV p27 antigens for ≥ 6 months after vaccination or challenge. "Transient viremia" refers to cats only positive once for this antigen. "Persistent antibody" refers to cats repeatedly ELISA-positive for FeLV antibody for ≥ 6 months after vaccination or challenge, whereas "transient antibody" refers to cats with only a short-lived (≤ 2 months) detectable antibody response. All cats that had to be euthanized for illness or were found dead were necropsied and studied microscopically.

RESULTS

The results of the FeLV novel vaccine trials in 6 groups of cats are shown in Tab. 2.

Cats Challenged with FeLV-A Without Vaccination (Group 1)

As a positive control, 22 kittens, 9 to 12 weeks of age, were inoculated with molecularly cloned, infectious wild-type FeLV-A. About 60% (14 of 22) recipients became persistently viremic, and all but 2 recipients made a persistent antibody response. However, 6 of these antibody-positive cats were nonviremic and thus had apparently been immunized by the FeLV-A inoculum, and 2 cats were apparently not infected by the inoculum. Six cats died from 9 to 18 months after inoculation; all were viremic and suffered from FeLV-related aplastic anemia and/or lymphoma. These results indicated that this strain of molecularly cloned FeLV-A was only moderately virulent, but could be suitable for FeLV vaccine trials. However, evidence of vaccine protection against challenge with this virus strain would have to demonstrate a dramatic decrease in the incidence of persistent viremia and related deaths accompanied in the same animals by a persistent antibody response.

Cats Immunized with AΔ and Challenged with FeLV-A (Groups 2 and 6)

Seven of 9 kittens (Group 2) inoculated with the AΔ virus alone made a detectable antibody response, but most of these same cats became persistently viremic and eventually died of FeLV-related disease. Following FeLV-A challenge of another group of 5 kittens inoculated 2 weeks before with AΔ at a one-tenth dose (Group 6), 3 made a detectable antibody response, but one of these became persistently viremic. One of these cats died of anemia. These results indicated that the reduced AΔ dose was still infectious in vivo and capable of inducing persistent viremia and associated disease. Moreover, AΔ failed to protect cats against persistent viremia following challenge with FeLV-A. Thus, the genetic manipulation to construct this AΔ virus did not cause any apparent attenuation of virulence and, used at either dosage, it would therefore not be suitable for vaccination.

Cats Immunized with CΔ and Challenged with FeLV-A (Groups 3 and 4)

Three of 8 kittens inoculated with CΔ (Group 3) made a detectable antibody response and remained nonviremic and well after 11 months. None of the antibody-negative cats were viremic. Following FeLV-A challenge of another group of 8 CΔ immunized kittens (Group 4), 5 became persistently viremic. Four of those eventually died of FeLV-associated disease. Thus, as administered, CΔ was apparently attenuated insofar as it caused no detectable viremia, but it was also not sufficiently antigenic to protect against challenge with FeLV-A.

Cats Immunized with A-env and AΔ and Challenged with FeLV-A (Group 5)

Four of 10 kittens inoculated with yeast-produced A-env followed by AΔ made a detectable antibody response, and all 10 were nonviremic 2 weeks later. The AΔ dose used was the same as that given Group 6. These 10 kittens were challenged 2 weeks later with FeLV-A. All 10 of these animals made an antibody response, and 9 remained nonviremic and healthy after 8 months. However, one cat that had a transient early viremia became viremic shortly before death which was due to anemia and an apparently unrelated cardiomyopathy.

DISCUSSION

Ideally, immunization for FeLV must confer lasting immunity without persistent viremia because the latter is inevitably associated with increased morbidity and mortality from FeLV-related diseases, mainly nonregenerative anemia, immunosuppression, and lymphoma. As with other vaccines, it is difficult to achieve this lasting immunity without some degree of virus replication occurring even transiently, to present a sufficient antigenic challenge to the host immune system (13). The goal of these immunization studies with the genetically altered FeLV was to achieve a low level of virus replication, perhaps only transiently, but sufficient to induce immunity and prevent persistent viremia. However, our results indicated that the modified live virus produced by envelope recombination (placing an FeLV subgroup A or C envelope on the FeLV subgroup B genome) and LTR enhancer deletion either did not adequately attenuate the virus

Tab. 2. Feline leukemia virus (FeLV) novel immunization approaches.

Group #	"Immunization"	Challenge[2]	# Recipients	FeLV Status[3]				Health status[4]		Current age of cats (mo)[1]
				V+ AB-	V+ AB+	V- AB+	V- AB-	Alive	Dead	
1.	--	FeLV-A	22	5	9	6	2	16	6	12-18
2.	AΔ[5]	--	9	2	5	2	0	6	3	18
3.	CΔ[5]	--	8	0	0	3	5	8	-	14
4.	CΔ[5,7]	FeLV-A	8	0	5	3	0	3	5	14
5.	A-env[6,7] +AΔ	--	10	0	0	4	6	9	1	11-14
		FeLV-A		0	1	9	0			
6.	AΔ[5,7,8]	FeLV-A	5	2	1	2	0	4	1	11

1 Recipient kittens were 8-9 weeks old when vaccinated, and 12 weeks old when challenged with FeLV. All cats were initially free of FeLV antigen and antibody.

2 Standard FeLV-A challenge was 1 x 10^5 infectious units/ml and cats were given 1.0 ml i.p. and 1.0 ml s.q.

3 V+, AB- = persistent viremia, no detectable antibody (serum 1:200); V+, AB+ = persistent viremia and antibody; V-, AB+ = no detectable viremia, but persistent antibody; and V-, AB- = no detectable viremia or antibody.

4 Most of the cats that died were viremic before death.

5 AΔ and CΔ = FeLV subgroups -A or -C, respectively, genetically engineered to be depleted of enhancing sequences in the 5' LTR and 3' LTR and having gag and pol genes derived from FeLV-B (see "Methods").

6 A-env = FeLV subgroup A envelope protein produced in yeast. Cats in Group 5 received an inoculation (320 μg) of A-env with complete Freund's adjuvant as a first dose.

7 Interval between vaccination and FeLV challenge = 2 weeks.

8 AΔ diluted 10-fold (1 x 10^4 infectious units/ml).

(AΔ) or did not confer protective immunity (AΔ and CΔ). It may yet prove possible to engineer an attenuated FeLV by intragenic recombination within the env gene of different subgroups. A–env protein followed by modified live virus AΔ induced antibody without detectable viremia in 4 of 10 cats, and, 8 months after FeLV–A challenge, 90% (9 of 10) of the cats immunized this way remained nonviremic and antibody-positive. Thus, A–env protein with Freund's adjuvant appears to prime the immune response and to protect against FeLV–A challenge. Cats immunized with A–env protein alone made antiviral antibodies as measured in an ELISA system (Parkes et al., unpubl. results); the degree of protection from challenge with live virus is under current investigation.

Current efforts are directed at devising the means to enhance the immunogenicity of the yeast-synthesized A-envelope protein, perhaps by application of new or improved adjuvants. Monitoring of cellular immunity to FeLV and determination of both neutralizing and cytotoxic antibody titers are also desirable for future vaccine studies.

Our results, presented in this chapter, indicate that development of immunization protocols for retrovirus infections and diseases using novel recombinant DNA technology is indeed possible but by no means an easy task. Yeast expression vectors provide a system for readily producing large quantities of viral envelope polypeptides that can be rigorously purified. The main challenge to be faced in immunization against the AIDS retrovirus will be achieving sufficient immunogenicity for lasting protection. Attainment of this goal will be guided by the work with retrovirus infection in the cat model as described here for FeLV and in other animal models such as simian species with an acquired immunodeficiency syndrome caused by a type D retrovirus (6).

SUMMARY

We have utilized 2 recombinant DNA strategies for immunization against FeLV in cats: (a) modified live virus was attenuated by mutation and recombination, and (b) an immunogen, consisting of subunit envelope protein, was prepared in genetically engineered yeast. Results indicated that the genetically manipulated live virus preparations were not protective against FeLV challenge because they were either not attenuated in virulence or were not sufficiently antigenic. Immunization with yeast-synthesized FeLV envelope protein followed by the modified live virus gave protective immunity in cats under experimental conditions. Future immunization attempts will concentrate on enhancing the immunogenic potency of the yeast- synthesized FeLV envelope protein.

REFERENCES

1. Barré-Sinoussi, F., J.C. Chermann, F. Rey, M.T. Nugeyre, S. Chamaret, J. Gruest, C. Dauguet, C. Axler-Blin, F. Ve'zinet-Brun, C. Rouzious, W. Rosenbaum, and L. Montagnier (1983) Isolation of a T-lymphotropic retrovirus from a patient at risk for acquired immune deficiency syndrome (AIDS). Science 220:868-871.
2. Coffin, J. (1982) Structure of the retroviral genome. In RNA Tumor Viruses, R. Weiss, N. Teich, H. Varmus, and J. Coffin, eds. Cold Spring Harbor Laboratory, Cold Spring Harbor, New York, pp. 261-268.
3. Elder, J.H., and J.I. Mullins (1983) Nucleotide sequence of the envelope gene of Gardner-Arnstein feline leukemia virus B reveals unique sequence homologies with a murine mink cell focus-forming virus. J. Virol. 46:871-880.

4. Emini, E.A., B.A. Jameston, A.J. Lewis, and E. Wimmer (1983) Priming and induction of anti-poliovirus neutralizing antibodies by synthetic peptides. Nature 304:699-702.

5. Gallo, R.C., S.Z. Salahuddin, M. Popovic, G.M. Shearer, M. Kaplan, B.F. Haynes, T.J. Palker, R. Redfield, J. Oleske, B. Safai, G. White, P. Foster, and P.D. Markham (1984) Frequent detection and isolation of cytopathic retroviruses (HTLV-III) from patients with AIDS and at risk for AIDS. Science 224:500-502.

6. Gardner, M.B., and P. Marx (1985) Simian acquired immunodeficiency syndrome. Adv. Viral Onc. 5:57-81.

7. Gardner, M.B., N. Pedersen, P. Marx, R. Henrickson, P. Luciw, and R. Gilden (1985) Vaccination against virally induced animal tumors. In Immunity To Cancer, A.E. Reif and M.S. Mitchell, eds. Academic Press, New York, pp. 605-617.

8. Hardy, Jr., W.D., P.W. Hess, E.G. MacEwen, A.J. McClelland, E.E. Zuckerman, M. Essex, S.M. Cotter, and O. Jarrett (1976) Biology of feline leukemia virus in the natural environment. Canc. Res. 36:582-588.

9. Jarrett, O. (1974) Feline leukemia virus subgroups. In Feline Leukemia Virus, W.D. Hardy and A.J. McClelland, eds. Elsevier/North Holland, Amsterdam, pp. 473-479.

10. Jarrett, W., L. Mackey, O. Jarrett, H. Laird, and C. Hood (1974) Antibody response and virus survival in cats vaccinated against leukemia. Nature New Biology 248:230-232.

11. Lutz, H., N.C. Petersen, and G.H. Theilen (1983) Course of feline leukemia virus infection and its detection by enzyme-linked immunosorbent assay and monoclonal antibodies. Am. J. Vet. Med. 44(11):2054-2059.

12. Mathes, L.E., M.G. Lewis, and R.G. Olsen (1980) Immunoprevention of feline leukemia: Efficacy testing and antigenic analysis of soluble tumor-cell antigen vaccine. In Feline Leukemic Virus, W.D. Hardy and A.J. McClelland, eds. Elsevier/North Holland, Amsterdam, pp. 211-216.

13. Norrby, E. (1983) Viral vaccines: The use of currently available products and future developments. Archives of Virology 75:163-177.

14. Poiesz, B.J., R.W. Ruscetti, A.F. Gazdar, P.A. Bunn, J.D. Minna, and R.C. Gallo (1980) Detection and isolation of type C virus particles from fresh and cultured lymphocytes of a patient with cutaneous T-cell lymphoma. Proc. Natl. Acad. Sci., USA 77:7415-7419.

15. Sarma, P.S., and T. Log (1971) Viral interference in feline leukemia-sarcoma complex. Virology 44:352-358.

16. Valenzuela, P., A. Medina, W.J. Rutter, G. Ammerer, and B.D. Hall (1982) Synthesis and assembly of hepatitis B virus surface antigen particles in yeast. Nature 298:347-350.

MOLECULAR APPROACHES TO VACCINES

Howard L. Bachrach

Plum Island Animal Disease Center
Agricultural Research Service
U.S. Department of Agriculture
Greenport, New York   11944

INTRODUCTION

Recombinant DNA (rDNA) and monoclonal antibody technologies are being used to develop new types of vaccines that are expected eventually to supplement or supplant conventional whole-agent vaccines for both animals and humans. Impetus for the new vaccines stems from findings that the surface proteins of most infectious agents possess immunogenic activity and that the corresponding genes can be manipulated to produce these products in quantity devoid of the infectious agents themselves. In addition, conventional vaccines that employ killed and attenuated viruses have certain limitations. For example, standard methods have failed to produce effective vaccines for many important diseases, including hepatitis A and B, and for most diseases caused by retroviruses, herpes viruses, and parasites, including those that cause malaria, trypanosomiasis, and anaplasmosis. Also, attenuated whole-agent vaccines can cause acute or slowly progressive diseases in immunodeficient individuals or in alternate animal hosts. Live poliovirus vaccine produces a few cases each year of paralytic polio in vaccinees or their contacts (91). Even so-called killed virus vaccines can contain infectious particles. This was the case with some of the early batches of commercially produced, inactivated poliovirus vaccine, and is still the case for foot-and-mouth disease (FMD) virus vaccines in wide use today. At least 44% of the outbreaks of FMD in Europe from 1968 to 1981 were caused by incomplete inactivation procedures or the escape of virus from vaccine manufacturing facilities (43). Whole-agent vaccines can also induce postvaccinal sequelae, including pyrogenic and allergic reactions, and Guillain-Barré and other neurological syndromes as well as immune enhancement of disease, e.g., dengue-2 virus infections of young children (62).

At least 5 strategies are being used to develop the new vaccines: (a) rDNA production of the immunogenic surface proteins of infectious agents, (b) chemical synthesis of peptides corresponding to the immunogenic sites of the surface proteins, (c) construction of recombinant viruses having guest genes expressing immunogenic surface proteins of infectious agents, (d) use of hybridoma technology to produce monoclonal anti-idiotype antibodies with sites that mimic immunodominant epitopes on surface proteins of infectious agents, and (e) genetic engineering of nonpathogenic mutants of infectious agents.

217

The new vaccines should circumvent many of the problems of conventional vaccines. The presence of only one or 2 essential proteins or immunogenic segments thereof in some of the new vaccines will eliminate the formation of antibodies against adventitious antigens that contaminate crude whole-agent vaccines, and thereby reduce the incidence of adverse reactions. It will also be possible to freeze-dry the new protein vaccines so as to ensure their potency in areas of the world where refrigeration is lacking. Moreover, the protein vaccines will be incapable of producing disease, because they lack nucleic acid and will be produced in facilities that are free of infectious agents.

Due in part to the length of government regulatory approval processes for new biological substances, the commercial introduction of the new vaccines will proceed slowly. The subunit vaccines presently in use include: (a) detergent- or ether-split influenza A and B virus vaccines, (b) a hepatitis B vaccine comprised of the 22-nm antigen particles purified from the serum of human carriers, and (c) an enterotoxigenic Escherichia coli vaccine that contains cloned pilus protein and a nontoxic segment of the bacterial toxin. The subunit vaccines for influenza produce fewer side effects than whole-virus vaccines, but 2 doses are needed for unprimed recipients (81). No whole-virus vaccine exists as an alternative to the 22-nm particle vaccine for hepatitis B; however, a number of investigators are engaged in developing cloned protein, synthetic peptide, and recombinant virus vaccines for this disease. Many bacteria and parasites also have surface proteins that are candidates for subunit vaccines. This chapter will describe the progress being made on new vaccine development by the approaches mentioned above--recombinant DNA, chemical synthesis, recombinant viruses, monoclonal anti-idiotype antibodies, and nonpathogenic mutants. It seems necessary to describe progress on both animal and human vaccines, particularly for zoonoses and because some new approaches are being applied to human pathogens prior to their use with animal disease agents.

RECOMBINANT DNA SUBUNIT VACCINES

The biosynthesis of surface proteins of infectious agents from rDNA cloned in single cells or animal hosts is at the forefront of the new approaches to vaccine development. In rDNA technology, a gene encoding a known protein is spliced into a vector, such as a double-stranded DNA virus or bacterial plasmid, for transfer into an alternate host cell for its replication and expression. This procedure permits genes (including those of RNA viruses after transcription into DNA using reverse transcriptase) to be transferred between plants, animals, and microorganisms. Escherichia coli, yeasts, and continuous mammalian cell cultures are commonly used hosts for the replication and expression of the chimeric DNA vectors. While E. coli is widely used, the expressed proteins are subject to cleavages by endogenous proteases and to contamination with E. coli toxins. Yeasts and mammalian cells are being used increasingly, because they are hospitable to eukaryotic genes and can carry out glycosylations that are needed to simulate native glycoproteins.

Recombinant DNA Protein Vaccines for Viral Diseases

Viruses that have surface proteins that are potentially useful as subunit vaccines are listed in Tab. 1; a few such vaccines are already in use. Those listed have one or more of the following activities: (a) reactivity with neutralizing antibody, (b) induction of neutralizing or precipitating antibodies, and (c) induction of protective immunity in recipients. The table also indicates whether or not the genes for these proteins have been cloned in alternate hosts. Protein vaccines that are in the most advanced stages of development include those for FMD; hepatitis B; feline, porcine, and mink parvovirus diseases; bovine papillomas (warts); herpes simplex 1 and 2; Epstein-Barr virus-associated diseases and cancers; infectious bovine rhinotracheitis; pseudorabies; Rift Valley fever; rabies and vesicular stomatitis; transmissible gastroenteritis of swine; and feline and bovine leukemias. These and others are discussed below, grouped into the replication classes (7) shown in Tab. 1, col. 1. Class VII is introduced provisionally to accommodate the hepatitis B group of DNA viruses.

Class I: Double-stranded DNA viruses. Class I viruses (Papova-, Adeno-, and Herpetoviridae) are distinguished by the fact that their isolated double-stranded DNAs can initiate infection that produces viral progeny. For bovine papilloma (wart) virus, 2 large open-reading frame genes, L1 and L2, in the nontransforming region of the genome have been cloned and expressed in E. coli for use of the protein products as vaccines (138). The hexons and penton fibers of adenoviruses have been shown to induce the formation of neutralizing antibodies (55), but little interest has been shown to date in cloning them for use as vaccines.

Considerable progress has been made in the development of rDNA protein vaccines for herpes diseases. It was demonstrated several years ago that 2 injections of 2 proteins from plasma membranes of cells infected with herpesvirus of turkeys would protect chickens against Marek's disease (84). Subsequently, 3 glycoproteins (ca gp115/110, gp63, and gp51) from herpesvirus of turkeys and Marek's disease virus, crossreactive with monoclonal antibodies, were found to elicit neutralizing antibodies in chickens and rabbits (77). Also, a 79-kd protein in herpesvirus of turkeys and Marek's disease virus-infected cells has epitopes common to these 2 viruses (163). Under low-stringency hybridization, the cloned DNAs of these 2 viruses display approximately 75% homology extending over 90-95% of their genomes (54). The surface glycoproteins of infectious bovine rhinotracheitis (103) and pseudorabies virions have been reported to induce neutralizing antibodies and protective immunity in cattle and swine, respectively. The experimental pseudorabies vaccine consisted of proteins and glycoproteins extracted from virus and emulsified in Freund's incomplete adjuvant (139). Pseudorabies virions are reported to have 4 major glycoproteins. Three of these (gp125, gp74, and gp50-58) are disulfide-linked and show extensive sequence homology; the fourth (gp58) is not complexed to other proteins. Monoclonal antibodies reactive with the major gp98 and gp50-58 moieties show complement-independent neutralization of virus (63,191). Progress is being made on the cloning and expression of the genes for the glycoproteins of infectious bovine rhinotracheitis (108) and pseudorabies viruses. A β-galactoside-pseudorabies glycoprotein fusion product from E. coli is reported to protect mice against challenge with pseudorabies virus (145). Feline rhinotracheitis virus has 3 prominent glycoproteins that may be candidate immunogens (105). A protein vaccine may also be possible for chickenpox and dermatomal zoster, because gp118 of varicella zoster elicits complement-independent neutralizing antibodies (46). Work is progressing toward subunit vaccines for cytomegaloviruses. A monoclonal antibody specific for a major gp64 of human cytomegalovirus neutralizes virus and binds to the plasma membranes of infected cells (27). Equine cytomegalovirus (equine herpesvirus-2) has been reported to have 3 major (gp83, gp87, and gp73) and 4 minor glycoproteins (25). Glycoprotein gp300/350 from Epstein-Barr virus has been reported to elicit both neutralizing antibodies and antibody-mediated cellular toxicity in experimental animals (143). This same protein, called gp340 by Epstein, also appears to prevent tumor development in cotton-top tamarin monkeys challenged with Epstein-Barr virus, and its gene is being cloned for expression in E. coli (119, 121). The greatest progress in cloning subunit protein vaccines for herpesviruses has, however, been made with herpes simplex viruses (HSVs). Glycoprotein gD from either HSV-1 or HSV-2 has been reported to protect mice against lethal challenge with homologous and heterologous virus types

Tab. 1. Candidate subunit vaccines for viral diseases.

Viral class	Family	Virus	Subviral immunogen	Cloned
I_{DNA}	Papova	bovine papilloma	L1, L2	+
	Adeno	adeno	hexons, fibers	+
	Herpes	Marek's disease, turkey	gp115, gp63, gp51	−
		infectious bovine rhinotracheitis	glycoproteins	+
		feline rhinotracheitis	glycoproteins	−
		pseudorabies	gp98, gp50−58	+
		simplex 1 & 2	gC, gD	+
		varicella	gp118	−
		Epstein−Barr	gp340	+
		cytomegalovirus	gp64	−
		equine cytomegalovirus	gp83, gp78, gp73	−
I_{DNP}	Orthopox	rabbitpox	5 proteins, HA	−
		vaccinia	5 proteins, HA	−
II	Parvo	cat, mink, pig	capsid proteins	+
III	Reo	reo	σ1, σ3, λ2	+
	Orbi	bluetongue	P2/t10, P3/t17	+
	Rota	simian, human	proteins	+
		bovine	VP7.2	−
	(Biseg)	infectious pancreatic necrosis	VP54	+
IV	Picorna	foot-and-mouth disease	VP_1	+
		polio	VP_1, VP_2, VP_3	+
		Coxsackie B3	VP_2	−
		rhino 14	VP_1, VP_3	−
		hepatitis A	VP_1, VP_3	+
	Calici	vesicular exanthema virus of swine	p61	−

* M. Essex, personal communication.
** Ref. 75a.
Note: Abbreviations not in text: HA, hemagglutinin; t, type;
STA, soluble tumor antigen; S/V, split virus. p15E of
feline leukemia virus is immunosuppressive. Class VII
is provisional.

Viral class	Family	Virus	Subviral immunogen	Cloned
IV cont.				
Toga:	Alpha	Semliki Forest	S/V, E_1, E_2, E_3	–
	Flavi	tick-borne encephalitis	gpV_3	–
	Pesti	hog cholera, bovine viral diarrhea	S/V, E_1, E_2	–
	Rubi	rubella	S/V, E_1, E_2	–
	Corona	human	13 kd	–
		transmissible gastro-enteritis	gp195, gp25	+
		infectious bronchitis	spike	+
V_{seg}	Orthomyxo	influenza	S/V, HA	+
		fowl plague	S/V, HA	+
	Bunya	Rift Valley fever	G_1, G_2	+
	Arena	Pichinde	pGP-C	+
V	Paramyxo	Sendai, Newcastle, parainfluenza-3	F, HN	–
		simian virus 5	F, HN	+
		measles, distemper	F, H	+(H)
		mumps	HN	–
		respiratory syncytial	F, G	+
	Rhabdo	rabies, vesicular stomatitis, infectious hematopoietic necrosis	G	+
VI	Retro	visna, caprine arthritis-encephalitis	glycoproteins	–
		human T-cell lymphotrophic virus-3/lymphadenopathy	gp120*	+
		human T-cell lymphotrophic virus-1	gp62	+
		Friend murine leukemia	gp71	+
		feline leukemia	STA, gp70	+
		bovine leukemia	gp51/60	+
		equine infectious anemia	gp90	+
		avian leukosis	gp85**	
VII	Hepadna	human hepatitis B	HBsAg, pre-S	+
		duck hepatitis B	DHBsAg	+

(102). In accord with this report, a gD-related polypeptide expressed in
E. coli from HSV-1-specific DNA elicits neutralizing antibodies to both
HSV-1 and HSV-2 (192).

Class I: Double-stranded deoxynucleoprotein viruses. Class I double-
stranded deoxynucleoprotein (DNP) viruses include the pox- and iridovirus-
es. Unlike Class I DNA viruses, the isolated DNAs of double-stranded DNP
viruses are not infectious; nearly intact virions containing core and capsid enzymes are required for the initiation of infection. To this author's
knowledge, there are no reports that the surface proteins, individually or
in combinations, of Class I double-stranded DNP viruses are capable of inducing immunity to infection. Orthopoxviruses have 5 or more surface antigens (14), some of which, from intracellular rabbitpox or vaccinia viruses,
will induce neutralizing antibodies to intracellular virus but not to extracellular virus (3,11). Nevertheless, animals with these antibodies are
not protected against infection. The complete extracellular virus with a
late hemagglutinin in its envelope appears to be necessary for the induction of protective immunity. In the case of African swine fever virus, an
iridovirus, even the complete virus is an inefficient inducer of neutralizing antibodies. Ninety-eight percent of its genome has now been cloned
(100).

Although subunit protein vaccines have not been reported for orthopoxviruses, outstanding progress has been made using vaccinia virus as a vector for the genes of the immunogenic proteins of other viruses (e.g., herpes, hepatitis B, influenza, rabies, vesicular stomatitis, Friend murine
leukemia, and porcine transmissible gastroenteritis viruses). In animals
inoculated with the recombinant vaccinia viruses, the guest genes are expressed and induce protection against challenge by virus from which the
genes were obtained (see section "Recombinant Virus Vaccines" below).

Class II: Single-stranded DNA viruses. Class II viruses are comprised solely of parvoviruses that contain separately encapsidated positive
or negative single-stranded DNA genomes. Upon extraction from virus, the
complementary DNA strands anneal and are infectious. Theoretically, either
strand alone should be infectious because their 3' ends self-prime the synthesis of double-stranded DNA intermediates. Surface proteins from both
H-1 and canine parvoviruses have been reported to elicit neutralizing antibodies in guinea pigs (S. Rhode, III, pers. comm.). Porcine parvovirus DNA
codes for 3 major proteins (90 kd, 65 kd, and 60 kd), with the smaller moieties being subsets of the largest protein. Antisera to the individual
SDS-denatured proteins neutralize virus infectivity (115). Aleutian disease virus, G strain, is reported to have 2 major structural proteins (85
kd and 75 kd) with extensive sequence homology. During infection in mink,
however, the Aleutian disease virus proteins are degraded proteolytically
to smaller, highly antigenic polypeptides (1). Porcine, feline panleucopenia and Aleutian disease virus (107) surface protein genes are being
cloned in E. coli, and tests of the immunogenicity of the proteins expressed by the porcine and feline panleucopenia genes are underway (24,45).
A recent report makes a presumptive connection between a human parvovirus
and rheumatoid arthritis (165). If confirmed, the experience being gained
on cloning animal parvovirus surface proteins should help to expedite the
development of a vaccine against arthritis.

Class III: Segmented, double-stranded ribonucleoprotein viruses.
Class III viruses include the double-stranded ribonucleoprotein (RNP) reo-,
orbi-, and rotaviruses, which have similar structures and replication
strategies, but different host ranges. The genomes of reo- and orbiviruses
are divided into 10 segments, and those of rotaviruses, into 11. The RNA
polymerases coded by and present in Class III virions are essential for
replication. Therefore, Class III viruses can be considered RNP and not

RNA viruses (7). This distinction also applies to Class V$_{seg}$ and Class V negative-stranded RNP viruses described below. For reoviruses, surface protein σ1 is a vaccine candidate because it evokes neutralizing antibodies in animals. For certain reovirus serotypes, σ3 and λ2 proteins also possess this activity. While most of λ2 resides in the reovirus core, the chain extends through to the viral surface (82).

For type 10 bluetongue virus, an orbivirus, protein P2 induces both neutralizing antibodies and protective immunity in sheep (75). For type 17 bluetongue virus, the neutralization-specific antigen is apparently P3, coded for by genome segment 2 (59). The cloning of genome segment 3 of type 10 bluetongue virus has been reported (141).

Rotaviruses also have proteins that are vaccine candidates. Unreduced p26 and gp34 of simian SA11 virus induce type-specific, neutralizing antibodies (13). Using monospecific polyclonal antisera, the principal neutralization-specific epitopes of calf rotavirus have been located in a minor protein, VP7.2, of the outer shell of the virus (92). Cloning of the surface protein genes of reo-, orbi-, and rotaviruses—particularly of human and bovine rotaviruses—in E. coli (44) is providing the quantities of protein that are needed for use in experimental vaccines (29).

Progress is being made on cloning and expressing capsid protein VP54 of infectious pancreatic necrosis virus, which is a bisegmented double-stranded RNP virus of fish. This protein has been shown to induce neutralizing antibodies in fish (98).

Class IV: Single-stranded RNA viruses. Class IV viruses, which contain messenger-sense, infectious, single-stranded RNA genomes, are composed of 4 families—the Picorna-, Calici-, Toga-, and Coronaviridae. While these families have markedly different architectures and compositions, their replication strategies are quite similar.

For picornaviruses, the prototype subunit vaccine was prepared from FMD virus (8) and, later, with protein cloned in E. coli (Fig. 1) (93). A 24-kd surface protein VP$_1$ (213 amino acid residues long) from FMD virus subtype A$_{12}$, or its CNBr-derived internal 13-kd fragment (amino acids 55-179) (9,10), and a 44-kd LE'-VP$_1$ fusion protein (amino acids 7-211) cloned in E. coli have been shown to elicit protective immunity in livestock. The cloned protein vaccine was characterized by U.S. Secretary of Agriculture John R. Block as "... the first effective vaccine produced by gene splicing for any disease of animals or humans." Because it stimulates immunity in both swine and cattle, the new vaccine promises to provide greater control of FMD, thereby increasing the world's production of meat and other animal products. Although the original cloned protein vaccine was for type A$_{12}$ FMD, those for FMD types A$_{24}$, A$_{27}$, A$_{79}$, and C$_3$ have now been cloned and shown to be as effective in cattle as the original type A$_{12}$ (P.D. McKercher, pers. comm.).

Subunit type O$_1$ FMD vaccine is still under development. Goats (67) and cattle given repeated doses of cloned type O$_1$ VP$_1$ vaccines developed high levels of neutralizing antibody, but their resistance to challenge with virus was less than satisfactory. Also, type O$_1$ VP$_1$ from virus evokes neutralizing antibodies in guinea pigs but only partial protection against challenge infection (83). Similar results have been obtained in cattle vaccinated with chemically synthesized, type O$_1$ VP$_1$-specific peptides (196) (see section "Chemically Synthesized Vaccines" below). Some evidence indicates that an effective type O$_1$ subunit vaccine will require the construction of a putative discontinuous epitope formed between amino acid residues in the 141-160 and 200-213 regions of the VP$_1$ chain (134).

223

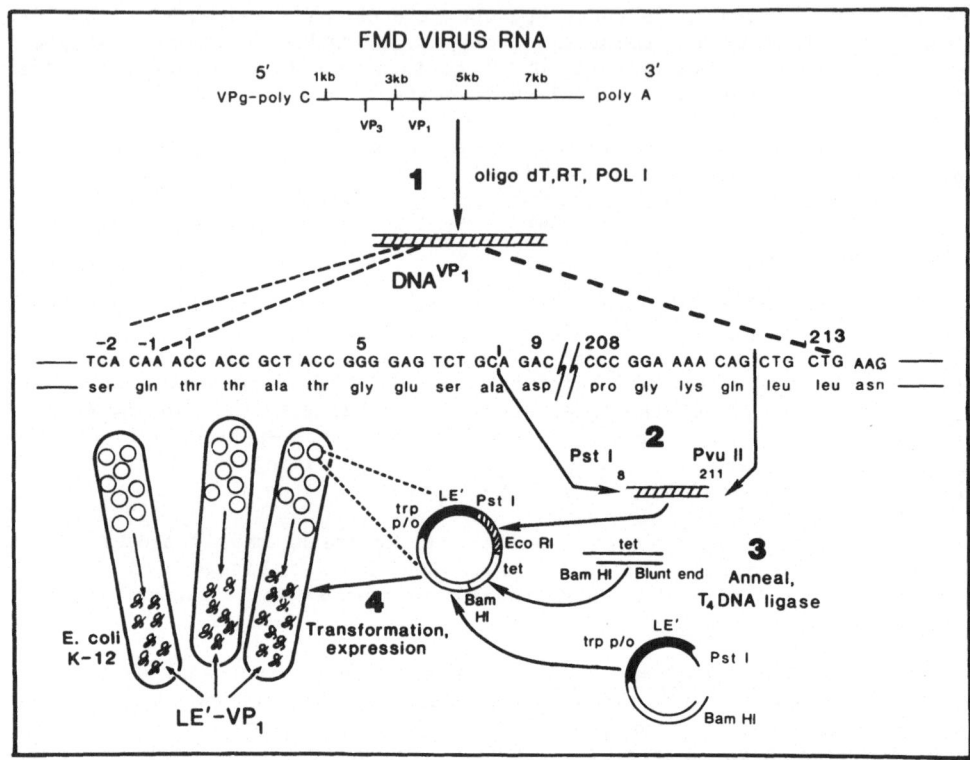

Fig. 1.   The strategy used to clone and express FMD VP$_1$ in E. coli. Using
oligo dT as the primer, purified genome RNA was transcribed into
DNA with reverse transcriptase and DNA polymerase I (step 1).
One of the DNA transcripts, amplified in a pBR322 vector in E.
coli, was shown by DNA sequencing to contain coding sequences for
the C-terminus of VP$_3$ and all of VP$_1$ including its 13-kd internal
region.   In addition, codons designated -2 and -1 at the C-terminus of VP$_3$ were contiguous with those for the known N-terminal
threonine codon of VP$_1$, showing that proteolytic cleavage of the
-gln-thr- juncture in their VP$_3$VP$_1$ precursor is not followed by
trimmings with exopeptidases.   In step 2, restriction endonucleases PstI and PvuII excised a fragment encoding amino acids 8
through 211 of VP$_1$, which in step 3 was spliced into a pBR322
plasmid having a tryptophan synthetase (trp) promoter-operator-
LE' leader and a tetracycline-resistance gene (tet) marker.
After annealing the complementary sticky ends and connecting the
blunt ends with T4 ligase, the chimeric plasmid was used to
transform E. coli (step 4).   After growing the transformants to
high concentration and increasing the copy number of the plasmid,
tryptophan was depleted from the growth medium to promote expression of 1 to 2 million copies per cell of a 44-kd fusion protein
LE-VP$_1$.   The fusion protein reacted with VP$_1$-specific antibody,
and 2 250-μg doses of the purified fusion protein in a modified
Freund's incomplete oil adjuvant elicited high levels of neutralizing antibody and protective immunity in swine and cattle.   In
contrast, unvaccinated swine and cattle developed generalized FMD
upon exposure to virus.   [Photo from New Approaches to Vaccines,
(1985); permission granted from Academic Press, Inc.]

224

Other picornaviruses have surface proteins that can elicit neutralizing antibodies in experimental animals. Capsid protein VP_2 from Coxsackie
virus B3 (15) and VP_1, VP_2, and VP_3 from poliovirus (19,26,197) display
this activity. However, the antibody responses to the poliovirus proteins
are generally quite weak. Using monoclonal antibodies, human rhinovirus 14
appears to have at least 2 dominant neutralization-specific epitopes: one
located on VP_1 and the other on VP_3 (161). In addition, virus-neutralizing
antisera have been produced in rats immunized with 3 to 4 doses of VP_1 or
VP_3 from hepatitis A virus (74). A prototype-inactivated hepatitis A virus
vaccine prepared from virus grown in cell cultures produces protective immunity in monkeys (190). DNA complementary to hepatitis A virus RNA has
been cloned and characterized, but the expression of viral proteins has not
been reported (178).

For caliciviruses, the single major capsid protein p61 of vesicular
exanthema virus of swine may be a vaccine candidate, because 2 vaccinations
of swine with purified p61 elicit neutralizing antibodies (Dardiri and
Bachrach, unpubl. data). The major capsid proteins of feline and sea lion
caliciviruses could be expected to have analogous activities in their
hosts.

Subunit vaccines have been prepared for several togavirus diseases.
For the _Alpha_ genus, multimeric micelles of the glycoprotein of Semliki
Forest virus (164) induce protective immunity, and glycoprotein VP_3 complexes from tick-borne encephalitis virus (65), a _Flavi_ virus, evoke hemagglutination-inhibiting and neutralizing antibodies and a protective immune
response. For the _Pesti_ genus, vaccinations with glycoproteins presumed to
be E_1 and E_2 from detergent-split hog cholera virus or from infected cells
induce higher levels of neutralizing antibody and protective immunity in
swine than detergent-split bovine viral diarrhea virus (31). Bovine viral
diarrhea virus, which is related antigenically to hog cholera virus, is reported to have 2 major proteins (p115 and p80) that are structurally related, and a major unrelated glycoprotein, gp55 (140). For the _Rubi_ genus,
rubella virus (which contains envelope glycoproteins E_1 and E_2; Ref. 85)
split with detergent (23) appears to have greater vaccine potency than either hemagglutinin rosettes or virosome preparations (182).

The coronaviruses have immunogenic surface proteins, and the genes for
some of them have been cloned and expressed. Coronaviruses possess the
longest-known, infectious RNA genomes (20,000 or more nucleotide residues).
A 13-kd protein from human coronavirus has been shown to evoke neutralizing
antibody against one viral strain and both neutralizing and hemagglutination-inhibiting antibodies against another strain (155). Multiple doses of
a transmissible gastroenteritis virus 25-kd glycoprotein in either aluminum
hydroxide gel or Freund's adjuvant induce neutralizing antibodies (apparently including the IgA class) and protective immunity in weanling pigs
(56). The gene for gp195 of transmissible gastroenteritis virus has been
cloned and expressed in _E. coli_, and while the protein product had little
immunogenic activity in mice, better results were obtained when the cloned
gene was first inserted into vaccinia virus (72,73) (see section "Recombinant Virus Vaccines" below). Similarly, the gene for the spike protein of
avian infectious bronchitis virus has been cloned in a pBR322/_E. coli_ system. Amino acid sequences of the spike protein have been deduced from the
corresponding cloned DNA sequence. This step represents progress toward a
new infectious bronchitis virus vaccine, because the spike is involved in
the induction of immunity against this disease (17).

Class V$_{seg}$: Segmented, single-stranded ribonucleoprotein viruses.
These RNP viruses are composed of members of the Orthomyxo- (8 segments),
Bunya- (3 segments), and Arenaviridae (2 segments). Influenza virus sub-
unit vaccines that are made from virus particles degraded with detergents

or ether are in wide use. They effect seroconversions and produce fewer side reactions than whole-virus vaccines; however, 2 doses are required for unprimed recipients (81). It has been suggested that whole virus should be added to the subunit vaccine in amounts sufficient to enhance the immune response of seronegative recipients without inducing adverse reactions (193). The hemagglutinin gene of fowl plague virus has been cloned and ex-pressed in E. coli (42), and that of wild-type human influenza virus has been cloned and expressed in both E. coli (32,64) and cultured eukaryotic cells (52). The eukaryotes express approximately 10^8 cell-associated hemagglutinin molecules that adsorb specifically to red blood cells and to viral antibodies. In addition, a truncated hemagglutinin expressed in the eukaryotes from an anchorless hemagglutinin gene is secreted into the ex-tracellular fluid (53). Nevertheless, cloned hemagglutinin molecules have not proven to be effective as vaccines for influenza, which is also true for chemically synthesized segments of the hemagglutinin molecule (see sec-tion "Chemically Synthesized Vaccines" below). Either the native conforma-tion of the hemagglutinin is not formed or epitopes are masked in the cloned and synthetic products. Inclusions of the envelope neuraminidase may help, because an experimental vaccine with both the hemagglutinin and neuraminidase of influenza virus has shown promising results in mice (118).

The middle-sized RNA segment of bunyaviruses codes for surface glyco-proteins G_1 and G_2 (49). In Rift Valley fever virus, the genes for these proteins, which map G_2G_1, have been cloned and partially expressed (USAMRDC contract, 1982; J. Dalrymple, pers. comm.). Since the expressed protein segments elicit only low levels of neutralizing antibodies and partial im-munity in mice, the central unexpressed region of the G_2G_1 genes may encode immunodominant epitopes of Rift Valley fever virus. Glycoprotein pGP-C, encoded by the small genome segment of Pichinde arenavirus, appears to have immunogenic activity.

Class V: Nonsegmented, single-stranded ribonucleoprotein viruses. This class includes paramyxo- and rhabdoviruses. The paramyxoviruses have 2 surface glycoproteins: one has either hemagglutinin (H) activity alone or both hemagglutinin and neuraminidase (HN) activities, and the other (the F glycoprotein) has cell fusion and hemolytic activities. The H protein from measles virus (16) and the HN from Sendai, Newcastle disease, simian-5, and parainfluenza-3 viruses induce the formation of neutralizing anti-bodies (69,111). By contrast, F glycoproteins elicit antibodies that inhibit the cell-to-cell spread of virus (111). The H protein, nucleocap-sid-associated phosphoprotein (P), and nucleocapsid (NP), matrix, and F proteins of measles virus are reported to stimulate lymphocyte prolifera-tion in high-responder individuals, but not in low-responder individuals (150). An oil-adjuvanted micellar mixture of HN and F glycoproteins from parainfluenza-3 virus is reported to elicit serum antibodies and protective immunity in lambs against parainfluenza-3 pneumonia (117). Newcastle di-sease virus HN has been shown by the use of monoclonal antibodies to have at least 4 epitopes on its surface (79). Cloning and characterization have been reported of complementary DNAs (cDNAs) that encode the following: the H and matrix proteins of measles virus (152); the H, matrix, and NP pro-teins of canine distemper virus (151); the P, NP, and matrix proteins of Sendai virus (60); and the F and G and other proteins of respiratory syncy-tial virus (39). The predicted primary sequences of simian virus-5 F and HN proteins determined by sequencing cloned cDNAs to the respective mRNAs are providing the basis for expression and examination of the immunogen-icity of these two surface glycoproteins (135).

The rhabdoviruses have a surface glycoprotein G that possesses both hemagglutinin and immunogenic activities. Rabies virus subunits and puri-fied glycoprotein G have been shown to elicit neutralizing antibodies and immunity in animals. Rabies and vesicular stomatitis virus G genes have been cloned and expressed in E. coli (149,199). The cloned vesicular sto-matitis virus G gene product is precipitated by virus-specific antibody, demonstrating a G-like epitope. However, much higher yields are obtained by cloning the G gene in an simian virus-40/eukaryote system (148). The eukaryotes have also been programmed to secrete an anchorless form of the vesicular stomatitis virus G protein. Cattle inoculated intramuscularly with a single 100-µg dose of G protein from vesicular stomatitis virus in Freund's complete adjuvant have resisted intradermolingual challenge with 400 infectious doses of virus (200). Restriction endonuclease maps of the cloned genome of infectious hematopoietic necrosis virus, a rhabdovirus of fish, have been used to identify the gene that codes for the immunogenic protein of the virus. A vaccine is needed to control infectious hematopoi-etic necrosis, which destroyed 15 million hatchery fish in the Columbia River basin in 1983 alone (95).

Class VI: Ribonucleoprotein retroviruses. Retroviruses, which repli-cate by an RNA-DNA-RNA pathway using the cell's DNA as the vector, possess immunogenic envelope glycoproteins. The first retrovirus glycoprotein shown to have this activity was gp71 from the envelope of Friend murine leukemia virus (76). Vaccines are being developed against several retro-virus diseases, e.g., avian, feline, and bovine leukemias, equine infec-tious anemia, and acquired immunodeficiency disease syndrome. A vaccine comprised of subunits and soluble tumor antigens from feline leukemia virus is reported to be about 80% effective in providing immunity against viral challenge (99). The vaccine must be free of feline leukemia virus protein p15E, which, if present, suppresses the cat's immune system (128). In a unique but generally applicable procedure, screenings with a monoclonal an-tibody of recombinant phages expressing short random feline leukemia virus DNA fragments were successful in mapping a virus-neutralizing epitope to a 14-amino-acid region in the N-terminal half of gp70 (127). Neutralizing antibodies have been induced with a crude bovine leukemia virus subunit vaccine, and also with glycoprotein isolated from virions (J.F. Ferrer, pers. comm.). Sequence analysis on gp60 and p30 proteins from bovine leu-kemia virus indicates that they are a surface glycoprotein and transmem-brane protein, respectively (156). The genome of bovine leukemia virus has been cloned using a lambda-phage vector (87), and its complete nucleotide sequence determined (154). Use of a cloned DNA probe has shown that a 150-kd protein present in the plasma of cattle with bovine leukemia blocks the transcription of the viral genome in infected lymphocytes in vivo (61). Because acquired immunodeficiency disease syndrome is reputed to be caused by retroviruses (lymphadenopathy virus and human T-cell lymphotropic virus-3) (12,47), the envelope glycoproteins of these viruses are under active investigation for immunogenic activity. In equine infectious anemia, both the protein and carbohydrate moieties of the viral envelope gp90 appear to evoke antibodies (116), whereas antigenicity has been attributed exclusive-ly to protein in Friend murine leukemia virus and to carbohydrate in bovine leukemia virus. The genetic relatedness of visna and caprine arthritis-encephalitis virus is reported to be confined principally to 5' gag-pol se-quences, indicating that their serological similarities and differences are due to core antigens and to envelope proteins, respectively (142). The ab-sence of neutralizing antibodies to caprine arthritis-encephalitis virus in goats can be overcome by vaccination with virus and large amounts of inactivated Mycobacterium tuberculosis (125). A physical map of caprine arthritis-encephalitis virus proviral DNA has been developed using 13 re-striction endonucleases (146).

Class VII (provisional): Partially double-stranded DNA viruses. This provisional class is composed of a small group of hepatitis B viruses (HBVs). One of the DNA strands in the viral core of HBV is incomplete, but a DNA polymerase in the core completes this strand after the virus is un-coated. The replication characteristics of HBV resemble in many respects those of retroviruses: (a) a DNA minus strand is transcribed from RNA before the DNA plus strand is synthesized (175), (b) HBV genomes and proviral DNA of retroviruses are organized similarly, and (c) a portion of the hepatitis B genome can integrate into host DNA and transform hepatic cells in vivo (106).

Hepatitis B surface antigen (HBsAg) 22-nm particles purified from human carrier serum are in use as a vaccine (66). Before approval for human use, the vaccine was shown to protect chimpanzees against viral challenge (35). Owing to the vaccine's high cost, however, its use is restricted primarily to medical, dental, and other health care personnel who are at high risk of infection in the workplace. The commercial vaccine elicits neutralizing antibodies that crossreact with all HBVs that have an "a" surface antigen (177). A vaccine that contains HBsAg from cultures of human hepatoma cells has been tested in chimpanzees and human volunteers (109).

The gene for HBsAg has been cloned and expressed in E. coli by several investigators, and the protein product has been shown to react with antibody to natural HBsAg (38). Cloning in yeasts (186), in an simian virus-40/transformed monkey kidney cell system (97), or in a continuous Chinese hamster ovary cell line (136), produces glycosylated HBsAg molecules that assemble into 22-nm particles that are similar structurally and antigenically to their natural counterparts. The rDNA subunit vaccine from yeast produces predominantly "a" epitope-specific antibodies in human volunteers (157). More than 1,500 people in several countries have received yeast-derived vaccine (20). These tests are reported to be the first exposure of humans to a vaccine prepared by rDNA technology. It has been reported recently that an epitope(s) programmed by the pre-S region of the hepatitis genome contributes significantly to the immune response to hepatitis B. An experimental vaccine obtained by cloning both pre-S and S epitopes in Chinese hamster ovary cell cultures is as potent in guinea pigs as the commercial HBsAg vaccine derived from human serum (113). By means of rDNA techniques, the surface proteins of other viruses can be incorporated into the HBsAg particle. Thus, a hybrid HBsAg-HSVgD particle produced in yeast, which includes the product of the pre-S region of HBV DNA reacts with both relevant antibodies (185). The efficacy of the hybrid particle is being tested as a bivalent vaccine. In a survey, approximately 10% of German-bred Pekin ducks were found to be chronically infected with duck HBV. In preparation for subunit vaccine studies, duck HBV DNA was cloned and shown to be infectious for Pekin ducks (172).

Recombinant DNA Protein Vaccines for Bacterial Diseases

The first work on rDNA bacterial protein vaccines involved the cloning of somatic pili of enterotoxigenic E. coli (ETEC) for use in preventing diarrhea in livestock. The pili of ETEC are composed of bundles of 14- to 22-kd protein molecules called pilin, which cause the bacteria to adhere to and colonize the mucosal surfaces of the small intestines. Several immunologically distinct pili have been isolated from ETEC: K88 and 987P from swine; K99 from cattle, sheep, and swine; CFA/1 and CFA/2 from humans; and type I and other common pili. Some pilin genes (type I and probably 987P) appear to be encoded by bacterial chromosomes, and others (K88, K99, CFA/1, and CFA/2) appear to be encoded by plasmids (80). Cloned vaccines for ETEC in swine have been developed in Western Europe (181) and in the United States. The cloned vaccine approved by the U.S. Department of Agriculture (USDA) contains 2 pilus antigens that evoke antibodies that block adhesion by ETEC and one antigen that induces the formation of an antitoxin (48). When administered to sows shortly before parturition, their colostrum and milk will contain protective antibodies for the newborn piglets. A recombinant plasmid/lambda lysogenic E. coli system has been used to express and externalize high yields of the B (adhesion) subunit of the heat-labile enterotoxin of E. coli (114). K99 monoclonal antibodies produced in hybridomas and amplified in mice are also used to control ETEC (160). Administered orally to calves, the monoclonal antibodies reduce both the severity and mortality of scours. First tested in Canada, these monoclonal antibodies have been approved by the USDA for use in the United States.

Progress has been reported on a pilus vaccine cloned in E. coli that may protect humans against penicillin-resistant strains of Neisseria gonorrhoeae (112,188). A pili-specific monoclonal antibody was used to detect expression of gonococcal DNA in the E. coli. Antibodies to gonorrheal pili are generally strain-specific, however, and they provide protection only against the homologous gonococcal strain (180).

The inability to culture the syphilis organism Treponema pallidum has been a continual barrier to vaccine development. Using rDNA procedures, at least 3 groups have cloned and expressed the genes of the envelope antigens of this spirochete in E. coli (173). This work may lead to the development of vaccines against syphilis and other spirochetes.

Molecular approaches are being made to vaccines for brucellosis and anaplasmosis of cattle. Porin proteins that bridge the outer membranes of Brucella abortus have immunogenic activity, but their genes have not yet been cloned (198). Neutralizing antisera have been used to identify 5 Anaplasma marginale initial body proteins (31 kd to 105 kd) that are potentially protective immunogens of cattle (129). The production of effective cloned protein vaccines against B. abortus and A. marginale could help to reduce the incidence of brucellosis and anaplasmosis in cattle in the United States.

The gene for the protective antigen of the tripartite protein toxin of Bacillus anthracis has been cloned and expressed in an E. coli/pBR322 system for the purpose of developing a more efficacious anthrax vaccine (187a).

Streptococcus mutans, which is the prime etiological agent of dental caries, possesses major antigens that have been cloned and expressed in a lambda-phage/E. coli system (120). Using a monoclonal antibody to S. mutans ribosomes, a 65-kd protein has been identified that is present on the surface of all variants of S. mutans. This protein is considered to be a candidate for vaccine tests.

Recombinant DNA Protein Vaccines for Parasitic Diseases

Work is in progress on the development of rDNA protein vaccines against parasitic diseases such as avian coccidiosis (183), malaria, East Coast fever, anaplasmosis, and trypanosomiasis. The gene for the surface protein of the sporozoite form of Plasmodium knowlesi has been cloned and expressed in E. coli for potential use in an animal model vaccine for malaria (40,201). Similarly, cloning of merozoite-stage, surface protein genes has yielded a protein that reacts with antisera to a 200-kd protein. This protein, in turn, has been shown to induce protective immunity in monkeys against Plasmodium falciparum (110). A monoclonal antibody that neutralizes different stocks of Theileria parva sporozoites has been used to identify a major 68-kd surface protein (34). This finding may lead to the development through rDNA technology of a broad spectrum vaccine against East Coast fever, a serious disease of cattle in East and Central Africa.

Recombinant DNA technology may also help to control African sleeping sickness in humans and trypanosomiasis of livestock. However, because of a unique defensive strategy employed by trypanosomes, vaccine developments are not likely to succeed. A single trypanosome is genetically programmed to coat itself with one or another of 20 to 100 variant surface glycoproteins. In active trypanosomiasis, wave-like alternations occur between immunity to organisms coated with one type of variant surface glycoprotein and emergence of organisms coated with another type of variant surface glycoprotein, until the host dies. DNA sequences have been cloned that are complementary to mRNAs that encode some variant surface glycoproteins of Trypanosoma brucei (137). Nevertheless, the large number of possible variant surface glycoproteins suggests that other methods should be explored for controlling trypanosomiasis. For example, it may be possible to exploit the finding that nonvariable phospholipids in the membrane of T. rhodesiense evoke 2 types of antibodies (189). Also, if the trypanotolerance of certain strains of cattle and sheep (78) resides in a few genes, these genes could be cloned and inserted into the germplasm of susceptible strains by the inoculation of the pronuclei in fertilized ova. The trypanotolerant genes acquired by the animal in this manner should then be transmitted to succeeding progeny through normal sexual breedings, as occurred in mice that were endowed with cloned human growth hormone genes (130).

A rDNA vaccine is being developed for the control of hookworm in animals and humans. The genes for the histolytic proteolytic enzyme that allows hookworms to bleed their hosts has been cloned in a lambda-phage vector. Antibodies to the putative histolytic proteolytic enzyme vaccine are expected to inactivate the histolytic proteolytic enzyme secreted by the hookworm and thereby prevent feeding on the host's blood (70).

CHEMICALLY SYNTHESIZED VACCINES

The chemical synthesis of experimental peptide vaccines is not new. Many years ago, a synthetic hexapeptide corresponding to the C-terminus of tobacco mosaic virus protein, crosslinked to BSA (2), and a synthetic peptide identical to the P2 region of MS-2 bacteriophage coat protein (96) were shown to evoke neutralizing antibodies in rabbits. Also, a synthetic peptide corresponding to the conformational antigenic determinant of egg-white lysozyme was shown to induce antibody to native lysozyme (4).

The recent work concerns principally synthetic peptides that mimic immunodominant epitopes of animal viruses and pathogenic bacteria. For example, a synthetic pentadecapeptide corresponding to a nucleotide sequence at the 3' end of the RNA of Moloney leukemia virus is reported to elicit antibody to previously undetected, viral-specific proteins present in infected cells (176).

Several peptides have been synthesized to mimic epitopes on HBV: (a) peptides corresponding to HBsAg residues 110-137 (Ref. 50) and 138-149, and one (120-145) (Ref. 126) that is covalently linked to liposomes, and (b) 2 related cyclic peptides (cyclized residues 124-137 with side chains containing residues 117-123 or 122-123) that are active without a protein carrier (36). These synthetic peptides provoke antibodies to HBV or HBsAg, or to both, but their ability to induce protective immunity in animals remains to be demonstrated conclusively (90).

Synthetic peptides mimicking regions of the hemagglutinin moiety of influenza viruses have been investigated in several laboratories. A synthetic peptide corresponding to hemagglutinin residues 91-108 of type A H_3N_2 virus linked to tetanus toxoid in Freund's adjuvant was found to stimulate neutralizing and hemagglutinin-inhibiting antibodies and induce partial protective immunity in mice (124). The 91-108 sequence is known to occupy an exposed corner of the folded hemagglutinin molecule (195). In addition, antibodies to 18 synthetic peptides that mimic 75% of the hemagglutinin-1 (HA1) portion of the hemagglutinin molecule (57) and to other peptides comprising the first 11 amino acids at the N-terminus in the cell fusion region of HA2 (5) bind to hemagglutinin and to virus. Nevertheless, these antipeptide antibodies do not neutralize virus. These findings emphasize the importance of conformation in epitope formation.

A synthetic hexadecapeptide corresponding to amino acid residues 8-23 of HSV-1 gD is reported to react with an anti-gD monoclonal antibody that neutralizes both HSV-1 and HSV-2. Also, this peptide, linked to carrier protein in Freund's complete adjuvant, induces antisera that react with native gD and neutralize both HSV-1 and HSV-2 (28). The protective effect of the peptide is being examined.

For picornavirus diseases, investigation of synthetic peptides was preceded by the demonstration that livestock could be immunized with a CNBr-derived 13-kd segment (amino acids 55-179) of VP_1 from type A_{12} FMD virus (9). A single dose of a synthetic type O_1 icosapeptide (amino acids 141-160 of VP1) linked to carrier protein was shown to induce neutralizing antibodies in guinea pigs and cattle and to protect guinea pigs, but not cattle, against viral challenge (18). Protection of the most important natural hosts for FMD--cattle and swine--has not yet been achieved using chemically synthesized peptides (196).* Capsid protein VP_1 from poliovirus induces low levels of neutralizing antibodies in rabbits (19,26). Five small synthetic peptides corresponding to hydrophilic domains of VP_1 were shown to react with poliovirus neutralizing antibodies (41); however, only one of these peptides and a VP_2-specific peptide (197) were able to elicit neutralizing antibodies in rabbits. Nevertheless, each of the 5 VP_1-specific peptides linked to BSA in Freund's complete adjuvant could prime rabbits for an anamnestic-like, virus-neutralizing antibody response to a later inoculation of poliovirus. Capsid protein VP_2 of Coxsackie virus B3 has been shown to induce neutralizing antibodies (15), but no synthetic peptide work has been initiated.

Synthetic peptide vaccines are being developed for bacterial diseases such as diphtheria, cholera, and ETEC-induced diarrheas. A hexadecapeptide corresponding to amino acid residues 186-201 of the A chain of diphtheria toxin, covalently linked to muramyl dipeptide adjuvant and multichain poly-DL-alanine carrier, has been shown to elicit protective antitoxic immunity in guinea pigs (6). This vaccine is reported to be the first fully synthetic vaccine with built-in adjuvant and carrier. For cholera, 2 antibodies are reported to be protective independently and synergistically for animals and humans (68). One antibody is directed against cholera lipopolysaccharide and the other against the cholera toxin. Because antibodies to the toxin's 103 amino acid-long B subunit prevent binding of the toxin to mucosal receptors, several synthetic B-subunit-specific peptides linked to tetanus toxoid or to multichain poly-DL-alanine are being tested as potential cholera vaccines (158). Synthetic peptides corresponding to epitopes on both heat-labile and heat-stable toxins of ETEC are reported to reduce the severity of ETEC-induced diarrhea in experimental animals (71).

Synthetic peptides are also being tested for use in vaccines against parasitic diseases. The immunodominant epitope on the sporozoite _Plasmodium falciparum_ has been shown to lie within the synthetic dodecapeptide

---

* Note added in proof: Concurrent with the editing of this chapter, a report of the immunization of cattle with a chemically synthesized peptide was submitted for publication by other workers.

$(Asn-Ala-Asn-Pro)_3$. This peptide is being considered for incorporation into a malarial vaccine, because it induces antibodies that neutralize P. falciparum sporozoites (202).

RECOMBINANT VIRUS VACCINES

Recombinant viruses that contain genes for the immunodominant proteins of unrelated infectious agents are being constructed for use as vaccines. The guest genes are inserted into a virus vector by site-directed homologous recombination. When the recombinant virus replicates in the recipient animal, it expresses the proteins that are encoded by the foreign genes. Vaccinia virus is the principal vector being used for the gene insertions; however, simian virus 40, bovine papilloma virus, adenoviruses, herpesviruses, and retroviruses are also being used. Vaccinia virus has certain advantages over other vectors: (a) it is not oncogenic, (b) it has a wide range, and (c) its large genome (187,000 bp) can accomodate up to 25,000 base pairs of foreign DNA and still retain infectivity (170). Vaccinia virus with different foreign genes inserted in tandem has the potential to induce immunity to several diseases concurrently (133).

Experimental recombinant vaccinia virus vaccines have been constructed for hepatitis B, herpes simplex, influenza, transmissible gastroenteritis of swine, rabies, vesicular stomatitis, Friend murine virus leukemia, and malaria. Rabbits inoculated with a vaccinia virus/HBsAg gene recombinant develop antibodies to HBsAg (169). Vaccinia virus that expresses HSV-1 gD is reported to induce HSV-gD and HSV-neutralizing antibodies in rabbits as well as protective immunity in mice against challenge with either HSV-1 or HSV-2 (30,132). Recombinant vaccinia viruses containing cloned influenza virus hemagglutinin genes induce both hemagglutination-inhibiting and influenza virus-neutralizing antibodies in rabbits and hamsters, and protect the hamsters against respiratory challenge with influenza virus (131,171). Vaccination of mice by the intradermal route protects only the lungs, whereas nasal vaccination prevents both nasal infection and pneumonia (166). A vaccinia virus/porcine transmissible gastroenteritis virus gp195 gene recombinant evokes transmissible gastroenteritis virus-neutralizing antibodies in mice (73). A vaccinia virus/rabies glycoprotein gene (V-RG) recombinant induces rabies virus-neutralizing antibodies and protective immunity in rabbits and mice (194). The V-RG recombinant has to have proline rather than leucine at position 8 in the rabies glycoprotein. Beta-propiolactone-inactivated V-RGpro8 was also protective for mice, indicating for the first time that vaccinia virus recombinants can incorporate the expressed foreign proteins into their surface membranes. A vaccinia virus/-VSV G gene recombinant is reported to produce vesicular stomatitis virus (VSV)-neutralizing antibody and protective immunity in mice. It also imparts degrees of protection to cattle against intradermolingual injection of VSV that correlate with the levels of neutralizing antibody in the animals' sera (104). A vaccinia virus recombinant has also been constructed that contains the gene for the surface glycoprotein gp71 of Friend leukemia virus (37).

A vaccinia virus/Plasmodium knowlesi sporozoite surface protein gene recombinant spawns 53- and 56-kd proteins that react with monoclonal antibody to sporozoite surface protein and evoke antibodies in rabbits that bind to sporozoites (168). To obtain protective immunity, however, it may be necessary to enhance the expression of the sporozoite protein gene and to add genes for merozoite surface proteins.

The safety of recombinant vaccinia virus is unresolved. While U.S. military recruits continue to be vaccinated against smallpox, it is known that the vaccine produces some cases of disseminated vaccinitis and encephalitis, and, in rare instances, even death. It has been reported, however, that vaccinia virus/foreign gene recombinants are considerably less virulent in mice than wild-type vaccinia virus (22). In more natural hosts such as cattle, the recombinants could be quite virulent. Conversely, in humans who have been vaccinated previously against smallpox, vaccinia virus recombinants may not replicate well enough to produce the immunity desired. If so, alternate virus vectors may be required. In this regard, the gene for HBsAg has been cloned and expressed in permissive monkey cells using an simian virus-40-based vector (101), in mouse cells using a bovine papilloma virus vector (33), and in Vero cells using HSV-1 as the vector (162).

GENETICALLY ENGINEERED REASSORTANTS AND DELETION MUTANTS

Reassortment of viral gene segments and gene deletion techniques are being used to attenuate viruses for potential use as vaccines. Selected HSV-1 x HSV-2 recombinants and HSV mutants of low virulence lacking genes expressed early in infection are reported to induce protective immunity in mice (147). These HSV recombinant and deletion mutants are being engineered to minimize the occurrence of back mutations to virulence.

Influenza virus reassortants with potential use as vaccines are being produced by the coinfection of cells or animals with wild-type virus, and either avirulent, temperature-sensitive (ts) mutants or avian strains that replicate poorly in mammals. Reassortants are selected from the coinfections that possess wild-type hemagglutinin and NA genes and the 6 internal genes of the other parental virus. Human volunteers vaccinated with reassortants from wild-type virus and ts mutants have shown lower rates of infection and less shedding of virus upon exposure to wild-type virus than individuals who have been vaccinated with inactivated influenza virus vaccine (122, 123). Similarly, a wild-type x avian influenza A reassorted virus possessed the immunogenic character of wild-type virus and was not transmissible between individuals. Analogous equine influenza virus reassortants are being constructed for use in horses (21).

Rotaviruses, which produce diarrhea in young animals and humans, have 11 gene segments that can reassort during mixed infections. Reassortants that have the gene for human rotavirus surface antigen VP7 and their other genes from bovine or rhesus rotaviruses display reduced growth in humans and appear to have potential use in immunizing agents (58). Because of its high degree of crossprotection, an attenuated bovine rotavirus vaccine (RIT 4237) is being considered as a vaccine for humans (187).

Bioengineered mutants of pathogenic bacteria also appear to have potential applications as vaccines. A gal E strain of Salmonella typhi, Ty21a, which is deficient in UDP-galactose-4-epimerase activity and in 2 other enzymes, replicates well enough following oral administration to confer immunity against wild strains of S. typhi (51). Similarly, Salmonella strains defective in aromatic biosynthesis (they require amino- and dihydroxybenzoate not present in host tissues), when administered intramuscularly or orally to calves in 2 doses, induce resistance to oral challenge with more than 10^{11} virulent S. typhi organisms (167).

A Salmonella typhi/Shigella sonnei hybrid is being examined as a vaccine for both typhoid fever and dysentery. Transfer by conjugation of a plasmid encoded with genes of S. sonnei into the attenuated Ty21a strain of S. typhi results in a hybrid bacterium that elicits antibodies to both typhoid and dysentery antigens (179).

Vibrio cholera genes are being manipulated in order to produce a vaccine. In one study, genes for the A and B subunits of cholera toxin were first deleted from the El Tor O_1 strain, and the B subunit gene was then restored by genetic transformation using a plasmid (86). Nine of 10 human volunteers vaccinated with the deletion mutant resisted oral challenge with 10^6 virulent El Tor Inaba organisms, whereas 7 of 8 unvaccinated recipients developed severe cholera symptoms. The mutant, however, still produces low-grade diarrheas and will require further attenuation.

ANTI-IDIOTYPE ANTIBODY VACCINES

Anti-idiotype antibodies have potential use as vaccines. However, they have shown generally only the ability to prime animals for anamnestic-like responses to infectious agents. The response of an animal to a foreign antigen gives rise to a cascade of idiotype, anti-idiotype, anti-anti-idiotype, and higher antibody forms. The anti-idiotype antibody of particular interest is the anti-paratope antibody. The paratope is the binding site in the Fab crevice of the antibody, and antibody to it will be in the image of the epitope of the original antigen. Some investigators consider anti-idiotype antibody vaccines to be a realistic goal, because anti-idiotype antibodies can now be produced in quantity in hybridomas or by cloning DNA that has been transcribed from mRNAs of anti-idiotype antibodies.

Anti-idiotype antibodies that have been examined as potential immunogens include those mimicking epitopes on HBsAg, poliovirus, reovirus, rabies virus, bacteria, trypanosomes, and cancer antigens. The injection of mice with anti-idiotype antibody to HBsAg primes a strong anamnestic-like antibody response to a later injection of either HBsAg (88) or a cyclic synthetic peptide containing amino acid residues 122-137 of HBsAg (90). The antibody induced by the anti-idiotype antibody recognizes a group-specific epitope on HBsAg, indicating that anti-idiotype antibodies may be useful as vaccines or, at least, as primers for more conventional vaccines (89). For polio, syngeneic monoclonal anti-idiotype antibody to type 2 poliovirus has been used to induce neutralizing antibodies in mice (184). Similarly, reovirus monoclonal anti-idiotype antibodies can induce cellular immunity in syngeneic mice (159). Two monoclonal anti-idiotype antibodies to rabies glycoprotein G have been produced that elicit virus-neutralizing antibody responses (144). For bacterial infections, the administration to newborn mice of idiotype or anti-idiotype antibodies to the polysaccharide capsular antigen of E. coli K13 primes the mice for protection against this bacterium (174). Interestingly, anti-idiotype antibodies have been reported that can actually induce protective immunity in mice against African trypanosomiasis (153). Analogous results, if attainable in livestock, would represent a step toward the control of this intractable disease. Cancer patients that have been shown to develop anti-idiotype antibodies following injection with mouse monoclonal antibody are reported to improve clinically and to enter periods of remission (94).

CONCLUSION

Several molecular approaches are being made to develop new vaccines. The intention, in general, is to present the recipient with immunodominant epitopes (or images thereof) located on purified proteins, on synthetic peptides, or on organisms that have been freed of pathogenic constituents. In most instances, however, considerable developmental work remains to be done to enhance the potency of the new vaccines and to perfect practical adjuvants and carriers. Also, the new vaccines will have to be proven safe and effective to gain approval of governmental regulatory agencies, and will have to be cost effective to satisfy users. Some of the new vaccines are expected to be superior to existing whole-agent vaccines, and others will provide control diseases for which vaccines have not been available previously.

ACKNOWLEDGEMENTS

The author acknowledges the contribution of the Plum Island Animal Disease Center's librarian, Mr. Stephen E. Perlman, for computer searches of databases. Owing to the size of the literature on the subject and to space limitations on this chapter, not all of the recent developments could be included.

REFERENCES

1. Aasted, B., R.E. Race, and M.E. Bloom (1984) *J. Virol.* 51:7-13.
2. Anderer, F.A. (1963) *Biochim. Biophys. Acta* 71:246-248.
3. Appleyard, G., and C. Andrews (1974) *J. Gen. Virol.* 23:197-200.
4. Arnon, R., E. Maron, M. Sela, and C.B. Anfinsen (1971) *Proc. Natl. Acad. Sci., USA* 68:1450-1455.
5. Atassi, M.Z., and R.G. Webster (1983) *Proc. Natl. Acad. Sci., USA* 80:840-844.
6. Audibert, F., M. Jolivet, L. Chedid, R. Arnon, and M. Sela (1982) *Proc. Natl. Acad. Sci., USA* 79:5042-5046.
7. Bachrach, H.L. (1978) In *Advances in Virus Research*, Vol. 22, M.A. Lauffer, F.B. Bang, K. Maramorosch, and K.M. Smith, eds. Academic Press, New York, pp. 163-166.
8. Bachrach, H.L., D.M. Moore, P.D. McKercher, and J. Polatnick (1975) *J. Immunol.* 115:1636-1641.
9. Bachrach, H.L., D.O. Morgan, P.D. McKercher, D.M. Moore, and B.H. Robertson (1982) *Vet. Microbiol.* 7:85-96.
10. Bachrach, H.L., D.O. Morgan, and D.M. Moore (1979) *Intervirology* 12:65-72.
11. Balachandran, N., P. Seth, and L.N. Mohapatra (1980) *Infect. Immun.* 29:846-852.
12. Barré-Sinoussi, F., J.C. Chermann, F. Rey, M.T. Nugeyre, S. Chamaret, J. Gruest, C. Dauguet, C. Axler-Blin, F. Vezinet-Brun, C. Rouzioux, W. Rozenbaum, and L. Montagnier (1983) *Science* 220:868-871.
13. Bastardo, J.W., J.L. McKimm-Breschkin, S. Sonza, L.D. Mercer, and I.H. Holmes (1981) *Infect. Immun.* 34:641-647.
14. Baxby, D. (1982) *J. Gen. Virol.* 58:251-262.
15. Beatrice, S.T., M.G. Katze, B.A. Zajac, and R.L. Crowell (1980) *Virology* 104:426-438.
16. Bellini, W.J., D.E. McFarlin, G.D. Silver, E.S. Mingioli, and H.F. McFarland (1981) *Infect. Immun.* 32:1051-1057.
17. Binns, M.M., M.E.G. Boursnell, D. Cavanagh, D.J.C. Pappin, and T.D.K. Brown (1985) *J. Gen. Virol.* 66:719-726.
18. Bittle, J.L., R.A. Houghten, H. Alexander, T.M. Shinnick, J.G. Sutcliffe, R.A. Lerner, D.J. Rowland, and F. Brown (1982) *Nature* 298:30-33.
19. Blondel, B., R. Crainic, and F. Horodinceanu (1982) *C.R. Acad. Sci. Paris* 294:91-94.
20. Brief report of the World Health Organization: Production of hepatitis B vaccine from yeast. (1985) *J. Med. Virol.* 15:211-212.
21. Brundage-Anguish, L.T., D.F. Holmes, N.T. Hosier, B.R. Murphy, J.G. Massicott, G. Appleyard, and L. Coggins (1982) *Am. J. Vet. Res.* 43:869-874.
22. Buller, R.M., G.L. Smith, B. Moss, K. Cremer, and A.L. Notkins (1985) In *Vaccines 85*, R.A. Lerner, R.M. Chanock, and F. Brown, eds. Cold Spring Harbor Laboratory, New York, pp. 163-167.

23. Cappel, R., and F. Decuyper (1976) <u>Arch. Virol.</u> 50:207-213.

24. Carlson, J., I. Maxwell, F. Maxwell, A. McNab, K. Rushlow, M. Mild-brand, Y. Teramoto, and S. Winston (1984) In <u>Modern Approaches to Vaccines</u>, R.M. Chanock and R.A. Lerner, eds. Cold Spring Harbor Laboratory, New York, pp. 195-201.

25. Caughman, G.B., J. Staczek, and D.J. O'Callaghan (1984) <u>Virology</u> 134:184-195.

26. Chow, M., and D. Baltimore (1982) <u>Proc. Natl. Acad. Sci., USA</u> 79:7518-7521.

27. Clark, B.R., J.A. Zaia, L. Balce-Directo, and Y-P. Ting (1984) <u>J. Virol.</u> 49:279-282.

28. Cohen, G.H., B. Dietzschold, M. Ponce de Leon, D. Long, E. Golub, A. Varrichio, L. Pereira, and R.J. Eisenberg (1984) <u>J. Virol.</u> 49:102-108.

29. Compans, R.W., and D.H.L. Bishop, eds. (1983) <u>Double Stranded RNA Viruses</u>, Elsevier Biomedical, New York, 505 pp.

30. Cremer, K.J., M. Mackett, C. Wolenberg, A.L. Notkins, and B. Moss (1985) <u>Science</u> 228:737-739.

31. Dalsgaard, K., and E. Overby (1976) <u>Acta Vet. Scand.</u> 17:465-474.

32. Davis, A.R., D.P. Nayak, M. Ueda, A.L. Hiti, D. Dowbenko, and D.G. Kleid (1981) <u>Proc. Natl. Acad. Sci., USA</u> 78:5376-5380.

33. Denniston, K.J., T. Yoneyama, B. Hoyer, and J. Gerin (1984) <u>Gene</u> 32:357-368.

34. Dobbelaere, D.A.E., S.Z. Shapiro, and P. Webster (1985) <u>Proc. Natl. Acad. Sci., USA</u> 82:1771-1775.

35. Dreesman, G.R., F.B. Hollinger, Y. Sanchez, P. Oefinger, and J.L. Melnick (1981) <u>Infect. Immun.</u> 32:62-67.

36. Dreesman, G.R., Y. Sanchez, I. Ionescu-Matiu, J.T. Sparrow, H.R. Six, D.L. Peterson, F.B. Hollinger, and J.L. Melnick (1982) <u>Nature</u> 295:158-160.

37. Earl, P., and B. Moss (1985) In <u>Program of the American Society of Virology Annual Meeting</u>, p. 14.

38. Edman, J.C., R.A. Hallewell, P. Valenzuela, H.M. Goodman, and W.J. Rutter (1981) <u>Nature</u> 291:503-506.

39. Elango, N., M. Satake, J.E. Coligan, E. Norrby, E. Camargo, and S. Venkatesan (1985) <u>Nucl. Acids Res.</u> 13:1559-1574.

40. Ellis, J., L.S. Ozaki, R.W. Gwadz, A.H. Cochrane, V. Nussenzweig, R.S. Nussenzweig, and G.N. Godson (1983) <u>Nature</u> 302:536-538.

41. Emini, E.A., B.A. Jameson, and E. Wimmer (1983) <u>Nature</u> 304:699-703.

42. Emtage, J.S., W.C.A. Tacon, G.H. Catlin, B. Jenkins, A.G. Perter, and N.H. Corey (1980) <u>Nature</u> 238:171-174.

43. European Commission Report for the Control of Foot-and-Mouth Disease, 24th Session, Food and Agriculture Organization (1981); and Outbreak, Isle of Wight (1981).

44. Flores, J., M. Sereno, A. Kalica, J. Keith, A. Kapikian, and R. Chanock (1984) In <u>Modern Approaches to Vaccines</u>, R.M. Chanock and R.A. Lerner, eds. Cold Spring Harbor Laboratory, New York, pp. 159-164.

45. Fox, G.M., D. Langley, and S. Hu (1984) In <u>Modern Approaches to Vac-cines</u>, R.M. Chanock and R.A. Lerner, eds. Cold Spring Harbor Laboratory, New York, p. 447.

46. Friedrichs, W.E., and C. Grose (1984) <u>J. Virol.</u> 49:992-996.

47. Gallo, R.C., S.Z. Salahuddin, M. Popovic, G.M. Shearer, M. Kaplan, B.F. Haynes, T.J. Palker, R. Redfield, J. Oleske, B. Safai, G. White, P. Foster, and P.D. Markham (1984) <u>Science</u> 224:500-502.

48. <u>Genetic Engineering News</u> (1983) Mary Ann Liebert, Inc., New York, Vol. 3, no. 3, p. 41.

49. Gentsch, J.R., E.J. Rozhon, R.A. Klimas, L.M. El Said, R.E. Shope, and D.H.L. Bishop (1980) <u>Virology</u> 102:190-204.

50. Gerin, J.R., R.H. Purcell, and R.A. Lerner (1985) In <u>Vaccines 85</u>, R.A. Lerner, R.M. Chanock, and F. Brown, eds. Cold Spring Harbor Laboratory, New York, pp. 235-239.

51. Germanier, R., and E. Furer (1983) In Devel. Biol. Standard, S. Karger, Basel, Vol. 53, pp. 3-7.
52. Gething, M.J., and J. Sambrook (1981) Nature 293:620-625.
53. Gething, M.J., and J. Sambrook (1982) Nature 300:598-603.
54. Gibbs, C.P., K. Nazerian, L.F. Velicer, and H-J. Kung (1984) Proc. Natl. Acad. Sci., USA 81:3365-3369.
55. Ginsberg, H.S. (1975) In Viral Immunology and Immunopathology, A.L. Notkins, ed. Academic Press, New York, pp. 317-326.
56. Gough, P.M., C.H. Ellis, C.J. Frank, and C.J. Johnson (1983) Antiviral. Res. 3:211-221.
57. Green, N., H. Alexander, A. Olson, S. Alexander, T.M. Shinnick, J.G. Sutcliffe, and R.A. Lerner (1982) Cell 28:477-487.
58. Greenberg, H., K. Midthun, R. Wyatt, J. Flores, Y. Hoshino, R. Chanock, and A. Kapikian (1984) In Modern Approaches to Vaccines, R.M. Chanock and R.A. Lerner, eds. Cold Spring Harbor Laboratory, New York, pp. 319-327.
59. Grubman, M.J., J.A. Appleton, and G.L. Letchworth, III (1983) Virology 131:335-366.
60. Gupta, K.C., E.M. Morgan, G. Kitchingman, and D.W. Kingsbury (1983) J. Gen. Virol. 64:1679-1688.
61. Gupta, P., S.V.S. Kashmiri, and J.F. Ferrer (1984) J. Virol. 50:267-270.
62. Halstead, S.B. (1982) In Progress in Allergy, P. Kallos, ed. S. Karger, Basel, Vol. 31, pp. 301-364.
63. Hampl, H., T. Ben-Porat, L. Ehrlicher, K-O. Habermehl, and A.S. Kaplan (1984) J. Virol. 52:583-590.
64. Heiland, I., and M.J. Gething (1981) Nature 292:851-852.
65. Heinz, F.X., W. Tuma, and C. Kunz (1981) Infect. Immun. 33:250-277.
66. Hilleman, M.R., E.B. Buynak, W.J. McAleer, A.A. McLean, P.J. Provost, and A.A. Tytell (1981) In Perspectives in Virology XI, M. Pollard, ed. A.R. Liss, Inc., New York, pp. 219-247.
67. Hofschneider, P.H., E. Burgelt, M. Kauzmann, M. Mussgay, R. Franze, R. Ahl, H. Bohm, K. Strohmaier, H. Kupper, and B. Otto (1981) In Munich Symposia on Microbiology, Biological Products for Viral Diseases, P.A. Bachmann, ed. Taylor and Francis, Ltd., London, pp. 105-113.
68. Holmgren, J., and A-M. Svennerholm (1982) Karger Gazette 44/45:9-11.
69. Hosaka, Y. (1980) Infect. Immun. 30:212-218.
70. Hotez, P.J. (1985) In Genetic Engineering News (article by S. Friedman), Mary Ann Liebert, Inc., New York, Vol. 5, no. 5, pp. 24-25.
71. Houghten, R.A., R.A. Lerner, S.R. Hoffmann, P.A. Worrell, P. Wright, and F.A. Klipstein (1985) In Vaccines 85, R.A. Lerner, R.M. Chanock, and F. Brown, eds. Cold Spring Harbor Laboratory, New York, pp. 91-94.
72. Hu, S., J. Bruszewski, T. Boone, and L. Sousa (1984) In Modern Approaches to Vaccines, R.M. Chanock and R.A. Lerner, eds. Cold Spring Harbor Laboratory, New York, pp. 219-223.
73. Hu, S., J. Bruszewski, R. Smalling, and J.K. Browne (1985) In Immunobiology of Proteins and Peptides-III: Viral and Bacterial Antigens, M.Z. Atassi and H.L. Bachrach, eds. Plenum Press, New York, pp. 63-82.
74. Hughes, J.V., C. Bennett, L. Stanton, D.L. Linemayer, and S.W. Mitra (1985) In Vaccines 85, R.A. Lerner, R.M. Chanock, and F. Brown, eds. Cold Spring Harbor Laboratory, New York, pp. 255-259.
75. Huismans, H., N.T. van der Walt, M. Cloete, and B.J. Erasmus (1983) In Double Stranded RNA Viruses, R.W. Compans and D.H.L. Bishop, eds. Elsevier Biomedical, New York, pp. 165-172.
75a. Hunter, E., E. Hill, M. Hardwick, D.E. Schwartz, and R. Tizard (1983) J. Virol. 46:920-936.
76. Ihle, J.N., J.C. Lee, J.J. Collins, P.J. Fischinger, N.H. Pazmino, V.

Moenning, W. Schafer, M.G. Hamia, Jr., and D.P. Bolognesi (1976) <u>Virology</u> 75:88–101.

77. Ikuta, K., S. Ueda, S. Kato, and K. Hirai (1984) <u>J. Virol.</u> 49:1014–1017.
78. <u>ILRAD Reports</u> (1983) No. 2, pp. 1–2.
79. Iorio, R.M., and M.A. Bratt (1984) <u>J. Virol.</u> 51:445–451.
80. Issacson, R.E. (1981) In <u>Proceedings of the International Symposium on Neonatal Diarrhea</u>, Vol. 3, pp. 213–236.
81. Jennings, R., T.L. Smith, R.C. Spencer, A.M. Mellresh, D. Edey, P. Fenton, and C.W. Potter (1984) <u>Vaccine</u> 1:75–80.
82. Joklik, W.K. (1981) <u>Microbiol. Rev.</u> 45:483–501.
83. Kaaden, O.R., K.H. Adam, and K. Strohmaier (1977) <u>J. Gen. Virol.</u> 34:397–400.
84. Kaaden, O.R., and B. Dietzschold (1974) <u>J. Gen. Virol.</u> 25:1–10.
85. Kalkkinen, N., C. Oker-Blom, and R.F. Pettersson (1984) <u>J. Gen. Virol.</u> 65:1549–1557.
86. Kaper, J.B., M.M. Levine, H.A. Lockman, M.M. Baldini, R.E. Black, M.L. Clements, and J.G. Morris (1985) In <u>Vaccines 85</u>, R.A. Lerner, R.M. Chanock, and F. Brown, eds. Cold Spring Harbor Laboratory, New York, pp. 107–111.
87. Kashmiri, S.V.S., R. Mehdi, and J.F. Ferrer (1984) <u>J. Virol.</u> 49:583–587.
88. Kennedy, R.C., K. Adler-Storthz, R.D. Henkel, Y. Sanchez, J.L. Melnick, and G.R. Dreesman (1983) <u>Science</u> 221:853–855.
89. Kennedy, R.C., J.L. Melnick, and G.R. Dreesman (1984) <u>Science</u> 223:930–931.
90. Kennedy, R.C., J.T. Sparrow, Y. Sanchez, J.L. Melnick, and G.R. Dreesman (1984) <u>Virology</u> 136:247–252.
91. Kew, O.M., B.K. Nottay, M.H. Natch, and J.F. Obijeski (1981) <u>J. Gen. Virol.</u> 56:337–347.
92. Killen, H.M., and N.J. Dimmock (1982) <u>J. Gen. Virol.</u> 62:297–311.
93. Kleid, D., D. Yansura, B. Small, D. Dowbenko, D.M. Moore, M.J. Grubman, P.D. McKercher, D.O. Morgan, B.H. Robertson, and H.L. Bachrach (1981) <u>Science</u> 214:1125–1129.
94. Koprowski, H., D. Herlyn, M. Lubeck, E. DeFreitas, and H.F. Sears (1984) <u>Proc. Natl. Acad. Sci., USA</u> 81:216–219.
95. Kurath, G., K.G. Ahern, G.D. Pearson, and J.C. Leong (1985) <u>J. Virol.</u> 53:469–476.
96. Langbeheim, H., R. Arnon, and M. Sela (1976) <u>Proc. Natl. Acad. Sci., USA</u> 73:4636–4670.
97. Laub, O., L.B. Rall, M. Truett, Y. Shaul, D.N. Strandring, P. Valenzuela, and W.J. Rutter (1983) <u>J. Virol.</u> 48:271–280.
98. Leong, J. (1985) In <u>Genetic Engineering News</u> (article by J. Cone), Mary Ann Liebert, Inc., New York, Vol. 5, no. 4, p. 12.
99. Lewis, M.G., L.E. Mathes, and R.G. Olsen (1981) <u>Infect. Immun.</u> 34:388–394.
100. Ley, V., J.M. Almendral, P. Carbonero, A. Beloso, E. Vinuela, and A. Talavera (1984) <u>Virology</u> 133:249–257.
101. Liu, C-C., D. Yansura, and A.D. Levinson (1982) <u>DNA</u> 1:213–221.
102. Long, D., T.J. Madara, M. Ponce de Leon, G.H. Cohen, P.C. Montgomery, and R.J. Eisenberg (1984) <u>Infect. Immun.</u> 37:761–764.
103. Lupton, H.W., and D.E. Reed (1980) <u>Am. J. Vet. Res.</u> 41:383–390.
104. Mackett, M., T. Yilma, J.K. Rose, and B. Moss (1985) <u>Science</u> 227:433–445.
105. Maes, R.K., S.L. Fritsch, L.L. Herr, and P.A. Rota (1984) <u>J. Virol.</u> 51:259–262.
106. Marx, J.L. (1982) <u>Science</u> 21:1021–1022.
107. Mayer, L.W., B. Aasted, C.F. Garon, and M.E. Bloom (1983) <u>J. Virol.</u> 48:573–579.
108. Mayfield, J.E., P.J. Good, H.J. VanOort, A.R. Campbell, and D.E. Reed (1983) <u>J. Virol.</u> 47:259–264.

109. McAleer, W.J., H.Z. Markus, D.E. Wampler, E.B. Bunyak, D.E. Miller, R.E. Weibel, A.A. McLean, and M.R. Hilleman (1984) Proc. Soc. Exptl. Biol. Med. 175:314-319.

110. McGarvey, M.J., B. Mach, and L.H. Perrin (1985) In Vaccines 85, R.A. Lerner, R.M. Chanock, and F. Brown, eds. Cold Spring Harbor Laboratory, New York, pp. 25-29.

111. Merz, D.C., A. Scheid, and P.W. Choppin (1980) J. Exp. Med. 151:274-283.

112. Meyer, T.F. (1982) Cell 30:45-52.

113. Michel, M-L., E. Sobczak, Y. Malpiece, P. Tiollais, and R.E. Streeck (1985) Biotechnology 3:561-566.

114. Miller, T.J., R. Peetz, A.P. Reed, T. Kost, A.L. Brown, J. Auerbach, and M. Rosenberg (1985) In Vaccines 85, R.A. Lerner, R.M. Chanock, and F. Brown, eds. Cold Spring Harbor Laboratory, New York, pp. 95-99.

115. Molitor, T.W., H.S. Joo, and M.S. Collett (1983) J. Virol. 45:842-854.

116. Montelaro, R.C., M. West, and C.J. Issel (1984) Virology 136:368-374.

117. Morein, B., M. Sharp, B. Sundquist, and K. Simons (1983) J. Gen. Virol. 64:1557-1569.

118. Morein, B., B. Sundquist, and S. Hoglund (1982) In Fourth International Conference on Comparative Virology, Banff, Alberta, Canada, S3-4, p. 87 (abstr.).

119. Morgan, A.J., M.A. Epstein, and J.R. North (1984) J. Med. Virol. 13:281-292.

120. Morrissey, P., G. Dougan, R.R.B. Russell, and M. Gilpin (1985) In Vaccines 85, R.A. Lerner, R.M. Chanock, and F. Brown, eds. Cold Spring Harbor Laboratory, New York, pp. 117-120.

121. Morse, G. (1984) Science News 125:354.

122. Murphy, B., A. Buckler-White, S-F. Tian, R. Chanock, M. Clements, H.F. Maasab, and W. London (1984) In Modern Approaches to Vaccines, R.M. Chanock and R.A. Lerner, eds. Cold Spring Harbor Laboratory, New York, pp. 329-337.

123. Murphy, B.R., D.L. Sly, E.L. Tierney, N.T. Hosier, J.D. Massicot, W.T. London, R.M. Chanock, R.G. Webster, and V.S. Hinshaw (1982) Science 218:1330-1332.

124. Muller, G.M., M. Shapira, and R. Arnon (1982) Proc. Natl. Acad. Sci., USA 79:569-573.

125. Narayan, O., D. Sheffer, D.E. Griffin, J. Clements, and J. Hess (1984) J. Virol. 49:349-355.

126. Neurath, A.R., S.B.H. Kent, and N. Strick (1984) Science 224:392-395.

127. Nunberg, J.H., G. Rodgers, J.H. Gilbert, and R.M. Snead (1984) Proc. Natl. Acad. Sci., USA 81:3675-3679.

128. Olsen, R.G. (1985) Vet. Med. 80:61-64.

129. Palmer, G.H., and T.C. McGuire (1984) J. Immunol. 133:1010-1015.

130. Palmiter, R.D., G. Norstedt, R.E. Gelinas, R.E. Hammer, and R.L. Brinster (1983) Science 222:809-814.

131. Panicali, D., S.W. Davis, R.L. Weinberg, and E. Paoletti (1983) Proc. Natl. Acad. Sci., USA 80:5364-5368.

132. Paoletti, E., B.R. Lipinskas, C. Samsonoff, S. Mercer, and D. Panicali (1984) Proc. Natl. Acad. Sci., USA 81:193-197.

133. Paoletti, E., M.E. Perkus, A. Piccini, S.M. Wos, B.R. Lipinskas, and S.R. Mercer (1985) In Vaccines 85, R.A. Lerner, R.M. Chanock, and F. Brown, eds. Cold Spring Harbor Laboratory, New York, pp. 147-150.

134. Parry, N.R., E.J. Ouldridge, P.V. Barnett, D.J. Rowlands, F. Brown, J.L. Bittle, R.A. Houghten, and R.A. Lerner (1985) In Vaccines 85, R.A. Lerner, R.M. Chanock, and F. Brown, eds. Cold Spring Harbor Laboratory, New York, pp. 211-216.

135. Paterson, R.G., S.W. Hiebert, and R.A. Lamb (1985) In Vaccines 85, R.A. Lerner, R.M. Chanock, and F. Brown, eds. Cold Spring Harbor Laboratory, New York, pp. 303-308.

136. Patzer, E.J., T.J. Gregory, G.R. Nakamura, C.C. Simonsen, A.D. Levinson, R.D. Hershberg, and J.W. Eichberg (1985) In Vaccines 85, R.A. Lerner, R.M. Chanock, and F. Brown, eds. Cold Spring Harbor Laboratory, New York, pp. 261-264.

137. Pays, E., M. Delronche, M. Lheureux, T. Vervoort, J. Bloch, F. Gannon, and M. Steinert (1981) Nucl. Acids Res. 8:5965-5981.

138. Pilacinski, W.P., D.L. Glassman, R.A. Krzyzek, and A.K. Robbins (1984) Biotechnology 2:356-360.

139. Platt, K.B. (1982) Vet. Microbiol. 7:515-534.

140. Purchio, A.F., R. Larson, and M.S. Collett (1984) J. Virol. 50:666-669.

141. Purdy, M., J. Petre, and P. Roy (1984) J. Virol. 51:754-759.

142. Pyper, J.M., J.E. Clements, S.M. Molineaux, and O. Narayan (1984) J. Virol. 51:713-721.

143. Qualtiere, L.F., R. Chase, and G.R. Pearson (1982) J. Immunol. 129:814-818.

144. Reagan, K.J., W.H. Wunner, T.J. Wiktor, and H. Koprowski (1983) J. Virol. 48:660-666.

145. Robbins, A.K., J.H. Weiss, L.W. Enquist, and R.J. Watson (1984) J. Mol. Appl. Genet. 2:485-496.

146. Roberson, S.M., and W.P. Cheevers (1984) Virology 134:489-492.

147. Roizman, B., J. Warren, C.A. Thuning, M.S. Fanshaw, B. Norrild, and B. Meignier (1982) In Devel. Biol. Standard, Vol. 52, S. Karger, Basel, pp. 287-304.

148. Rose, J.K., and J.E. Bergmann (1982) Cell 30:753-762.

149. Rose, J.K., and A. Shafferman (1981) Proc. Natl. Acad. Sci., USA 78:6670-6674.

150. Rose, J.W., W.J. Bellini, D.E. McFarlin, and H.F. McFarland (1984) J. Virol. 53:988-991.

151. Rozenblatt, S., O. Eizenberg, G. Englund, and W.J. Bellini (1985) J. Virol. 53:691-694.

152. Rozenblatt, S., C. Gesand, V. Lavie, and F.S. Neuman (1982) J. Virol. 42:790-797.

153. Sacks, D.L., and A. Sher (1983) J. Immunol. 131:1511-1515.

154. Sagata, N., T. Yasunaga, J. Tsuzuku-Kawamura, K. Ohishi, Y. Ogawa, and Y. Ikawa (1985) Proc. Natl. Acad. Sci., USA 82:677-681.

155. Schmidt, O.W., and G.E. Kenney (1982) Infect. Immun. 35:515-522.

156. Schultz, A.M., T.D. Copeland, and S. Oroszlan (1984) Virology 135:417-427.

157. Scolnick, E.M., A.A. McLean, D.J. West, W.J. McAleer, W.J. Miller, and E.B. Buynak (1984) J. Am. Med. Assoc. 251:2812-2815.

158. Sela, M., C.O. Jacob, and R. Arnon (1984) In Modern Approaches to Vaccines, R.M. Chanock and R.A. Lerner, eds. Cold Spring Harbor Laboratory, New York, pp. 87-92.

159. Sharpe, A.H., G.N. Gaulton, K.K. McDade, B.N. Fields, and M.I. Greene (1984) J. Exp. Med. 160:1195-1205.

160. Sherman, D.M., S.D. Acres, P.L. Sadowski, J.A. Springer, B. Bray, T.J.G. Raybould, and C.C. Muscoplat (1983) Infect. Immun. 42:653-658.

161. Sherry, B., and R. Rueckert (1985) J. Virol. 53:137-143.

162. Shih, M-F., M. Asrsenakis, B. Roizman, and P. Tiollais (1985) In Vaccines 85, R.A. Lerner, R.M. Chanock, and F. Brown, eds. Cold Spring Harbor Laboratory, New York, pp. 177-180.

163. Silva, R.F., and L.F. Lee (1984) Virology 136:307-320.

164. Simons, K., A. Helenius, B. Morein, J. Balcorova, and M. Sharp (1980) In New Developments with Human and Veterinary Vaccines, A. Mizrahi, I. Hertman, and M.A. Klingberg, eds. A.R. Liss, Inc., New York, pp. 217-228.

165. Simpson, R.W., L. McGinty, L. Simon, C.A. Smith, C.W. Godzeski, and R.J. Boyd (1984) Science 223:1425-1428.

166. Small, Jr., P.A., G.L. Smith, and B. Moss (1985) In Vaccines 85, R.A.

Lerner, R.M. Chanock, and F. Brown, eds. Cold Spring Harbor Laboratory, New York, pp. 175-176.

167. Smith, B.P., M. Reina-Guerra, S.K. Hoiseth, B.A.D. Stocker, and F. Habasha (1984) Am. J. Vet. Res. 45:59-66.

168. Smith, G.L., G.N. Godson, V. Nussenzweig, R.S. Nussenzweig, J. Barnwell, and B. Moss (1984) Science 224:397-399.

169. Smith, G.L., M. Mackett, and B. Moss (1983) Nature 302:490-495.

170. Smith, G.L., and B. Moss (1983) Gene 25:21-28.

171. Smith, G.L., B.R. Murphy, and B. Moss (1983) Proc. Natl. Acad. Sci., USA 80:7155-7159.

172. Sprengel, R., C. Kuhn, C. Mason, and H. Will (1984) J. Virol. 52:932-937.

173. Stamm, L.V., and P.J. Bassford, Jr. (1982) DNA 1:329-333.

174. Stein, K., and T. Soderstrom (1984) J. Exp. Med. 160:1001-1011.

175. Summers, J., and W.S. Mason (1982) Cell 29:403-415.

176. Sutcliffe, J.G., T.M. Shinnick, N. Green, F-T. Liu, H.L. Niman, and R.A. Lerner (1980) Nature 287:801-805.

177. Szmuness, W., C.E. Stevens, E.J. Harley, E.A. Zang, H.J. Alter, P.E. Taylor, A. Devera, G.T.S. Chen, and A. Keller (1982) New Engl. J. Med. 307:1481-1486.

178. Ticehurst, J.R., V.R. Racaniello, B.M. Baroudy, D. Baltimore, R.H. Purcell, and S.M. Feinstone (1983) Proc. Natl. Acad. Sci., USA 80: 5885-5889.

179. Tramont, E.C., R. Chung, S. Berman, D. Keren, C. Kapfer, and S.B. Formal (1984) J. Infect. Dis. 149:133-136.

180. Tramont, E.C., J.C. Sadoff, J.W. Boslego, J. Ciak, D. McChesney, C.C. Brinton, Jr., S. Wood, and E. Takafuji (1981) J. Clin. Invest. 68: 881-888.

181. Trevis, J., and A. Bertelsen (1982) Feedstuffs 54(5):32-37.

182. Trudel, M., and P. Payment (1982) In Fourth International Conference on Comparative Virology, Banff, Alberta, Canada, S3-6, p. 88 (abstr.).

183. U.S. Department of Agriculture (1985) Agricultural Res. 33(3):13.

184. Uytdehaag, F.G.C.M., and A.D.M.E. Osterhaus (1985) J. Immunol. 134: 1225-1229.

185. Valenzuela, P., D. Coit, M.A. Medina-Selby, C.H. Kuo, G. Van Nest, R.L. Burke, P. Bull, M.S. Urdea, and P.V. Graves (1985) Biotechnology 3:323-326.

186. Valenzuela, P., A. Medina, W.J. Rutter, G. Ammerer, and B.D. Hall (1982) Nature 298:347-350.

187. Vesikari, T., E. Isolauri, F.E. Andre, E. d'Hondt, A. Delem, G. Zissis, G. Beards, and T.H. Flewett (1985) In Vaccines 85, R.A. Lerner, R.M. Chanock, and F. Brown, eds. Cold Spring Harbor Laboratory, New York, pp. 369-372.

187a. Vodkin, M.H., and S.H. Leppla (1983) Cell 34:693-697.

188. Walter Reed Army Institute Research (1982) Rept. Div. Commun. Dis. and Immunol. 3-4, 1-2.

189. Walter Reed Army Institute Research (1982) Rept. Div. Biochem. 3-4, 4-5.

190. Walter Reed Army Institute Research (1985) WRAIR Res. Rept. 6-1, p. 1.

191. Wathen, M.W., and L.M.K. Wathen (1984) J. Virol. 51:57-62.

192. Watson, R.J., J.H. Weis, J.S. Salstrom, and L.W. Enquist (1982) Science 218:381-384.

193. Webster, R.G., W.P. Glezen, C. Hannoun, and W.G. Laver (1977) J. Immunol. 119:2073-2077.

194. Wiktor, T.J., R.I. MacFarlan, K.J. Reagan, B. Dietzschold, P.J. Curtis, W.H. Wunner, M-P. Kieny, R. Lathe, J-P. Lecocq, M. Mackett, B. Moss, and H. Koprowski (1984) Proc. Natl. Acad. Sci., USA 81:7194-7198.

195. Wiley, D.C., I.A. Wilson, and J.J. Skehel (1981) Nature 289:373-378.

196. Wilson, T. (1984) Biotechnology 2:29-39.
197. Wimmer, E., B.A. Jameson, and E.A. Emini (1984) Nature 308:19.
198. Winter, A.J., D.R. Verstreate, C.E. Hall, R.H. Jacobson, W.L. Castleman, M.P. Meredith, and C.A. McLaughlin (1983) Infect. Immun. 42: 1159-1167.
199. Yelverton, E., S. Norton, J.F. Obijeski, and D.V. Goeddel (1983) Science 219:614-620.
200. Yilma, T., R.G. Breeze, S. Ristow, J.R. Gorham, and S.R. Leib (1985) In Immunobiology of Proteins and Peptides-III: Viral and Bacterial Antigens, M.Z. Atassi and H.L. Bachrach, eds. Plenum Press, New York, pp. 101-115.
201. Young, J.F., W.T. Hockmeyer, M. Gross, W.R. Ballou, R.A. Wirtz, J.H. Trosper, R.L. Beaudoin, M.R. Hollingdale, L.H. Miller, C.L. Diggs, and M. Rosenberg (1985) Science 228:958-962.
202. Zavala, F., J.P. Tam, M.R. Hollingdale, A.H. Cochrane, I. Quakyi, R.S. Nussenzweig, and V. Nussenzweig (1985) Science 228:1436-1440.

IDENTIFICATION OF SEX IN MAMMALIAN EMBRYOS

G.B. Anderson

Department of Animal Science
University of California
Davis, California 95616

INTRODUCTION

One of the uncertainties in reproduction of farm animals, regardless of whether one considers natural reproduction or reproduction assisted by artificial insemination or embryo transfer, is sex of the resulting off-spring. This uncertainty can be of significant consequence when the economic value of one sex is considerably greater than that of the other. Little progress has been made despite long-term interest in controlling the sex ratio, or at least in exerting a measure of control. Recorded in folk-lore, the popular press, and to some extent in the scientific literature are time-honored but unsubstantiated procedures for affecting the sex ratio in animals and humans (see Ref. 2 and 17 for historical reviews). Perhaps the ideal method of controlling the sex ratio is by separation of X-bearing and Y-bearing spermatozoa. Although some progress has been reported in conjunction with human artificial insemination, these procedures have not been successfully extended to agriculturally important animals. Despite the existence of procedures that are commercially available for sexing of semen, researchers have generally concluded from results of controlled studies that no such procedures can be used to consistently control the sex ratio (1).

An alternate approach to affecting the sex ratio is identification of the sex of preimplantation embryos after they have been flushed from the reproductive tract of a female and before they are used for embryo transfer procedures. A decision can be made based on the sex of the embryo whether or not it will be transferred to a recipient for development to term. One approach to identification of embryonic sex has included cytological meth-ods by which the embryo is determined to be XX or XY by examination of the sex chromosomes or chromatin. Immunological methods have been used to identify sex-specific antigens on embryos in an effort to distinguish male from female embryos. Quantification of differences in metabolic activity of male and female embryos has also been proposed as a method for distin-guishing embryos on the basis of their sex (20). Williams (35) recently reported that, by using a colorimetric test for glucose 6-phosphate-dehy-drogenase activity, he was able to correctly identify the sex of 64% of the mouse embryos in his study. He observed a slight decrease in viability of the embryos with this procedure. Among the newer techniques being proposed is DNA hybridization for identification of male embryos. With this method cells are removed from an embryo and a probe for sequences specific to the

Y chromosome is hybridized with DNA from the biopsied cells. Although each of these procedures has shown promise for identification of embryonic sex, only for cytological and immunological methods do sufficient data exist to evaluate efficacy. In this chapter, data that provide a basis for evaluation of these approaches to identification of embryonic sex will be reviewed and summarized.

APPROACHES TO SEX IDENTIFICATION

Cytological Methods

Betteridge et al. (3) and King (11) have reviewed the scientific literature regarding identification of embryonic sex by examination of sex chromosomes. Details on various procedures are included in these reviews and will not be repeated here.

Identification of the inactivated X chromosome. The first documented use of cytological methods for identification of sex in a living embryo was reported by Gardner and Edwards (7). Several hundred cells were removed from day-5.5 rabbit blastocysts and prepared for examination of the Barr body in female cells. Although the procedure was highly accurate, only a relatively small proportion of the embryos could be sexed. X-inactivation is incomplete in early mammalian embryos, which requires that a relatively large number of cells be removed for analysis. This procedure has not been extended to embryos of the livestock species because the granular nature of their egg cytoplasm obscures nuclear material.

Biopsy of trophoblasts from elongated blastocysts. Hare et al. (10) described a procedure whereby trophoblastic cells were removed from day-14 to day-15 bovine embryos and prepared for examination of sex chromosomes. Betteridge et al. (3) summarized results from 4 laboratories that used this procedure on day-12 to day-15 bovine embryos and concluded that only approximately two-thirds of the embryos could be sexed with certainty. Absence of metaphase spreads and presence of only incomplete spreads after processing the biopsied cells were some resultant problems (37). Embryos at day 14 are approximately 1 week older than those commonly collected and transferred by nonsurgical means. They cannot currently be stored for more than a few hours and, upon transfer to recipients, only approximately a third of the embryos result in pregnancy. Despite the promise that this procedure showed and the enthusiasm it generated when first described, it has been largely abandoned for commercial use.

Biopsy of cells from morulae. A refinement of the procedure, whereby a large number of cells are removed from an elongated blastocyst-stage embryo, is the removal of relatively few cells from a morula-stage embryo. The advantages of using embryos at this stage of development are their higher survival after nonsurgical transfer to recipient females and their ability to survive freezing and thawing. Variable success has been reported in accurate identification of sex by removal of cells from bovine morulae. Moustafa et al. (14) reported that when 8 to 10 cells were removed from bovine morulae, metaphase chromosomes were obtained from 75% of the embryos. In 63% of the embryos, sex was accurately identified. Using similar procedures, but removing 15 to 17 cells per embryo, Singh and Hare (23) were able to identify sex in only approximately a third of the embryos tested. These researchers also incubated morulae from cattle, rabbits, and mice in Colcemid for 4 to 6 hours and determined that the mitotic index for morulae from these 3 species was only approximately 10%, which is probably too low for routine examination of metaphase chromosomes from a small number of cells. Rottmann (21) confirmed the observation of a low mitot-

244

ic index in rabbit embryos and was able to identify sex in only 14% of the embryos from which 6 to 7 cells per embryo were removed. It would seem that identification of embryonic sex by analysis of sex chromosomes from a small number of cells will require a higher mitotic index than normally exists in early mammalian embryos, long-term culture of the biopsied cells to increase the number of cells available for analysis, or an improved method for identification of sex chromosomes. Such procedures have not yet been reported as available for use in identification of embryonic sex.

Use of half embryos. The ability to bisect morulae and blastocysts and produce offspring from each of the halves is being used increasingly in bovine embryo transfer. Pregnancy rates from transfer of half embryos that are only slightly lower than can be obtained from intact embryos have been reported (16,36). Half embryos can be frozen and thawed and transferred nonsurgically in cattle. The use of half embryos for identification of embryonic sex is a modification of procedures for removal of a small number of cells from morula-stage embryos. Because a larger number of cells are removed, however, metaphase chromosomes can be obtained from a larger proportion of embryos. Picard et al. (18) reported that sex could be identified in approximately 60% of bovine half embryos after culture for 4 hours in Colcemid. Only slightly fewer than 30% of bovine half embryos could be sexed by other researchers (cited in Ref. 11). Even if these procedures can be improved to the point where a high proportion of embryos can be sexed, a major disadvantage of this procedure remains: a potential offspring is lost. Decisions must be made on an individual basis whether one offspring of known sex is more desirable than 2 offspring of unknown sex until birth. A less invasive approach to sex identification would allow the viability of half embryos to be utilized in the production of genetically identical twins of known sex. Such an approach to sex identification is described in the next section.

Identification of Sex-specific Antigens

The existence of a male-specific antigen was first proposed by Eichwald and Silmser (5) when it was observed that female mice from a highly inbred strain rejected skin grafts from males of the same strain. All other sex combinations were accepted: male to male, female to female, and female to male. To explain this phenomenon, a male-specific transplantation antigen was proposed. This male-specific antigen was later given the name H-Y antigen, because its gene was thought to be located on the Y chromosome (4). The H-Y antigen has been proposed as the testicular-organizing factor responsible for differentiation of the indifferent gonad into a testis (reviewed by Ref. 26). The antigen is highly conserved across species (28) and is present in the heterogametic sex, except in certain transitional systems (15). A great deal of research has been conducted in regard to the expression and role of H-Y antigen in sexual differentiation and recent reviews on the subject are available (26,27).

Controversy continues to exist over both the location of the gene for H-Y antigen on the chromosome and the role that H-Y antigen plays in normal development. Recently, questions were raised over whether the male-specific antigen that is detected serologically is indeed H-Y antigen, which was originally described as a tissue-transplantation antigen. Silvers et al. (22) have proposed that the male-specific antigen that is detected by serological means be referred to as serologically detectable male-specific antigen or SDM antigen. Other researchers have argued that H-Y antigen and SDM antigen are either identical or highly cross-reactive (27). In this discussion, the term H-Y antigen will be used with the understanding that there may be one or more male-specific antigens, and the antigen detected on preimplantation embryos may be different than the antigen originally described by Eichwald and Silmser.

The H-Y antigen has been reported to be detectable on mouse embryos as early as at the 8-cell stage (12). This expression of a paternal allele confirmed that the embryonic genome is activated during cleavage. This observation also suggested that detection of H-Y antigen may be useful for identification of male embryos. Epstein et al. (6) repeated the experiments of Krco and Goldberg and, in addition, examined the sex chromosomes of embryos presumed to be female by their lack of H-Y antigen. In the experiments of Epstein and co-workers, 92% of the 8-cell embryos that continued to develop after incubation in H-Y antiserum and guinea-pig complement were confirmed to be female. These authors concluded, however, that this procedure was not effective in producing pure populations of embryos of a particular sex, because accuracy was not 100%.

Mouse embryos. A series of experiments were initiated in this laboratory in which the expression of H-Y antigen on embryos of various species was studied and possibilities for using such an approach for identification of embryonic sex were examined. Antisera to H-Y antigen were produced by injection of spleen cells from C57/Bl male mice into females of the same strain (32). In an initial experiment, 2,000 8-cell mouse embryos were incubated in either medium containing antiserum and guinea-pig serum as a complement source or in various control media. Embryos that continued to develop in culture were classified as having been "unaffected" by culture in antiserum and complement and were presumed to be female. Embryos that showed evidence of cytolysis in individual blastomeres or failed to develop to the blastocyst stage were classified as having been "affected" by culture and were presumed to be male. Unaffected embryos were transferred to the reproductive tracts of pseudopregnant females and allowed to develop to term. Eighty-six percent of the pups born from embryos presumed to be female were female; approximately 50% of the pups that developed from embryos cultured in control media were female (32). The results of this experiment confirmed that H-Y antigen is expressed on early preimplantation mouse embryos and that its detection may be useful for identification of embryonic sex.

Because 86% rather than 100% of the embryos classified as female were indeed female, a second experiment was conducted in which the polyclonal antiserum used in the first experiment was replaced with an anti-H-Y monoclonal antibody. In addition, an indirect immunofluorescence assay for detection of H-Y antigen was compared with the cytolytic assay that results in destruction of male embryos. FITC-labeled anti-mouse IgM was used as a second antibody in the immunofluorescence assay. Both fluorescing (H-Y positive) and nonfluorescing (H-Y negative) embryos were transferred to the reproductive tracts of pseudopregnant females. Only unaffected (H-Y negative) embryos from the cytolytic assay were transferred to recipients. H-Y positive and H-Y negative embryos from the indirect immunofluorescence assay survived equally well after transfer; 78% of the pups that developed from fluorescing embryos were male and 83% of the pups that developed from nonfluorescing embryos were female. Eighty-one percent of the pups that developed from unaffected embryos from the cytolytic assay were female (33). It was demonstrated that both male and female embryos could be sexed and their viability maintained with the immunofluorescence assay and, therefore, this procedure replaced the cytolytic assay in subsequent experiments. Because no improvement in accuracy was observed with the monoclonal antibody over results obtained with the polyclonal antiserum, use of the polyclonal antiserum was retained.

The H-Y antigen is a weak antigen and the response to immunization is unpredictable. Furthermore, antiserum of high anti-H-Y titer is difficult to identify. A bioassay that has been used to identify anti-H-Y activity is the ability of the antiserum and complement to kill spermatozoa (9). We observed a significant correlation of 0.86 between the ability of an antiserum to kill spermatozoa and its ability to detect embryonic H-Y antigen (19). The correlation between specific sperm cytotoxicity of an antiserum and the accuracy with which nonfluorescing embryos were classified female was also 0.86. The corresponding correlation between cytotoxicity and accuracy with which fluorescing embryos were correctly classified male was 0.78, which suggested a greater accuracy of classifying nonfluorescing embryos as female than fluorescing embryos as male. A similar observation has been made in other experiments and will be mentioned again below.

The indirect immunofluorescence assay that was developed with murine antiserum and for use with murine embryos has since been modified for use with embryos of other species. The mouse anti-H-Y antiserum is absorbed against female white blood cells from the appropriate species, a step that removes antibodies that bind to an embryo because it is of a different species. Morula-stage embryos are then incubated in anti-H-Y antiserum for approximately 45 min. After several washes to remove unbound antibodies, the embryos are incubated for approximately 45 min in a secondary antibody, usually FITC-labeled goat anti-mouse IgG. After several washes to remove unbound second antibody, embryos are examined individually under a fluorescence microscope.

Although the indirect immunofluorescence assay has been shown to be a reasonably accurate method for identification of embryonic sex of murine embryos, the assessment of fluorescence is highly subjective. Embryo quality appears to be a factor in the degree and type of fluorescence that is observed. Poor-quality embryos with lysed or dead cells often show a reasonable degree of fluorescence that is unrelated to expression of H-Y antigen. Extruded blastomeres, for example, are often observed to be fluorescent. Some embryos tend to show fluorescence in the zona pellucida that must be ignored. Diffuse fluorescence in the perivitelline space or fluorescence that does not appear to be associated with an individual blastomere must also be ignored. We have also observed that detection of H-Y antigen is stage-specific (19,31,34; and unpubl. data). H-Y antigen is readily detectable at the 8-cell stage of development. It can be detected throughout cleavage and at the morula stage, but for reasons that are yet unknown, it becomes more difficult to detect on blastocysts. Cell-specific fluorescence that is observed in blastocysts appears to be associated primarily with cells of the inner cell mass, but not the trophoblast. H-2 antigens, which are the major histocompatibility antigens of the mouse, are synthesized by and expressed on preimplantation murine embryos (8,30), including on blastocysts. In one report, however, paternal H-2 antigens were described as being expressed at the morula stage, but not at the blastocyst stage (13). Although H-Y antigen is not part of the major histocompatibility complex in mice, perhaps an analogous situation exists where H-Y antigen is more readily detectable on morulae than on blastocysts. It is not known whether the antigen is not expressed on trophoblastic cells or whether its expression is masked, but it is interesting to speculate that its lack of detection on cells in direct contact with the endometrium may contribute to survival of the embryo.

Bovine embryos. Experiments were designed to learn at which point during bovine embryo development H-Y antigen can be detected and at which stages of development sex can be accurately identified. Bovine embryos from the 4-cell to the blastocyst stages were examined for expression of H-Y antigen. As had been originally reported for mouse embryos (12), H-Y antigen was detected at the 8-cell, but not the 4-cell, stage. Furthermore, H-Y antigen was detected on approximately half of the embryos at the 8-cell through the blastocyst stages, which agrees with expectation if embryos were separated on the basis of sex. Karyotypes were prepared from embryos in order to determine the accuracy with which embryonic sex was identified by detection of H-Y antigen. Readable karyotypes were prepared from 125 of 149 sexed embryos. Eighty-nine percent of the nonfluorescing embryos were confirmed to be female, and 80% of the fluorescing embryos were male, giving an overall accuracy of 84% (31; and unpubl. data). As had been observed with mouse embryos, the accuracy with which nonfluorescing embryos were classified as female was greater than the accuracy with which fluorescing embryos were classified as male. Wachtel (27) has also reported that detection of H-Y antigen can be used to identify sex in bovine embryos.

Porcine embryos. In initial experiments with porcine embryos, difficulty was experienced in obtaining cell-specific fluorescence that has been observed on male embryos of other species. It was hypothesized that supernumerary spermatozoa on and in the zona pellucida of porcine embryos, apparently present because of an incomplete block to polyspermy at the level of the zona pellucida, were interfering with the assay. This interference appeared to be due in part to fluorescence of the spermatozoa themselves as well as to their absorption of primary antibody and a resulting reduction in antibody available to the embryo. When the zona pellucida was removed using standard micromanipulation procedures, it was observed that the denuded embryos reacted much like intact embryos of other species; only with porcine embryos was it found to be necessary to remove the zona pellucida. When 4-cell stage to hatched blastocyst-stage porcine embryos were treated with primary and then secondary antibody, it was observed that H-Y antigen was detected on few, if any, 4-cell embryos and hatched blastocysts. Approximately half of the 8-cell-stage through blastocyst-stage embryos were classified as H-Y positive. Karyotypes were prepared from embryos sexed by indirect immunofluorescence. Readable karyotypes were obtained from 59 embryos, of which 48 (81%) had been correctly sexed (34).

Ovine embryos. Ovine embryos have also been sexed by indirect immunofluorescence using anti-H-Y antiserum. After sexing they were either transferred to the reproductive tracts of synchronized ewes or prepared for analysis of sex chromosomes. Of 68 embryos from which either lambs were born at term or readable karyotypes were obtained, 58 (85%) had been correctly sexed (unpubl. data). Data from sheep embryos were very much in line with those obtained with embryos of mice, cattle, and pigs.

Embryos from other species. Limited data available for embryos of other species indicate that, here too, detection of H-Y antigen may be useful for identification of embryonic sex. Utsumi et al. (24) reported that, using a cytolytic assay and rat H-Y antiserum, 97% of rat pups born from unaffected embryos were female and 80% of pups born from affected embryos were male. These researchers have also reported that approximately half of the caprine embryos incubated in their cytolytic assay were arrested in their development and classified as H-Y positive (25). In our laboratory, a limited number of equine embryos have been subjected to the indirect immunofluorescence assay; approximately half were classified as H-Y positive and the remaining as H-Y negative. Consistent with previous observations of the conservation of H-Y antigen across species, H-Y antigen is readily detectable on embryos of various domestic species by murine antiserum. H-Y antigen is expressed in cleavage-stage embryos in each of the species studied thus far.

CONCLUSIONS

Male-specific antigen is detectable on male preimplantation mammalian embryos of various species and may provide the basis for identification of embryonic sex. Indirect immunofluorescence can be used as a noninvasive procedure for identification of embryonic sex, although at present the procedure is highly subjective and requires simplification for widespread use.

If soluble H-Y antigen is released by male embryos into the surrounding medium, perhaps an enzyme-immunoassay can be used to measure H-Y antigen after short-term culture. Enhancement of specific binding of H-Y antibody to male embryos, perhaps through the use of monoclonal antibody, or enhancement of fluorescence of male embryos may also improve the accuracy with which male-specific antigen can be detected. Those of us interested in production agriculture often measure the value of a procedure by its potential for practical application to field conditions. Procedures for identification of embryonic sex are not widely available except as experimental procedures. The accuracy with which young of a specified sex can be produced, however, is sufficient enough that additional research is warranted.

REFERENCES

1. Amann, R.P., and G.E. Seidel, Jr. (1982) Prospects for Sexing Mammalian Sperm, Colorado Associated Universities Press, Boulder, Colorado.
2. Betteridge, K.J. (1984) The folklore of sexing. Theriogenology 21: 3-6.
3. Betteridge, J., W.C.D. Hare, and E.L. Singh (1981) Approaches to sex selection in farm animals. In New Technologies in Animal Breeding, B.G. Brackett, G.E. Seidel, Jr., and S.M. Seidel, eds. Academic Press, New York, pp. 109-125.
4. Billingham, R.E., and W.K. Silvers (1960) Studies on tolerance of the Y-chromosome antigen. J. Immunol. 85:14-26.
5. Eichwald, E.J., and C.R. Silmser (1955) Untitled publication. Transplantation Bulletin 2:148-149.
6. Epstein, C.J., S. Smith, and B. Travis (1980) Expression of H-Y antigen on preimplantation mouse embryos. Tissue Antigens 15:63-68.
7. Gardner, R.L., and R.G. Edwards (1968) Control of the sex ratio at full term in the rabbit by transferring sexed blastocysts. Nature 218:346-348.
8. Goldbard, S.B., S.O. Gollnick, and C.M. Warner (1985) Synthesis of H-2 antigens by preimplantation mouse embryos. Biol. Reprod. 33:30-36.
9. Goldberg, E.H., E.A. Boyse, D. Bennett, M. Scheid, and E.A. Carswell (1971) Serological demonstration of H-Y (male) antigen on mouse sperm. Nature 232:478-480.
10. Hare, W.C.D., D. Mitchell, K.J. Betteridge, M.D. Eaglesome, and G.C.B. Randall (1976) Sexing two week old bovine embryos by chromosomal analysis prior to surgical transfer: Preliminary methods and results. Theriogenology 5:243-253.
11. King, W.A. (1984) Sexing embryos by cytological methods. Theriogenology 21:7-17.
12. Krco, C.J., and E.H. Goldberg (1976) Detection of H-Y (male) antigen on 8-cell mouse embryos. Science 193:1134-1135.
13. Lala, P.K., and P. Kim (1983) Are paternal type H-2K antigens expressed on murine preimplantation blastocysts? Proc. 2nd Intl. Congr. Reprod. Immun.
14. Moustafa, L.A., J. Hahn, and R. Roselius (1978) Versuche zur Geschlechtsbestimmung an 6 und 7 Tage alten Rinderembryonen. Berl. Munch. Tierarztl. Wschr. 91:236-238.
15. Nakamura, D., S.S. Wachtel, and K. Kallman (1984) H-Y antigen and the evolution of heterogamety. J. Hered. 75:353-358.
16. Ozil, J.P., Y. Heyman, and J.P. Renard (1982) Production of monozygotic twins by micromanipulation and cervical transfer in the cow. Vet. Rec. 110:126-127.
17. Parkes, A.S. (1971) Mythology of the human sex ratio. In Sex Ratio at Birth--Prospects for Control: A Symposium, C.A. Kiddy and H.D. Hafs, eds. American Society of Animal Science, Pennsylvania State University, University Park, Pennsylvania, pp. 38-42.

18. Picard, L., W.A. King, and K.J. Betteridge (1984) Cytological studies of bovine half-embryos. Theriogenology 21:252 (abstr.).

19. Piedrahita, J.A., and G.B. Anderson (1985) Investigation of sperm cytotoxicity as an indicator of ability of antisera to detect male-specific antigen on preimplantation mouse embryos. J. Reprod. Fertil. (in press).

20. Rieger, D. (1984) The measurement of metabolic activity as an approach to evaluating viability and diagnosing sex in early embryos. Theriogenology 21:138-149.

21. Rottmann, O.J. (1981) Chromosome preparation from single blastomeres after colcemid treatment and removal from rabbit morulae--Unsuitable for sexing in routine embryo transfer. Theriogenology 5:321-326.

22. Silvers, W.K., D.L. Gasser, and E.M. Eicher (1982) H-Y antigen, serologically detectable male antigen and sex determination. Cell 28: 439-440.

23. Singh, E.L., and W.C.D. Hare (1980) The feasibility of sexing bovine morula stage embryos prior to embryo transfer. Theriogenology 14: 421-427.

24. Utsumi, E., E. Satoh, and M. Yuhara (1983) Sexing of mammalian embryos exposed to H-Y antisera. Proc. 2nd Intl. Congr. Reprod. Immun.

25. Utsumi, K., E. Satoh, and M. Yuhara (1984) Sexing of goat and cow embryos by rat H-Y antibody. Proc. 10th Intl. Congr. Anim. Reprod. and A.I., University of Illinois, Urbana, Illinois.

26. Wachtel, S.S. (1983) H-Y Antigen and the Biology of Sex Determination, Grune and Stratton, New York.

27. Wachtel, S.S. (1984) H-Y antigen in the study of sex determination and control of sex ratio. Theriogenology 21:18-28.

28. Wachtel, S.S., G.C. Koo, and E.A. Boyse (1975) Evolutionary conservation of H-Y ('male') antigen. Nature 254:270-272.

29. Wachtel, G.M., S.S. Wachtel, D. Nakamura, C.A. Moreira-Filho, M. Brunner, and G.C. Koo (1984) H-Y antibodies recognize the H-Y transplantation antigen. Transplantation 37:8-13.

30. Warner, C.M., and D.J. Spannaus (1984) Demonstration of H-2 antigens on preimplantation mouse embryos using conventional antisera and monoclonal antibody. J. Exp. Zool. 230:37-52.

31. White, K.L., M.W. Bradbury, G.B. Anderson, and R.H. BonDurant (1984) Immunofluorescent detection of a male-specific factor on preimplantation bovine embryos. Theriogenology 21:275 (abstr.).

32. White, K.L., G.M. Lindner, G.B. Anderson, and R.H. BonDurant (1982) Survival after transfer of "sexed" mouse embryos exposed to H-Y antisera. Theriogenology 18:655-662.

33. White, K.L., G.M. Lindner, G.B. Anderson, and R.H. BonDurant (1983) Cytolytic and fluorescent detection of H-Y antigen on preimplantation mouse embryos. Theriogenology 19:701-705.

34. White, K.L., G.B. Anderson, P.J. Berger, R.H. BonDurant, and R.L. Pashen (1985) Expression of a male-specific factor (H-Y antigen) on preimplantation porcine embryos. Proc. Ann. Mtg. Am. Soc. Anim. Sci., University of Georgia, Athens, Georgia (abst.).

35. Williams, T.J. (1985) Embryo sexing with X-linked enzymes. Proc. 17th Mtg. Aust. Soc. for Reprod. Biol., Adelaide, Australia (abstr.).

36. Williams, T.J., R.P. Elsden, and G.E. Seidel, Jr. (1984) Pregnancy rates with bisected bovine embryos. Theriogenology 22:521-532.

37. Wintenberger-Torres, S., and P.C. Popescu (1980) Transfer of cow blastocysts after sexing. Theriogenology 14:309-318.

CRYOBIOLOGY: PRESERVATION OF MAMMALIAN EMBRYOS

S.P. Leibo

Rio Vista International, Inc.
Route 9, Box 242
San Antonio, Texas   78227

INTRODUCTION

The freezing of mammalian embryos has become a routine procedure. It can be reasonably estimated that tens of thousands of live young have been born from embryos that had been frozen and stored in liquid nitrogen at -196°C. The first reports of the successful freezing of mouse embryos were published in 1972 (56,61). Since that time, embryos of 9 other mammalian species have been "successfully" frozen, "successfully" in the sense that live young have been born. As shown in Tab. 1, these species include laboratory animals, domestic animals, wild animals, and even humans. It is somewhat ironic that the successful freezing of human embryos was accomplished before that of nonhuman primates, experimental species often touted as models for the human. There is every reason to believe that this list will grow with time, as an ever-increasing variety of species' embryos are subjected to this procedure. Parenthetically, only embryos of the swine have been reported as being so sensitive even to chilling to around 0°C that they have not been successfully frozen (35). Despite that, the reason for optimism is that the freezing of mammalian embryos has become not only a routine procedure, but one that rests on a firm mechanistic understanding of some of the factors responsible for cell injury and death caused by freezing and thawing. In many respects, the freezing of mammalian embryos has contributed to that understanding. The purpose of this chapter is to describe (a) how embryos are frozen routinely, (b) the mechanistic basis of successful embryo freezing, (c) some of the more novel methods reported recently, and (d) the degree of success that can be achieved by embryo freezing.

Procedural Aspects of Embryo Freezing

Briefly, the freezing of embryos can be considered as a unidirectional sequence of the following steps (shown diagrammatically in Fig. 1):

(a) Exposure of the embryo to cryoprotective solution. No embryos have been reported to survive freezing in the absence of a protective compound.

(b) Cooling of the sample to a temperature slightly below 0°C, and seeding. Concentrated solutions such as those used in freezing have freezing points a few degrees below 0°C. Seeding is used to

251

Tab. 1. Live offspring produced from frozen-thawed embryos of mammalian species.

Species	Year	First observation
Mouse	1972	Whittingham et al. (56)
Cow	1973	Wilmut and Rowson (62)
Rabbit	1974	Bank and Maurer (1)
Sheep	1974	Willadsen et al. (60)
Rat	1975	Whittingham (55)
Goat	1976	Bilton and Moore (2)
Horse	1982	Yamamoto et al. (65)
Antelope	1983	Kramer et al. (12)
Human	1983	Trounson and Mohr (54)
Baboon	1984	Pope et al. (36)

induce ice formation under controlled conditions just below the solution's freezing point. Although some of the latest, novel techniques for embryo freezing have eliminated the necessity of seeding, most procedures require it.

(c) Controlled cooling of a sample to an intermediate subzero temperature. Initially, successful embryo freezing required slow cooling to temperatures as low as $-80°$ to $-100°C$. It was subsequently shown that embryos can also be successfully frozen by cooling them slowly only to temperatures as high as $-20°$ to $-40°C$.

(d) Rapid cooling to $-196°C$ for long-term storage. Virtually all procedures for long-term storage require the embryos to be cooled to below about $-130°C$. The easiest and most practical way to accomplish this is to store frozen embryos in liquid nitrogen. It has been formally estimated that embryos stored at $-196°C$ will remain alive for hundreds, and probably thousands, of years (27).

(e) Warming and thawing of frozen embryos. In general, the specifics of the warming procedure depend upon the cooling conditions that preceded it. It is for this reason that embryo freezing must be considered a unidirectional process.

(f) Finally, dilution and washing of the embryo to remove the protective compound. For an embryo to function and develop into a living fetus, the multimolar solution used to protect it against freezing damage must be removed. Often, embryos have been successfully frozen and thawed, only to be killed during the dilution process.

Published procedures to freeze most species of embryos are all variations on this general scheme. Although various compounds have been shown to protect embryos against freezing damage, all procedures require some type of cryoprotectant, as these compounds have come to be called. Some procedures stipulate slow chilling to $0°C$; others specify rapid chilling. There seems to be little or no difference between them. With the exceptions to be described below, all procedures use seeding to induce ice formation in the sample. The reason is simply that embryos, like virtually all cells, respond osmotically to their suspending medium. Seeding induces

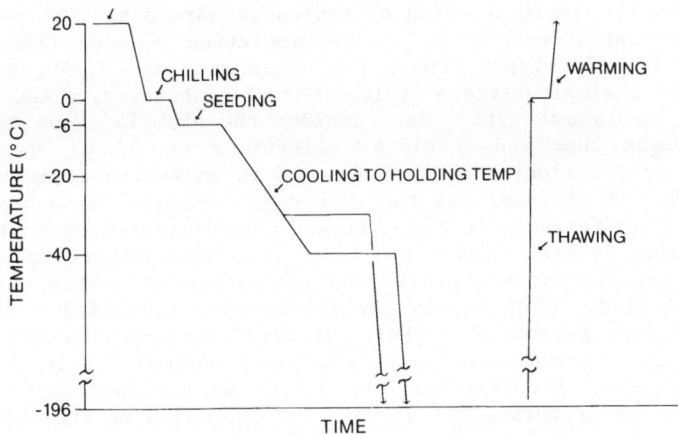

Fig. 1. Diagram of the steps of embryo freezing. This shows the temperatures to which embryos are exposed as a function of time in arbitrary units. See text for details.

a phase change of water to ice, with a concomitant increase in the concentration of the suspending solution. Once ice has formed, the sample is
cooled rather slowly at about 0.3° to 0.5°C/min to lower subzero temperatures. As will be described in more detail below, this slow cooling permits embryos to continue to respond osmotically. The slow cooling is often
continued to an intermediate temperature of about -20° to -40°C, at which
point the sample may be held briefly to permit further time for osmotic response of the embryo. Then, the sample is cooled abruptly in liquid nitrogen at -196°C. At that temperature, embryos can be stored for days, weeks,
or years with no decrease in viability. It has been explicitly demonstrated that the in vitro and in vivo survival of mouse embryos is the same after about 5 years storage at -196°C as after 24 hr storage (24). After
storage, the frozen embryos are thawed and warmed. As mentioned above, the
specifics of the thawing procedure depend upon the cooling that preceded
it. In general, if embryos have been cooled slowly to low subzero temperatures of -60°C or below, then they are warmed relatively slowly at about 5°
to 50°C/min. If they have been cooled slowly only to relatively high subzero temperatures of about -30°C, then they are warmed rapidly at rates of
about 200°C/min or greater. Finally, once the embryos have been thawed and
warmed, they are diluted out of the protective solution. This aspect will
be addressed in more detail below. The ultimate step of most embryo freezing procedures is to transfer the diluted embryos into recipient females,
in whom the embryos will develop into live young animals.

Reasons for Embryo Freezing

The reasons for freezing embryos are varied. Sometimes the purpose is
to preserve a specific mutant for future use. Examples of this are found
in the Mouse Embryo Banks that have been established at the Jackson Laboratory in the United States, or in the Laboratory Animals Centre at Carshal-
ton in the United Kingdom. Sometimes embryos are frozen so that they can
be collected in one place and shipped over long distances in the frozen
state to be transferred into recipients in another place. This is coming
to be a common practice by which different laboratories can exchange mouse
strains, for example. Sometimes embryos are frozen for purely logistical
purposes. An example of this is found in the bovine embryo transfer industry.

Cows, like females of many other mammalian species, can be induced to superovulate by the administration of exogenous hormones. The response of the cow to hormones, such as follicle-stimulating hormone (FSH), is extremely variable. It ranges from a few to as many as 50, 60, or even 75 ova ovulated in a single estrous cycle. Normally, however, a cow will ovulate about 12 to 15 ova (51). But consider the example shown in Fig. 2. These photographs show the complete collections of embryos from 2 cows whose embryos were collected on the same day in an embryo transfer clinic. One cow yielded 38 embryos, and the second yielded 42, for a total of 80 embryos to be transferred. It has been amply demonstrated that the likelihood of pregnancy is maximal when the donor from whom the embryos were collected is in estrous synchrony with the recipients into whom the embryos will be transferred. That is, if the embryos are collected 7 days after the donor exhibited estrus, then the recipients also ought to have exhibited estrus 7 days previously. Since the cow's estrous cycle is 21 days long, the chances of 2 cows exhibiting estrus on the same day is, therefore, 1/21. In other words, out of 21 cows, one will, on the average, exhibit estrus on a given day. For there to be synchronous recipients available on a given day for the 80 embryos shown in Fig. 1, a total herd of 21 x 80 = 1680 cows would be required. Even if one allows 24-hr asynchrony, i.e., a recipient exhibits estrus 24 hr before or after the donor's estrus,

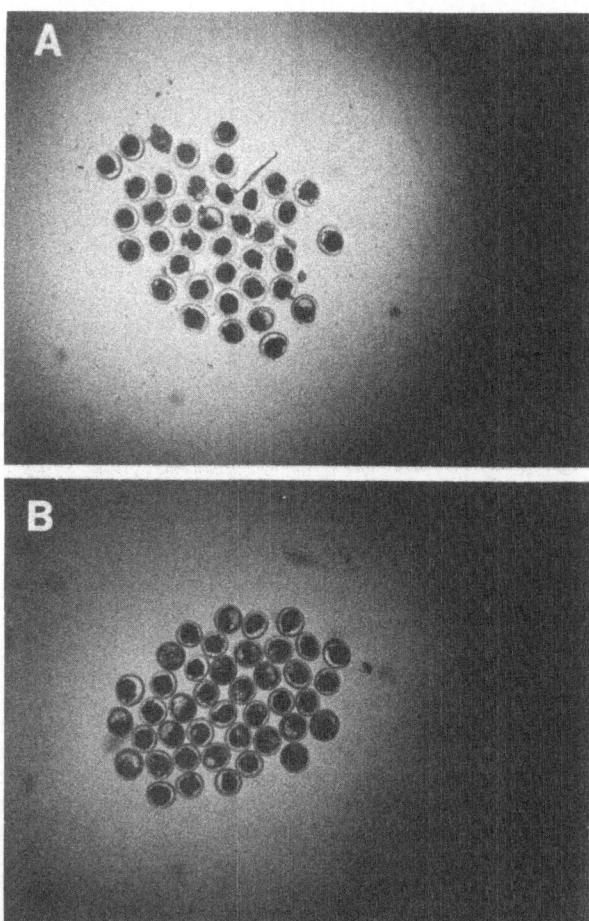

Fig. 2. Photographs of bovine embryos collected on the same day from 2 superovulated donor cows at a commercial embryo transfer clinic.

the chance of 2 cows exhibiting synchronous estrus is 3/21 or 1/7. Therefore, 7 x 80 = 560 cows would still be required as potential recipients.
Only 80 of those 560 cows would actually be used as recipients. The cost
of maintaining the other 480 cows becomes a major expense in the entire
procedure. The alternative is to freeze the embryos. Once frozen, each
embryo can then be thawed as a recipient becomes available.

In summary, embryos are frozen as a means of preservation, to aid in
their shipment, or for logistical reasons, as well as for other reasons.
This brings us to the mechanistic basis of successful embryo freezing.

FUNDAMENTAL ASPECTS OF EMBRYO FREEZING

To survive freezing, embryos, like most other mammalian cells, must be
suspended in an aqueous solution of a protective compound. Compounds that
have been demonstrated to protect embryos include dimethyl sulfoxide (DMSO)
and glycerol (56), various glycols (32,33,44,46), and methanol (40). All
of these compounds are low in molecular weight and permeate the cells of
the embryo.

Aqueous Solutions at Subzero Temperatures

When an aqueous solution is frozen to a temperature below its freezing
point (fixed by the initial concentration of the solute), water is removed in the form of ice. As the solution is cooled to lower and lower
temperatures, more ice forms, resulting in a concomitant increase in the
concentration of the solute. This behavior of solutions can be illustrated graphically as a phase diagram, such as the one for 1.5 M glycerol in

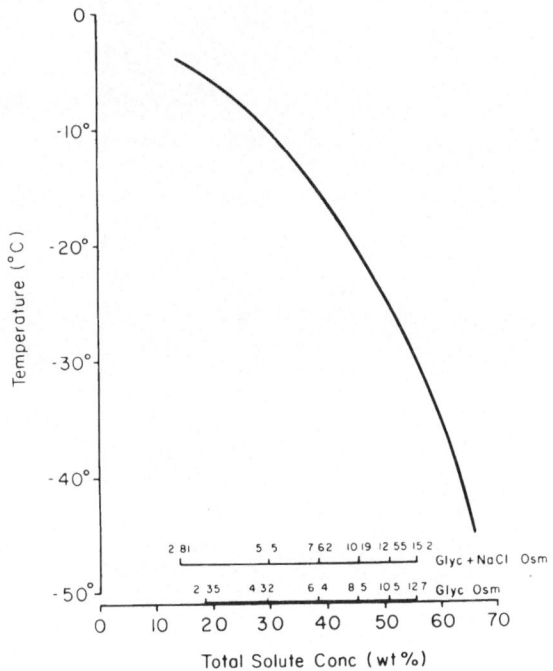

Fig. 3.   Phase diagram of the system glycerol-PBS (phosphate-buffered sa-
           line). The initial glycerol concentration is 1.5 M, and the
           ratio (R value) of glycerol to NaCl is 14.9. The graph was re-
           drawn from data in Rall (37).

phosphate-buffered saline (PBS) shown in Fig. 3. This graph shows that as the temperature of such a solution is lowered, the total solute concentration in the unfrozen portion increases from about 15 weight-percent at about −4°C to almost 70 weight-percent at about −45°C. The corresponding values for the osmolalities of glycerol alone and of glycerol plus NaCl are also shown in the graph. It is obvious that an embryo frozen in a 1.5 M glycerol solution is exposed to exceedingly high solute concentrations at low temperatures. Solutions of the other compounds listed above exhibit analogous behavior during freezing. Such behavior of solutions can, of course, be described mathematically.

Osmotic Behavior of Embryos

It was mentioned above that compounds such as DMSO, glycerol, and ethylene glycol permeate the cells of an embryo, a point to which we shall return below. But permeability is a highly temperature-dependent process. It has been shown that the permeation of glycerol into an ovum, for example, effectively ceases at about 0°C (9). It is reasonable to assume that, at low subzero temperatures, a solute such as glycerol acts effectively like a nonpermeable solute. The response of an embryo to a nonpermeating compound can be studied at temperatures above 0°C by exposing it to solutions of sucrose, since this solute does not permeate cells. Figure 4 illustrates the behavior of a bovine morula transferred from PBS (A) into

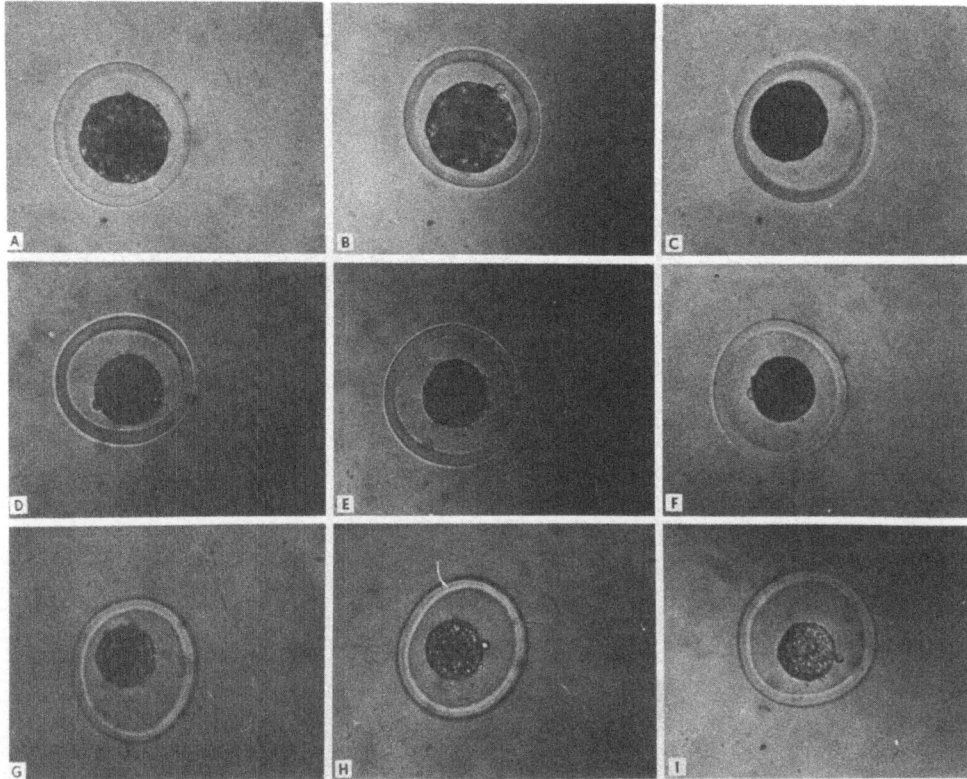

Fig. 4. Photographs of a bovine morula in phosphate-buffered saline (PBS) (A, B for 2 views) equilibrated successively at about 20°C in solutions of sucrose in PBS. The embryo was rinsed and then held for 15 min in sucrose solutions of 0.2, 0.4, 0.6, 0.8, 1.0, 1.25, and 1.5 M, respectively (C-I).

increasingly concentrated sucrose solutions (B to I). The embryo responds osmotically by losing water as the impermeant solute concentration is increased. By measuring the cross-sectional area, the volume of the embryo can be calculated and plotted as a function of the reciprocal of the osmotic pressure of the solution. Similar observations have been published for mouse ova (15,37). A comparison of the results for mouse ova and bovine embryos is shown as a Boyle–Van't Hoff plot in Fig. 5. Several interesting facts emerge from these data. First, mouse ova and bovine embryos behave as "perfect osmometers" in that their volumes are directly proportional to the reciprocal of the osmolality of their suspending solutions. Second, mouse ova and bovine embryos respond similarly. Thirdly, extrapolation of the lines to an infinitely concentrated solution (1/osmolality = 0) leads to the conclusion that about 15 to 20% of the volume of ova and embryos is cell solids. The corollary is that more than 80% of the volume of ova and embryos is cell water. Freezing is concerned with the state and disposition of cell water during cooling and warming.

A Boyle–Van't Hoff plot is an equilibrium measurement of cell volume at a constant solute concentration. However, the movement of water across a cell membrane is a kinetic process, expressed as the water permeability coefficient or hydraulic conductivity of the cell, L_p. It has been previously shown that the L_p of mouse ova has a value of 0.4 $\mu m^3/\mu m^2$-min-atm at 20°C, and that the activation energy of L_p is about 14 Kcal/mole (15). Compared to other cells, these values indicate that ova have a lower hydraulic conductivity than red blood cells, for example, with a higher acti-

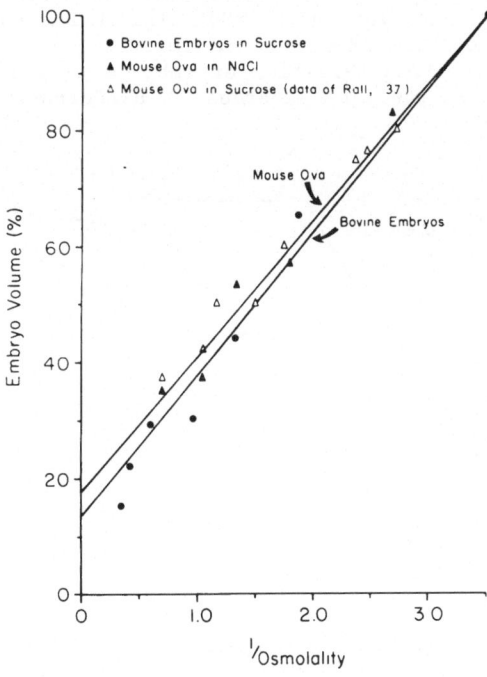

Fig. 5. Boyle–Van't Hoff plot of bovine embryos and mouse ova. Relative volumes of bovine morulae and mouse ova as a function of the reciprocal of the osmolality of the suspending solutions. The volumes for morulae were calculated from photographs such as those shown in Fig. 4. The data for mouse ova are from Rall (37) and from Leibo (15), who describes the experimental details of this analysis.

vation energy, but that they are similar to lymphocytes with respect to their water permeability. In other words, water moves across the membranes of ova quite slowly, and this water movement has a high temperature coefficient. This subject of the water permeability of ova and embryos compared to other types of cells has been discussed previously (14,16). The main point here is simply to recognize that the water permeability of ova and embryos can be described quantitatively.

It was mentioned above that embryos are frozen in solutions of permeating compounds, such as DMSO or glycerol. It is, therefore, important to consider the osmotic response of embryos exposed to such solutions. When a cell is first exposed to a hypertonic solution of a permeating compound, it shrinks by losing water. The reason is that water moves across the membrane faster than the solute. By water loss, the intracellular contents will quickly increase to the osmolality of the extracellular solution. At the same time, however, the solute has begun to enter the cell. As a consequence, water will begin to re-enter the cell, so as to maintain osmotic equilibrium between the intracellular and extracellular solutions. This continues until the concentration of the permeating solute has reached equilibrium. At that time, the cell will have returned to approximately its isotonic volume. The actual volume will be slightly larger than the isotonic volume due to the intracellular presence of the solute. By measuring the response to different solutes as a function of time at constant temperatures, the permeability characteristics of ova and embryos can be determined. A detailed study has been published for mouse ova in glycerol (9). Similar studies have been initiated for bovine embryos (18,49). An example of the responses of bovine embryos to DMSO and glycerol is shown in Fig. 6. The results in that figure demonstrate that bovine embryos equilibrate faster with glycerol than with DMSO. This means that glycerol enters bovine embryos faster than DMSO. It is important to recognize, however, that the permeability characteristics of embryos of different species may well be different. Furthermore, embryos of different stages of the same

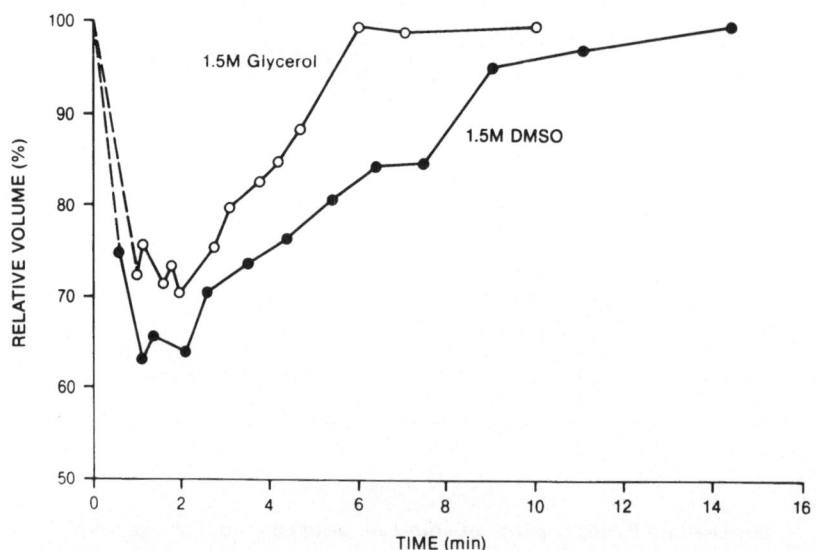

Fig. 6.   The change in volume of bovine blastocysts, relative to their isotonic volume, when suspended in each of the indicated solutions at 24°C. The figure is from Leibo (18), who describes the experimental method used. It is used with permission of the editor.

species have different permeability characteristics (31), and fertilization of an ovum also changes its permeability properties (9). Most importantly, the effect of temperature on permeability may well differ for different solutes, for different developmental stages of the same species, and for the same stage of different species.

These questions of solute permeability are important in the context of embryo freezing for several reasons. It appears as if embryos survive freezing only if some cryoprotective compound has permeated the cells prior to freezing. Although indirect, the evidence for this conclusion is shown in Fig. 7. This graph (redrawn from observations in Ref. 22 and 33) shows the survival of embryos exposed to different solutes for various times prior to being frozen and thawed. It is clear that embryo survival was independent of exposure time to ethylene glycol, was slightly dependent on exposure time to DMSO, and was highly dependent on exposure time to diethylene and triethylene glycol. Unpublished observations of embryo responses to these compounds performed in the manner discussed above for Fig. 6 demonstrate that the embryos are most permeable to ethylene glycol, are somewhat less permeable to DMSO, and are only slowly permeable to the other two glycols. In other words, if embryos are to be successfully frozen, sufficient time must be allowed for the protective compound to permeate the cells.

There is another equally important reason for understanding the permeability characteristics of embryos. In general, if embryos previously equilibrated with a concentrated solution of DMSO or glycerol are rapidly diluted with isotonic PBS, they will swell as water moves into the cells to restore osmotic equilibrium. As mentioned above, embryos have often been successfully frozen and thawed only to be killed by this osmotic shock during dilution. A standard method to avoid this osmotic shock is to dilute

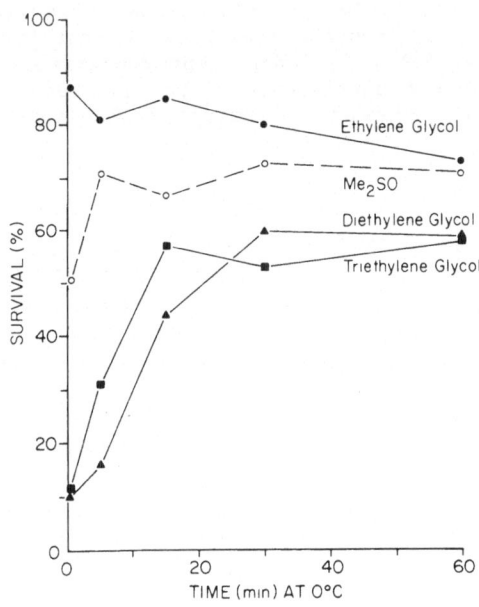

Fig. 7.   Survival of 8-cell mouse embryos frozen to -79°C or below in 1.2 M solutions of each of the indicated solutions. The data are from Miyamoto and Ishibashi (33), except for those data for Me$_2$SO, which are from Leibo et al. (22). The figure is from Leibo (16), and is used with permission of the publisher.

the protective compound gradually by a stepwise dilution. Schneider and Mazur (49) determined the permeability coefficient of bovine embryos to glycerol, and then calculated the osmotic consequences of a 6-step dilution procedure. Their calculations are shown in Fig. 8. The embryo exhibits volume spikes at each step of the dilution as water enters the embryo; as the glycerol diffuses out, the embryo volume returns to approximately its isotonic volume. As those authors pointed out, the embryo will survive as long as the volume increase does not exceed a critical volume of about 2.7 times the isotonic volume.

A number of years ago, an alternative method was suggested to dilute embryos out of protective solutions (20). In this method, an impermeant solute such as sucrose is used to act as an osmotic buffer to prevent swelling of the embryo during dilution. Figure 8 also illustrates the volume consequences of this one-step (versus the 6-step procedure described above) dilution method. When the embryo is transferred from glycerol or DMSO, for example, directly into sucrose, the embryo shrinks by water loss. The reason that it shrinks is that glycerol diffuses out of the embryo in response to the concentration gradient, i.e., little or no extracellular glycerol. As glycerol diffuses out, water also leaves the embryo to maintain osmotic equilibrium between the intracellular and extracellular solutions. The rationale of this one-step dilution method is illustrated diagrammatically in Fig. 9. Once the glycerol has left the embryo, it can be safely diluted with PBS, or can be transferred directly into the uterus of a recipient female.

Another advantage accrues from this one-step dilution method. Most embryo freezing-thawing methods require that the embryos be repeatedly handled with the aid of a stereomicroscope. A few years ago, it was shown that embryos can be frozen in small, plastic straws, and, upon thawing, can be diluted in one step with sucrose in the straw itself. This method, developed independently by 2 groups of investigators, permits bovine embryos to be thawed under field conditions, and to be transferred into recipient females in a manner quite analogous to artificial insemination of cattle (17,18,19,45,47,48). An independent comparison of the two methods has shown that both yield approximately the same results (4). Actual results obtained by this approach to embryo freezing will be described in

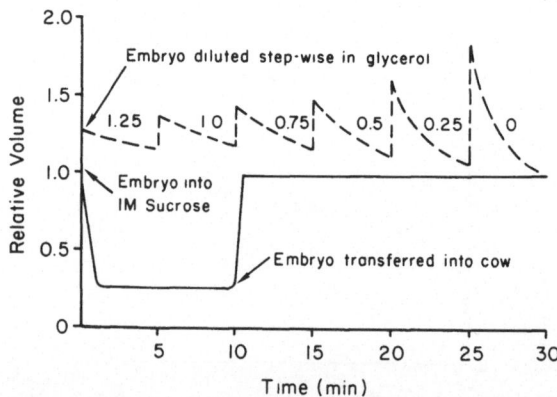

Fig. 8. Calculated volumes of bovine blastocysts previously equilibrated with 1.5 M glycerol in phosphate-buffered saline (PBS), and then diluted by either of the methods shown. The method of calculation and the figure itself are from Schneider and Mazur (49), and are used with permission of the publisher.

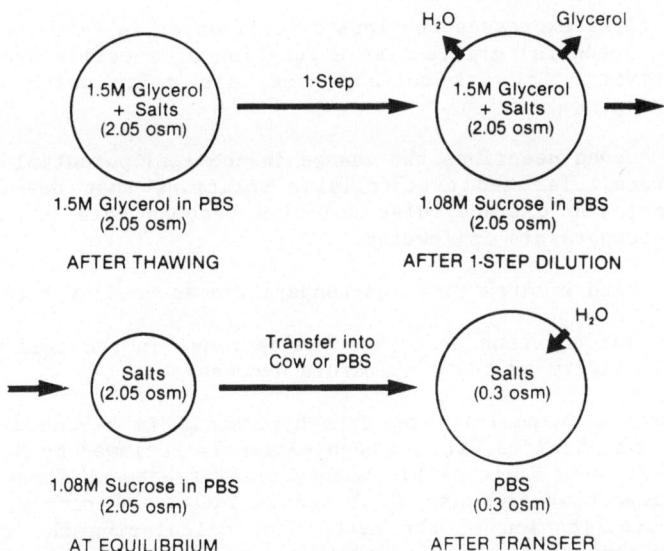

Fig. 9. Rationale of one-step dilution. Diagrammatic representation of the volume changes exhibited by a cell that was frozen and thawed in 1.5 M glycerol in phosphate-buffered saline (PBS), and then diluted in one step into 1.08 M sucrose in PBS. The figure is from Leibo (19), and is used with permission of the publisher.

more detail below. The fact is that a basic understanding of the osmotic behavior of embryos has led to a highly practical method to freeze, thaw, and dilute embryos.

Responses of Embryos to Freezing and Thawing

The first reports of the freezing of mammalian embryos emphasized what appeared to be unique requirements for the preservation of embryos: very low cooling rates to low subzero temperatures followed by low warming rates (22,56,61). In 1977, however, Willadsen (58) reported that survival of sheep and bovine embryos could be obtained when slow cooling was continued only to relatively high subzero temperatures as long as the frozen embryos were warmed rapidly. Continuation of these studies (57,59) quickly demonstrated that this shorter and more practical method of freezing embryos could yield higher survival. Extension of these findings yielded the even more striking observation that high survival of mouse embryos could be obtained by cooling embryos in 2 steps: from about -7° to -20°C, holding them briefly at the latter temperature, and then cooling them rather rapidly to -100°C, or below (10,63). Again, high survival required rapid warming. The response of embryos subjected to these various treatments has been analyzed in detail elsewhere (42). To understand the reasons why embryos can withstand this variety of treatments, and to appreciate the significance of the latest novel procedures of embryo freezing, however, require an examination of the basic aspects of embryo freezing.

As described above, when aqueous solutions are frozen, water is removed in the form of ice, resulting in an increase in the concentration of dissolved solutes. More than twenty years ago, Mazur (26) proposed a formal, mathematical hypothesis to describe the response of a cell to freezing. He derived a series of equations to quantify 4 elements of cell freezing, as follows:

(a) The first expresses the loss of cell water in response to a chemical potential gradient as a function of a cell's hydraulic conductivity, Lp, its surface area, its molar water volume, and temperature.

(b) The second describes the change in chemical potential between the intracellular and extracellular solutions that develops as ice forms; the extracellular solution becomes more concentrated as the temperature is lowered.

(c) The third relates time and temperature as cooling rate.

(d) The last equation describes the decrease in the cell's hydraulic conductivity as the temperature decreases.

A mathematical exposition of this hypothesis is beyond the scope of this chapter, and, besides, it has been elegantly reviewed by Mazur himself (28,29). Others have modified his mathematical treatment somewhat (8,50), although the important elements of it remain intact. Recently, Mazur (30) introduced a modification of the method for calculating the relationship between Lp and temperature. This modification permits the coupling between Lp and temperature to be calculated assuming an Arrhenius relationship rather than an exponential one. Although the calculated results are somewhat different, the overall conclusions are very similar to those presented previously (14,28,29).

The results of such a theoretical calculation for a mouse ovum frozen in 1 M DMSO are shown in Fig. 10. Briefly, theory argues that an ovum cooled infinitely slowly so as to remain in osmotic equilibrium (EQ) will

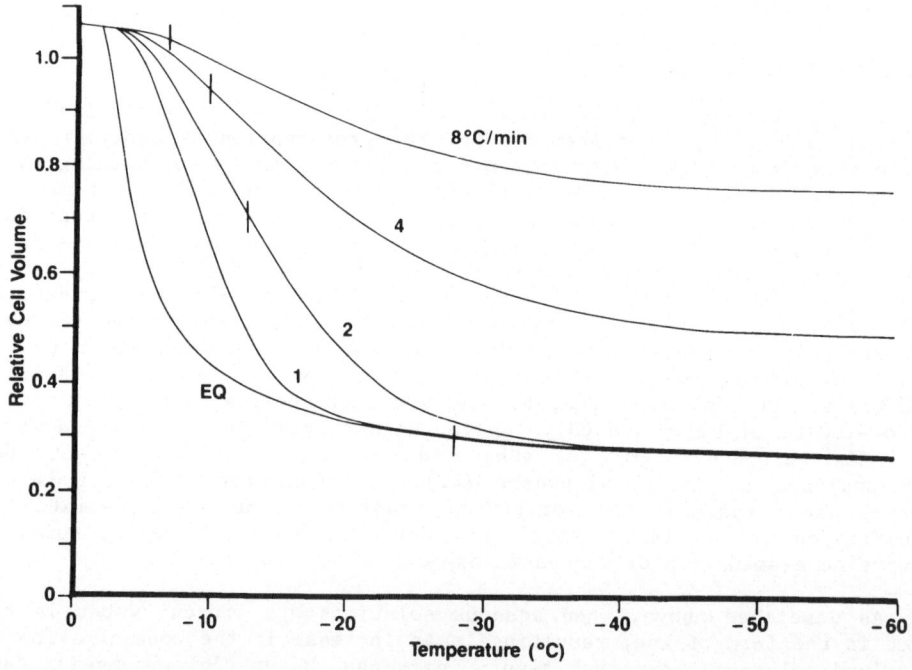

Fig. 10. Calculated change in volume of mouse ova in 1 M dimethyl sulfoxide (DMSO) during freezing at each of the indicated rates. The figure is from Mazur et al. (30), who describe the details of the calculation. It is used with permission of the publisher.

lose almost 60% of its volume by the time it is cooled to -10°C, and it will lose 70% by the time it is cooled to -25°C. An ovum cooled at finite rates of 1°,2°,4° or 8°C/min will have less time to lose cell water, and ova cooled at those rates will contain more water at higher temperatures. In other words, their volumes will be greater at any temperature as the cooling rate is increased.

More than 10 years ago, a cryomicroscope was designed and constructed to test Mazur's hypothesis directly (6,7). Since that time, many other versions of cryomicroscopes have been developed (5,11,43,52). Several years ago, a cryomicroscope was used to study the response of mouse ova during cooling (23). Briefly, it was found that few, if any, ova cooled at about 1°C/min froze intracellularly, whereas an increasing proportion of ova froze intracellularly as the cooling rate was increased. All ova cooled at 5°C/min or faster froze intracellularly. It was also observed that ova cooled at rates that yielded high survival had lost a considerable fraction of their original volume, whereas ova cooled at rates that yielded no survival had lost little of their original volume (14). Since that time, detailed cryomicroscopical studies of mouse embryos suspended in various solutions and cooled and warmed under many different conditions have been published (39,40,41,42). Similar studies have also been made of bovine embryos (13). These studies have contributed greatly to an understanding of the conditions under which ice does or does not form within embryos during cooling and warming. Suffice it to say that these latter studies by Rall and his colleagues have elucidated many basic features of intracellular ice formation and melting. Among other findings, they have demonstrated that embryos may be cooled in certain solutions in such a way that the intracellular contents of embryos do not crystallize, but rather form a glassy solid (i.e., the embryo vitrifies). We will return to this subject below.

Recently, further cryomicroscopical observations of bovine embryos have been made to define in more detail the conditions under which intracellular ice forms and melts (21). Within a group of embryos cooled at a given rate, intracellular ice nucleated, as well as melted, at different temperatures. However, the mean nucleation temperatures of embryos cooled at different rates increased dramatically with increasing rates. Moreover, the corresponding mean melting temperatures also increased with increasing rates. These results, summarized in Fig. 11, show that embryos cooled slowly enough so as to lose water (see Fig. 10) may still freeze intracellularly at very low temperatures. When ice has formed at low temperatures, e.g., about -45°C, it also melts at low temperatures, e.g., about -27°C. Embryos cooled more rapidly do not dehydrate as much and tend to freeze intracellularly at higher temperatures. One final item deserves note. The difference between the nucleation and melting temperatures of intracellular ice is a measure of the extent to which a given embryo has supercooled when intracellular ice forms. This fact has important implications with respect to theories of cell freezing. It indicates that intracellular ice formation is not invariant with cooling rate, and does not depend upon a fixed amount of cell water. These cryomicroscopical observations yielded another interesting finding. During cooling, some embryos shrank as spheres and were not distorted or obscured by ice crystals. This permitted rather precise measurements to be made of these embryos during the actual cooling. One example of such measurements of a bovine morula cooled at 1°C/min in 1.5 M glycerol is shown in Fig. 12. In that figure, the observed loss of embryo volume is compared to the calculated loss of volume for a mouse ovum in 1 M DMSO. This latter curve is the same as that shown in Fig. 10, labeled 1°C/min. As mentioned above, those calculations are from Mazur et al. (30). Given that a mouse ovum is not a bovine morula, and that 1 M DMSO is not 1.5 M glycerol, nevertheless the agreement between theory and observation of an embryo in a cryoprotective solution seems extremely good.

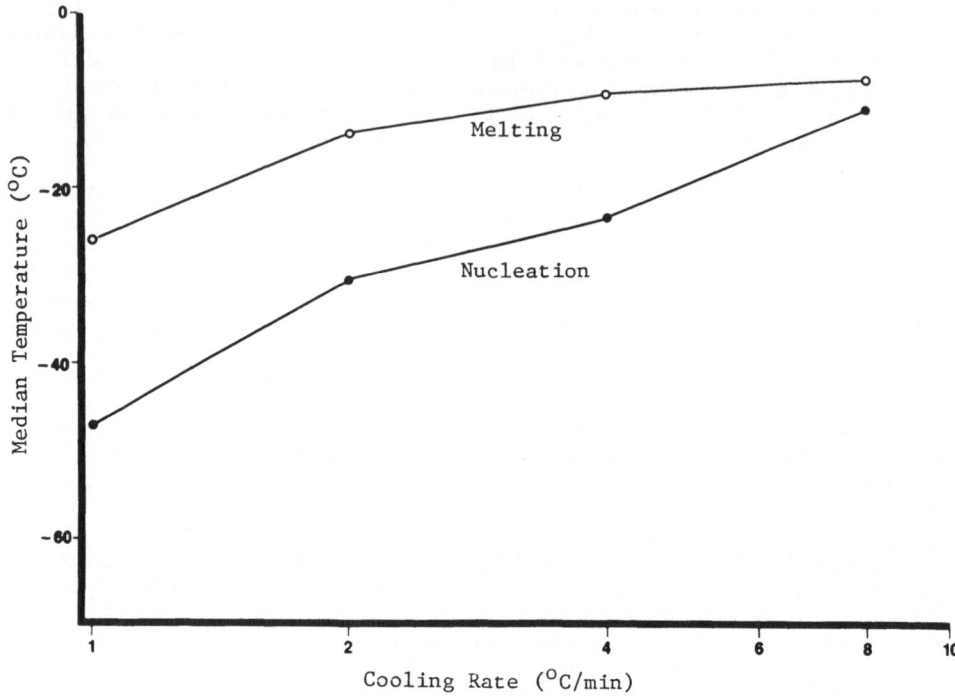

Fig. 11. The median temperatures at which intracellular ice was observed
to form (nucleation) and to melt in groups of 10 bovine embryos
in 1.5 M glycerol. Each embryo was observed during cooling at
one of 4 rates, and then during warming at 10°C/min. The data
are those of Leibo et al. (21).

In summary, the responses of embryos during freezing and thawing have
been studied in great detail. These studies have demonstrated that embryos
can survive freezing and thawing under a wide variety of conditions, and
that their responses can be interpreted and understood as "typical" of
cells in general. The study of embryo freezing is actually contributing to
a fuller understanding of cryobiology. Furthermore, because of the detail
with which embryo freezing has been studied, novel approaches to the successful freezing of embryos have recently been introduced.

NOVEL APPROACHES TO EMBRYO FREEZING

The literature on the freezing of mammalian embryos has grown rapidly
since the first publications in 1972. A bibliography compiled by Whittingham and Wood and published in May 1984 in the Bibliography of Reproduction
listed 220 full articles and 73 abstracts that had appeared between 1977
and 1983. The past year has seen that list grow even more. Among the latest publications are 5 that deserve special attention. The reasons are
several.

By application of the growing understanding that has emerged of the
mechanisms responsible for freezing damage of embryos, these authors have
developed efficient, rapid, and clever methods to freeze mammalian embryos.
It is only reasonable to expect that these novel methods will find wider
and wider application in the preservation of mammalian embryos. The common features of these procedures are that they permit embryos to be frozen

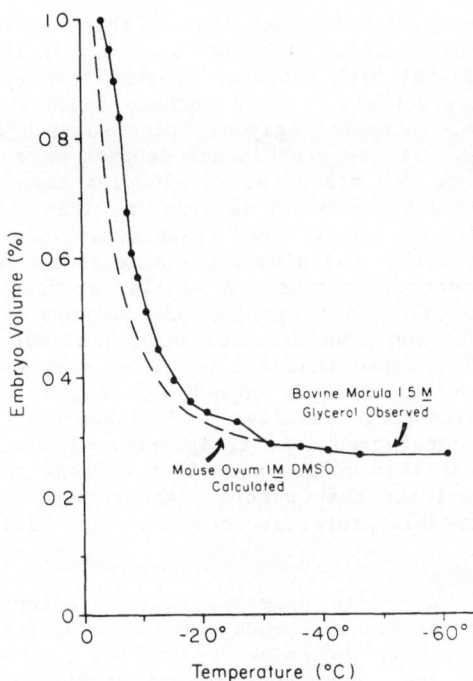

Fig. 12. Comparison of the calculated change in the volume of a mouse ovum
in 1 M dimethyl sulfoxide (DMSO) (redrawn from Fig. 10) with the
observed change of a bovine morula in 1.5 M glycerol being cooled
at 1°C/min. The latter data are from Leibo et al. (21).

rapidly with a minimum of equipment and to be thawed in an equally rapid
and efficient way.

Four of these novel methods are quite similar in approach, although
they differ in detail. The briefest report of this group (53) describes
the preservation of mouse embryos in plastic straws. It was found that
mouse embryos suspended in solutions of glycerol plus sucrose could be rapidly mixed with a sucrose solution within the straw, and then immediately
plunged in liquid nitrogen. When embryos in glycerol concentrations of 3
or 4 molar plus 0.5 M sucrose were mixed with 1.5 or 2 M sucrose just before freezing, high survivals were obtained. This method clearly demonstrates that mouse embryos can be frozen without seeding and without slow
cooling. As should be obvious from the previous discussion of the osmotic
behavior of embryos, this method rests on the principle that embryos can be
"preloaded" with glycerol and then dehydrated with sucrose prior to freezing. This obviates the necessity to cool embryos slowly to cause dehydration.

A related approach was used by Massip and van der Zwalmen (25) to
freeze bovine embryos. In this report, embryos were suspended in a mixture
of about 1.4 M glycerol plus 0.25 M sucrose. This method still required
slow cooling of the embryos at about 0.3°C/min to -25° or -35°C before
plunging the embryos into liquid nitrogen. The frozen embryos were thawed
rapidly and transferred into recipients without further dilution. Five
pregnancies resulted from embryos frozen in the glycerol-sucrose mixture.
Although, as the authors themselves acknowledge, the number of embryos frozen was very small, clearly the principle of the method was demonstrated.

Two much more comprehensive studies of this approach have recently been published by the same group of investigators. In the first, Renard et al. (46) demonstrated that high survival of rabbit embryos can be obtained by a 2-step freezing procedure. These authors found that pretreatment of embryos in propanediol (propylene glycol) plus sucrose caused the embryos to shrink osmotically. If the preshrunken embryos were cooled rather rapidly to -30°C for 30 to 240 min prior to plunging them into liquid nitrogen, about 80% of the embryos would develop in vitro. Furthermore, 21 of 81 embryos (26%) frozen in this way and transferred into recipients yielded live young. This percentage was almost the same as the 32% live young produced from unfrozen control embryos. A similar approach was then used to freeze bovine embryos (3). In this case, the authors found that high survival of bovine embryos could be obtained using a mixture of 1.5 M glycerol plus 1 M sucrose. They found that 44% of 52 embryos would hatch in vitro after having been cooled directly to -30°C, held for 30 min, and then plunged into liquid nitrogen. Finally, 7 of 21 embryos frozen in this way, thawed rapidly, and transferred into recipients, yielded pregnancies. The practical advantages of this method are clear. These several papers taken together obviously indicate that partial dehydration of embryos, together with the use of a permeable protective compound, can yield high embryo survival.

The final novel procedure deserves special attention. This method "relies on the ability of highly concentrated aqueous solutions of cryoprotective agents to supercool," becoming "so viscous that they solidify without the formation of ice, a process termed vitrification" (38). These authors devised a complex solution consisting of DMSO, acetamide, propylene glycol, and polyethylene glycol. The unusual property of this solution is that it vitrifies, i.e., it forms a glassy solid at low temperatures. Although this solution is toxic to embryos, the authors overcame this problem by exposing the embryos to it at about 4°C. After equilibration, the embryos were placed into plastic straws and the straws were plunged directly into liquid nitrogen at -196°C. Thawing consisted of simply transferring the vitrified straws from liquid nitrogen into a 0°C water bath. The authors also examined the effects of other cooling and warming procedures. In brief, 87% of 531 mouse embryos vitrified under optimum conditions developed in vitro into blastocysts. An additional 83% of 636 embryos vitrified by variations of the general method also survived and developed in culture. Very recently, Rall, Wood, Kirby, and Whittingham (unpubl. observ.) have produced a total of 98 normal fetuses and live young that developed from vitrified embryos. When extended to other species of embryos, as it surely will be, this vitrification procedure will have important practical, as well as fundamental, implications for embryo preservation.

DEGREES OF SUCCESS WITH EMBRYO FREEZING

As mentioned previously, freezing of mammalian embryos has become a routine procedure. Often, the purpose of the freezing is to study fundamental aspects of cryobiology. In other situations, the purpose is to demonstrate that a given species or a given embryonic stage can be successfully frozen. Alternatively, the purpose may be to examine an aspect of early embryology or reproductive biology. But underlying all of these studies there is the ultimate goal of producing living young animals by the transfer of frozen-thawed embryos. The most rigorous test of any embryo freezing procedure, then, must be the number of live young produced. In the case of laboratory species, this number can be rather easily related to the number of embryos originally frozen. In the case of the large domestic species, e.g., the bovine, the degree of success is usually reported as the percentage pregnancy. Sometimes, this percentage is calculated on the basis of the number of embryos actually transferred. This latter figure may often be somewhat less than the former, since not all thawed embryos are necessarily transferred. It is common practice, for example, to cull obviously damaged or degenerate embryos. This does color, to a certain extent, the measure of success of an embryo freezing procedure.

Rather than attempting to give an exhaustive review of all degrees of success reported, I shall here only survey some of the results that have been obtained from the transfer of frozen-thawed embryos. Furthermore, this survey will cite those examples in which large numbers of embryos have been transferred. "Success" in this survey is intended to mean that normal, living fetuses or live young were obtained, or that well-established pregnancies were produced (for example, in the cow, pregnancies of 90 days or more duration were established).

In the first report on embryo freezing in which transfers were performed (56), a total of 267 fetuses and live young were obtained from the transfer of 927 frozen-thawed mouse embryos. This percentage, 28.8%, has often been exceeded in later studies. For example, in a study of the long-term effects of background radiation on frozen mouse embryos, Lyon et al. (24) reported that the transfer of 521 embryos stored for times between 6 to 65 months at -196°C yielded a total of 258 normal fetuses and live mouse pups. In other words, about 50% of the frozen-thawed embryos developed normally in vivo. When embryos, especially of the mouse, are cultured in vitro after thawing, those that do not develop normally can be culled. This can improve the overall percentage of in vivo survival, since fewer embryos are transferred than were originally frozen and thawed. For example, Whittingham et al. (57) were able to obtain 64% in vivo development of 308 embryos that had been cultured for 24 hours after thawing. It has recently been reported that mouse embryos transferred directly after thawing yielded the same overall survival rate as those cultured for 24 or 48 hours after thawing (44). That is, Renard and Babinet obtained 124 live fetuses from a total of 188 frozen-thawed embryos (66%). These results clearly demonstrate that the success "rate" that can be achieved to produce live mice from frozen-thawed embryos can be extremely high.

The other species for which a large number of observations have been made with frozen-thawed embryos is the bovine. Some of these observations, although for relatively small numbers of embryos, have reported pregnancy rates almost as high as those obtained with mouse embryos. For example, the single-step dilution method of Renard et al. (47) has been reported to yield high survival. In one study (45), 16 pregnancies were produced from the direct transfer of 33 frozen-thawed embryos (48% pregnancy). An independent study (4) of the same procedure reported a 45% pregnancy percentage from the transfer of 47 thawed embryos.

Recently, 2 articles have been published that report on pregnancy results resulting from the transfer of large numbers of frozen-thawed bovine embryos. In one (34), the embryos were frozen by the method first described by Willadsen (58), and were surgically transferred into the recipients. Pettit achieved extremely good results: 54% of 550 thawed embryos yielded pregnancy. It would appear from that report, however, that only the best quality embryos were frozen. The second report (64) described the results for the nonsurgical transfer of a larger number of frozen-thawed embryos. Wright achieved an overall percentage pregnancy of 32% from the transfer of 832 embryos. On the other hand, he froze embryos of various stages and of varying quality. Nevertheless, he produced a total of 267 bovine pregnancies.

The effect of these various factors on the ultimate number of pregnancies can be illustrated by some recent results from my laboratory. Table 2 summarizes the pregnancy rates achieved from the direct nonsurgical trans-

Tab. 2. Pregnancies from frozen-thawed, one-step-diluted bovine embryos.

A.

Developmental stage	No. pregnant/No. transferred	% Pregnant
Late morula	100/222	45.1
Very early blastocyst	68/172	39.5
Expanded blastocyst	34/ 82	41.5
Total	202/476	42.4

B. Summary

Number of donors whose embryos were frozen	73
Mean number of embryos frozen/donor	6.5
Mean number of pregnancies/donor	2.8

fer of embryos of various stages of development. The embryos had been frozen, thawed, and diluted by the one-step sucrose dilution method described above (19). All embryos that were thawed were transferred without examination. The results in that table show that a total of 202 pregnancies were produced from the transfer of 476 embryos (42%). It should also be noted that an average of 6.5 embryos were frozen per donor, and that each donor's collection yielded an average of 2.8 pregnancies.

One final graph illustrates the measure of success that can be achieved with frozen-thawed bovine embryos diluted by the one-step method. The results in Fig. 13 compare the pregnancy results produced by the non-surgical transfer of fresh and frozen embryos as a function of their grade or quality. Embryo grade is a rather subjective judgement of the morphological quality of an embryo. Despite that, it can be seen that the likelihood of pregnancy exhibits a good correlation with the grades assigned to a large number of bovine embryos. The grades of the frozen embryos were assigned prior to their being frozen. This comparison demonstrates that, for the better quality embryos (grades 1 and 2), freezing can yield a pregnancy rate that is more than 80% of that produced by the transfer of fresh embryos.

CONCLUSIONS

The preservation of mammalian embryos by freezing has become a reliable, standard procedure. Thousands of live young animals of 10 species have been produced from frozen-thawed embryos. Thousands more embryos remain preserved in liquid nitrogen, awaiting thawing and transfer into recipient females years, decades, or perhaps centuries from now.

Because of the inherent practical value of embryo freezing, as well as its intellectually provocative nature, the cryobiology of mammalian embryos has been studied intensely. Therefore, there is a firm mechanistic understanding of the physiological and embryological consequences caused by freezing.

Given the advances in this discipline during the last 13 years, it seems reasonable to predict that embryo freezing will play a growing role in research, agriculture, and human and veterinary medicine.

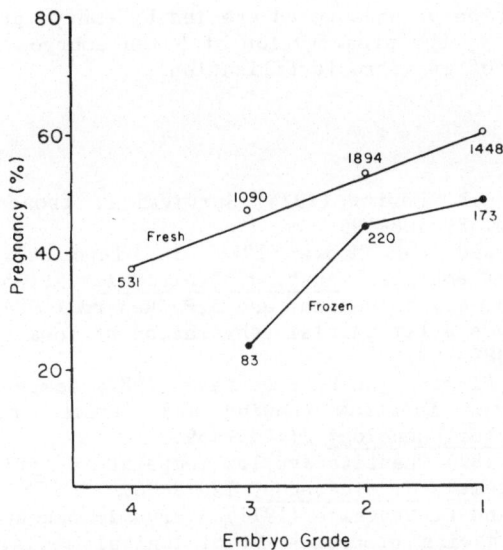

Fig. 13. Bovine pregnancies from nonsurgically transferred embryos. The pregnancy percentages obtained during a 12-month period at a commercial embryo transfer clinic. All embryos were transferred nonsurgically by standard methods. The frozen embryos had been frozen, thawed, and diluted by the one-step method described in Leibo (19).

SUMMARY

    The preservation of mammalian embryos has become a routine procedure. Thousands of live offspring have been produced from frozen-thawed embryos transferred into recipient foster mothers. Species whose embryos have been successfully preserved include mouse, rat, rabbit, sheep, goat, cattle, horse, antelope, baboon, and human. During the past few years, novel procedures have been introduced that permit embryos to be frozen and thawed rapidly, and to be transferred into recipients under field conditions almost immediately upon thawing. Thus, the transfer of frozen-thawed embryos of domestic animals is becoming almost as efficient as is artificial insemination using frozen-thawed semen.

    Because of both the inherent fundamental interest and the practical applications of embryo freezing, a substantial understanding of the mechanisms responsible for freezing damage of embryos has been achieved. To survive freezing, embryos must be exposed to protective compounds; to function after thawing, embryos must be washed free of these compounds. Based on fundamental physiology, efficient methods to accomplish such washing have been developed. Furthermore, to survive freezing, embryos must be cooled under conditions in which intracellular ice does not form. This can be accomplished either by pretreating the embryo or by cooling it in such a way as to cause it to dehydrate during freezing. Maximum survival of embryos appears to be achieved when intracellular water does not crystallize during cooling or during warming.

    As a result of the growing efficiency of embryo preservation, this method is being applied to a variety of practical situations. For example, large banks of frozen embryos of laboratory animals are being established to preserve valuable research resources. The freezing of cattle embryos is being used with increasing frequency as an adjunct to commercial embryo transfer. Preservation of endangered species by embryo preservation is beginning. And finally, the preservation of human embryos is finding application in the field of in vitro fertilization.

REFERENCES

1. Bank, H., and R.R. Maurer (1974) Survival of frozen rabbit embryos. Exptl. Cell Res. 89:188-196.
2. Bilton, R.J., and N.W. Moore (1976) In vitro culture, storage and transfer of goat embryos. Aust. J. Biol. Sci. 29:125-129.
3. Bui-Xuan-Nguyen, N., Y. Heyman, and J.P. Renard (1984) Direct freezing of cattle embryos after partial dehydration at room temperature. Theriogenology 22:389-399.
4. Chupin, D., B. Florin, and R. Procureur (1984) Comparison of two methods for one-step in-straw thawing and direct transfer of cattle blastocysts. Theriogenology 21:455-459.
5. Diller, K.R. (1982) Quantitative low temperature optical microscopy of biological systems. J. Microscopy 126:9-28.
6. Diller, K.R., and E. Cravalho (1971) A cryomicroscope for the study of freezing and thawing processes in biological cells. Cryobiology 7: 191-199.
7. Diller, K.R., E.G. Cravalho, and C.E. Huggins (1972) Intracellular freezing in biomaterials. Cryobiology 9:429-440.
8. Fahy, G.M. (1981) Simplified calculation of cell water content during freezing and thawing in nonideal solutions of cryoprotective agents and its possible application to the study of "solution effects" injury. Cryobiology 18:473-482.
9. Jackowski, S.C., S.P. Leibo, and P. Mazur (1980) Glycerol permeabilities of fertilized and unfertilized mouse ova. J. Exp. Zool. 212:329-341.
10. Kasai, M., K. Niwa, and A. Iritani (1980) Survival of mouse embryos frozen and thawed rapidly. J. Reprod. Fert. 58:51-56.
11. Körber, C., M-W. Scheiwe, and K. Wollhöver (1984) A cryomicroscope for the analysis of solute polarization during freezing. Cryobiology 21: 68-80.
12. Kramer, L., B.L. Dresser, C.E. Pope, R.D. Dalhausen, and R.D. Baker (1983) The nonsurgical transfer of frozen-thawed eland embryos. Proc. Am. Assoc. Zoo Veterinarians pp. 104-105.
13. Lehn-Jensen, H., and W.F. Rall (1983) Cryomicroscopic observations of cattle embryos during freezing and thawing. Theriogenology 19:263-277.
14. Leibo, S.P. (1977) Fundamental cryobiology of mouse ova and embryos. In The Freezing of Mammalian Embryos, Ciba Foundation Symposium 52 (new series), K. Elliott and J. Whelan, eds. Excerpta Medica, Amsterdam, pp. 69-92.
15. Leibo, S.P. (1980) Water permeability and its activation energy of fertilized and unfertilized mouse ova. J. Membrane Biol. 53:179-188.
16. Leibo, S.P. (1981) Physiological basis of the freezing of mammalian embryos. In Immunologic Defects in Laboratory Animals, Vol.2, M.E. Gershwin and B. Merchant, eds. Plenum Publishing Corporation, New York, pp. 353-379.
17. Leibo, S.P. (1982) A one step method for direct nonsurgical transfer of frozen-thawed bovine embryos. In International Congress: Embryo Transfer in Mammals, Annecy, France, p. 97.
18. Leibo, S.P. (1983) A one-step in situ dilution method for frozen-thawed bovine embryos. Cryo Letters 4:387-400.
19. Leibo, S.P. (1984) A one-step method for direct nonsurgical transfer of frozen-thawed bovine embryos. Theriogenology 21:767-790.
20. Leibo, S.P., and P. Mazur (1978) Methods for the preservation of mammalian embryos by freezing. In Methods in Mammalian Reproduction,

J.C. Daniel, Jr., ed. Academic Press, New York, pp. 179-201.

21. Leibo, S.P., M.F. Dowgert, and P. Steponkus (1984) Observation of intracellular ice formation and melting in bovine embryos cooled at various rates. Cryobiology 21:711.

22. Leibo, S.P., P. Mazur, and S.C. Jackowski (1974) Factors affecting survival of mouse embryos during freezing and thawing. Exptl. Cell Res. 89:79-88.

23. Leibo, S.P., J.J. McGrath, and E.G. Cravalho (1978) Microscopic observation of intracellular ice formation in unfertilized mouse ova as a function of cooling rate. Cryobiology 15:257-271.

24. Lyon, M.F., P.H. Glenister, and D.G. Whittingham (1981) Long-term viability of embryos stored under irradiation. In Frozen Storage of Laboratory Animals, G.H. Zeilmaker, ed. Gustav Fischer, Stuttgart, pp. 139-148.

25. Massip, A., and P. van der Zwalmen (1984) Direct transfer of frozen cow embryos in glycerol-sucrose. Vet. Rec. 115:327-328.

26. Mazur, P. (1963) Kinetics of water loss from cells at subzero temperatures and the likelihood of intracellular freezing. J. Gen. Physiol. 47:347-369.

27. Mazur, P. (1976) Freezing and low temperature storage of living cells. In Basic Aspects of Freeze Preservation of Mouse Strains, O. Mühlbock, ed. Gustav Fischer Verlag, Stuttgart, pp. 1-12.

28. Mazur, P. (1977) The role of intracellular freezing in the death of cells cooled at supraoptimal rates. Cryobiology 14:251-272.

29. Mazur, P. (1977) Slow freezing injury in mammalian cells. In The Freezing of Mammalian Embryos, Ciba Foundation Symposium 52 (new series), K. Elliott and J. Whelan, eds. Excerpta Medica, Amsterdam, pp. 19-48.

30. Mazur, P., W.F. Rall, and S.P. Leibo (1984) Kinetics of water loss and the likelihood of intracellular freezing in mouse ova. Cell Biophys. 6:197-213.

31. Mazur, P., N. Rigopoulos, S.C. Jackowski, and S.P. Leibo (1976) Preliminary estimates of the permeability of mouse ova and early embryos to glycerol. Biophys. J. 16:232a.

32. Miyamoto, H., and T. Ishibashi (1977) Survival of frozen-thawed mouse and rat embryos in the presence of ethylene glycol. J. Reprod. Fert. 50:373-375.

33. Miyamoto, H., and T. Ishibashi (1978) Protective action of glycols against freezing damage of mouse and rat embryos. J. Reprod. Fert. 54:427-432.

34. Pettit, W.H. (1985) Commercial freezing of bovine embryos in glass ampules. Theriogenology 23:13-16.

35. Polge, C., I. Wilmut, and L.E.A. Rowson (1974) The low temperature preservation of cow, sheep, and pig embryos. Cryobiology 11:560.

36. Pope, C.E., V.Z. Pope, and L.R. Beck (1984) Live birth following cryopreservation and transfer of a baboon embryo. Fertil. Steril. 42:143-145.

37. Rall, W.F. (1979) Physical-chemical aspects of cryoprotection of human erythrocytes and mouse embryos. Ph.D. Dissertation, University of Tennessee, Knoxville, Tennessee.

38. Rall, W.F., and G.M. Fahy (1985) Ice-free cryopreservation of mouse embryos at -196°C by vitrification. Nature 313:573-575.

39. Rall, W.F., and C. Polge (1984) Effect of warming rate on mouse embryos frozen and thawed in glycerol. J. Reprod. Fert. 70:285-292.

40. Rall, W.F., M. Czlonkowska, S.C. Barton, and C. Polge (1984) Cryoprotection of day 4 mouse embryos by methanol. J. Reprod. Fert. 70:293-300.

41. Rall, W.F., D.S. Reid, and J. Farrant (1980) Innocuous biological freezing during warming. Nature 286:511-514.

42. Rall, W.F., D.S. Reid, and C. Polge (1984) Analysis of slow-warming

injury of mouse embryos by cryomicroscopical and physio-chemical methods. Cryobiology 21:106-121.

43. Reid, D.S. (1978) A programmed controlled temperature microscope stage. J. Microscopy 114:241-248.

44. Renard, J-P., and C. Babinet (1984) High survival of mouse embryos after rapid freezing and thawing inside plastic straws with 1-2 propanediol as cryoprotectant. J. Exp. Zool. 230:443-448.

45. Renard, J-P., and Y. Heyman (1983) Effect du mode d'addition et de dilution du cryoprotecteur sur la viabilite des blastocystes de vache apres de congelation. In 16th Intl. Congr. Refrig. Comm. C1:69-75.

46. Renard, J-P., N. Bui-Xuan-Nguyen, and V. Garnier (1984) Two-step freezing of two-cell rabbit embryos after partial dehydration at room temperature. J. Reprod. Fert. 71:573-580.

47. Renard, J-P., Y. Heyman, and J-P. Ozil (1982) Congelation de l'embryon bovin: Une nouvelle methode de décongélation pour le transfert cervical des embryons conditionnés une seule fois en paillettes. Ann. Med. Vet. 126:23-32.

48. Renard, J-P., Y. Heyman, P. Leymonie, and J-C. Plat (1983) Sucrose dilution: A technique for field transfer of bovine embryos frozen in the straw. Theriogenology 19:145.

49. Schneider, U., and P. Mazur (1984) Osmotic consequences of cryoprotectant permeability and its relation to the survival of frozen-thawed embryos. Theriogenology 21:68-79.

50. Schweiwe, M., and C. Körber (1983) Basic investigations on the freezing of human lymphocytes. Cryobiology 20:257-273.

51. Seidel, G.E. (1981) Superovulation and embryo transfer in cattle. Science 211:351-358.

52. Steponkus, P.L., M.F. Dowgert, J.R. Ferguson, and R.L. Levin (1984) Cryomicroscopy of isolated plant protoplasts. Cryobiology 21:209-233.

53. Takeda, T., R.P. Elsden, and G.E. Seidel, Jr. (1984) Cryopreservation of mouse embryos by direct plunging into liquid nitrogen. Theriogenology 21:266.

54. Trounson, A.O., and L. Mohr (1983) Human pregnancy following cryopreservation thawing and transfer of an eight-cell embryo. Nature (London) 305:707-709.

55. Whittingham, D.G. (1975) Survival of rat embryos after freezing and thawing. J. Reprod. Fert. 43:575-578.

56. Whittingham, D.G., S.P. Leibo, and P. Mazur (1972) Survival of mouse embryos frozen to -196°C and -269°C. Science 178:411-414.

57. Whittingham, D.G., M. Wood, J. Farrant, H. Lee, and J.A. Halsey (1979) Survival of frozen mouse embryos after rapid thawing from -196°C. J. Reprod. Fert. 56:11-21.

58. Willadsen, S.M. (1977) Factors affecting the survival of sheep and cattle embryos during deep-freezing and thawing. In The Freezing of Mammalian Embryos, Ciba Foundation Symposium 52 (new series), K. Elliott and J. Whelan, eds. Excerpta Medica, Amsterdam, pp. 175-194.

59. Willadsen, S.M., C. Polge, and L.E.A. Rowson (1978) Viability of deep-frozen cow embryos. J. Reprod. Fert. 52:391-393.

60. Willadsen, S.M., C. Polge, L.E.A. Rowson, and R.M. Moor (1974) Preservation of sheep embryos in liquid nitrogen. Cryobiology 11:560.

61. Wilmut, I. (1972) The effect of cooling rate, warming rate, cryoprotective agent, and stage of development on survival of mouse embryos during freezing and thawing. Life Sci. 11:1071-1079.

62. Wilmut, I., and L.E.A. Rowson (1973) Experiments on the low-temperature preservation of cow embryos. Vet. Rec. 92:686-690.

63. Wood, M.J., and J. Farrant (1980) Preservation of mouse embryos by two-step freezing. Cryobiology 17:178-180.

64. Wright, J. (1985) Commercial freezing of bovine embryos in straws. Theriogenology 23:17-29.

65. Yamamoto, Y., N. Oguri, Y. Tsutsumi, and Y. Hachinohe (1982) Experiments in the freezing and storage of equine embryos. J. Reprod. Fert. Suppl. 32, pp. 399-403.

LEGAL AND REGULATORY ASPECTS OF GENETICALLY ENGINEERED ANIMALS

Daniel D. Jones*

Food Safety and Inspection Service
United States Department of Agriculture
300 12th Street, S.W.
Washington, D.C. 20250

INTRODUCTION

The commercialization of genetically engineered animals, particularly food animals, may pose a number of legal and regulatory questions that have not been encountered before. In addressing these questions, it is helpful to distinguish between the process and the products of animal genetic engineering. From a legal standpoint, the process of animal genetic engineering will probably be regulated in much the same manner as other currently used veterinary procedures and it is discussed briefly below.

On the other hand, the product of animal genetic engineering, namely, a genetically modified or transgenic animal, is the major focus of this chapter. The principal aspects that I would like to address are the inspection and labeling of novel food animals and food products prepared from them. Most people are familiar with traditional food animals such as cows and chickens. The criteria and procedures for the inspection of these traditional food animals and the labeling of their products have been established for many years (28). By the term "novel food animal" I mean any potential food animal for which there are no established criteria, policies, or procedures for inspection and labeling. Under this definition a novel food animal could be produced either by traditional animal breeding or by the newer methods of genetic engineering. In either case, there may be an initial period of time during which the health, safety, and environmental and economic aspects of the animal are examined closely to determine its suitability for commercial development as a source of food. As soon as that determination is made and the inspection and labeling criteria become established, whether by statute, regulation, policy, or case-by-case decision, the food animal ceases to be novel. Several examples are given below.

---

* The author is Chief of the Standards Branch, Standards and Labeling Division, Meat and Poultry Inspection Technical Services, Food Safety and Inspection Service (FSIS), U.S. Department of Agriculture (USDA), Washington, D.C. The views expressed are the author's and do not necessarily represent the policies or interpretations of the FSIS, the USDA, or any other government agency.

The developing fields of biotechnology and genetic engineering promise to confer great benefits on animal health and nutrition and on food products prepared from them. The other side of this coin is that these developments may also raise important questions about the safety of animal genetic engineering, its possible environmental effects, and the wholesomeness of engineered food animals as well as the labeling of food products prepared from them. The Food Safety and Inspection Service (FSIS) of the U.S. Department of Agriculture (USDA) has general statutory responsibility for the safety, wholesomeness, and labeling of food products made from animals, whether they are traditionally bred or genetically engineered. Other federal agencies may have statutory interests in the safety and effectiveness of certain tools of genetic engineering such as artificial vectors. Depending on the composition of the vector and the specific purpose for which it is used, a vector used in animal gene transfer may be regulated either as an animal drug (7) under the Federal Food, Drug and Cosmetic Act (FFDCA) (20), or as an animal biologic (27) under the Virus, Serum and Toxin Act (VSTA) (17). In general, if a vector used in animal genetic engineering is a virus, serum, toxin, or analogous substance and is used to diagnose, treat, or prevent an animal disease, then it would probably be regulated as an animal biologic. A vector of a different composition, such as a plasmid, that is used to diagnose, treat, or prevent animal disease or to affect a structure or function of the animal body would probably be regulated as an animal drug (10,11). The regulatory classification of vectors used in gene transfer as animal drugs or animal biologics is not necessarily straightforward and will probably be determined on the time-honored case-by-case basis. Toward this end, the Food and Drug Administration and the Animal and Plant Health Inspection Service of the USDA have issued a memorandum of understanding (8) and have formed a joint interagency committee. This committee or some similar mechanism may be used to determine how vectors used in animal gene transfer are to be regulated.

Once genetic material is successfully transferred into a host animal, replicated, and appropriately transcribed and translated into gene product(s), it becomes part of the host animal. As such, it may be subject to statutes that provide inspectional authority to determine the safety and wholesomeness of host animals that are slaughtered for human food. These statutes include the Federal Meat Inspection Act (FMIA) (19) and the Poultry Products Inspection Act (PPIA) (21).

Assuring the safety and wholesomeness of food animals can be approached in a number of different ways. One possible approach is to focus on the capability of the animal for use as human food without regard for the taxonomic identity of the animal. The FMIA defines the term "capable of use as human food" as applying to "any carcass, or part of a carcass, of any animal [emphasis added]" (19). Focusing on the capability of use as human food rather than on animal species identity would certainly simplify the slaughter and inspection of hybrid animals and other combinations that zoologists have been reluctant to confront taxonomically (9). But the original intent of Congress seems to have been to restrict eligibility for slaughter and mandatory meat inspection to a closed list of species and other taxonomic groupings. A number of definitions in the FMIA, including that for meat food product, specify cattle, sheep, swine, goats, and equines (19). The PPIA defines poultry somewhat more broadly as any domesticated bird (21), but the poultry product inspection regulations do specify chickens, turkeys, ducks, geese, or guinea fowl (31).

There are other reasons besides perceived Congressional intent for restricting eligibility for slaughter and mandatory inspection for wholesomeness to certain species. The very definitions of animal health and disease are species-dependent. Diseases of a nutritional, infectious, and parasitic nature frequently exhibit species specificity (3). Therefore,

Tab. 1. Four kinds of food animal slaughter and inspection.

- Mandatory inspection
- Voluntary inspection
- Conditional inspection
- Custom processing

---

there is an advantage to assessing the health or disease state of an animal of known species identity as opposed to one of either unclear or unknown species identity. A further incentive for knowing and maintaining the species identity of an animal food product is consumer preference and expectation. Many meat food products are perceived as being of higher quality or greater value if they are derived from a single species of food animal rather than from a mixture or from unspecified species. For example, products such as pork chops and lamb chops are probably preferred by consumers over a mixture such as pork and lamb chops or the more generic animal chops.

The following section describes several slaughter and inspection programs for food animals developed by the USDA under various food statutes.

FOOD ANIMAL SLAUGHTER AND INSPECTION

Food animals can be grouped into 4 classes depending on if, or how, they are subjected to federal inspection for wholesomeness. These 4 classes can be designated roughly as mandatory inspection, voluntary inspection, conditional inspection, and custom processing (Tab. 1).

Mandatory inspection (30,31) is authorized by the FMIA and the PPIA. These statutes require the USDA to inspect certain domestic animals that are slaughtered and sold as human food. The species and other taxonomic groupings inspected under these statutes are cattle, sheep, swine, goats, equines, and domestic poultry. These inspections are mandatory and the cost of inspection, except for overtime and holiday inspection, is required to be borne by the United States (Tab. 2).

Tab. 2. Mandatory meat and poultry inspection.

Statutory Authorities

Federal Meat Inspection Act (FMIA)

Poultry Products Inspection Act (PPIA)

Species

Cattle	Goats
Sheep	Equines
Swine	Domestic poultry

Financing

Tax-supported

Tab. 3.   Voluntary meat and poultry inspection.

Statutory Authority

Agricultural Marketing Act (AMA)

Species

Rabbits	Pigeons
Reindeer	Pheasants
Buffalo	Quail
Deer	Waterfowl

Financing

Fee-for-service reimbursement

A second class of inspection is voluntary inspection (33). Authority for the voluntary inspection program is provided by the Agricultural Marketing Act (AMA) (18). Animals inspected under this program include rabbits, domesticated reindeer, and buffalo and deer raised under conditions approaching those under which domestic food animals are raised. The USDA has recently proposed a number of options for the voluntary inspection of buffalo (34). Pigeons, pheasants, quail, and migratory waterfowl may also be inspected under the AMA. All these inspections are voluntary and reimbursable to the government on a fee-for-service basis (Tab. 3).

A third class of inspection is what might be called conditional inspection and it applies mainly to animals used for research (29,32). The meat and poultry inspection regulations state that no animal amenable to inspection that is also used in a research investigation involving an experimental biological product, drug, or chemical shall be eligible for slaughter at an official establishment unless certain conditions are met. These conditions include compliance with certain provisions of the VSTA (17), the FFDCA (20), and the Federal Insecticide, Fungicide and Rodenticide Act (FIFRA) (22). Before an animal used for research can be eligible for slaughter, the operator, investigator, or sponsor must show either that the experimental biologic, drug, or chemical was used in compliance with one of the above-mentioned statutes or that the food products prepared from the research animal are not otherwise adulterated. Based on its review of the data, the Meat and Poultry Inspection Program may approve the slaughter of such an animal for human food purposes (Tab. 4).

Tab. 4.   Conditional inspection.

Intended mainly for research or experimental animals.

Codified under mandatory inspection regulations.

Cross-referenced to:

Virus, Serum and Toxin Act (VSTA)

Federal Food, Drug and Cosmetic Act (FFDCA)

Federal Insecticide, Fungicide and
Rodenticide Act (FIFRA)

Tab. 5. Custom processing of food animals.

May be slaughtered for sole use of owner.

May not be inspected.

May not be sold.

Blends of game meat and inspected meat
  may be prepared for owner.

A fourth class of food animal slaughter is custom processing and it applies mainly to food animals that are intended solely for their owner's use. These may include FMIA-amenable animals that are not delivered to a USDA-inspected establishment because of distance, time, or cost considerations. Meat from such animals is not inspected for wholesomeness and cannot be used as an ingredient in meat food products. The USDA also has provisions under which game animals may be slaughtered for owners, provided adequate facilities are available and their handling does not create a health hazard (23). Custom products, consisting of meat from custom-processed animals mixed with pork, beef, or lamb, may also be prepared for the owners of custom-processed animals. Such products are not inspected and they are not sold (Tab. 5).

Figure 1 shows a schematic decision network for the slaughter and inspection of food animals. It is based largely on the USDA's previous experience with food animals inspected under both the mandatory and voluntary programs as well as on custom-slaughtered animals and animals used for

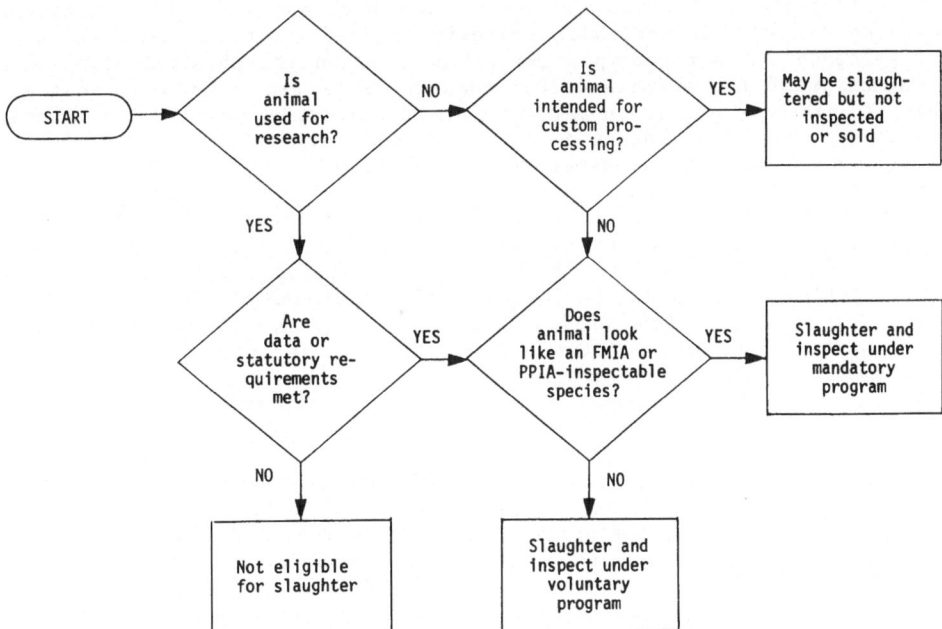

Fig. 1.   Food animal slaughter and inspection flowchart. (This flowchart is intended for conceptual purposes only and does not necessarily reflect procedural requirements of the Federal Meat and Poultry Inspection Program.)

research. The decision network also may be a useful guide for the possible slaughter and inspection of novel food animals, whether produced by traditional breeding methods or by genetic engineering. As a further guide to the slaughter and inspection of novel food animals, I would now like to describe the USDA's previous inspection experience with certain hybrid food animals.

"CATTALO" VS "BEEFALO"

The Federal Meat and Poultry Inspection Program has had prior experience with 2 hybrid crosses of cattle and buffalo. The purpose of these hybridizations was to combine the winter hardiness and range-foraging ability of buffalo with the beef production characteristics of domestic cattle. These hybrids were produced by traditional animal breeding methods and they provide an instructive example of how the inspection of some novel food animals has been handled in the past.

One hybrid, called "cattalo," resulted from direct crossbreeding of buffalo and cattle (24). The hybrid animals had the physical appearance of buffalo, but the males were usually sterile and the cows and calves suffered high mortality. The Federal Meat and Poultry Inspection Program determined that since cattalo have the physical appearance of buffalo, they are not amenable to the FMIA, but may be slaughtered under the voluntary, reimbursable inspection program.

A second hybrid, called "beefalo," is said to have resulted from a cross of three-eighths buffalo and five-eighths cattle (25). These hybrids have the general physical appearance of domestic cattle and they cannot be distinguished easily from cattle on antemortem inspection. For this reason, they are inspected as cattle under the mandatory program and the FMIA.

The example of cattalo and beefalo may have established an important precedent for the inspectional criteria applied to novel food animals. That precedent is a phenotypic criterion based on the physical appearance or phenotype of the animals rather than on their genetic makeup or genotype. In other words, if an animal looks like cattle, sheep, swine, goats, or equines, then it is inspected under the mandatory program as beefalo is. On the other hand, if an animal does not look like cattle, sheep, swine, goats, or equines, then it may be inspected under the voluntary program as cattalo is. If the number of novel food animals increases significantly in the future, then in order to provide for systematic slaughter and inspection, it may be necessary to develop improved criteria for animal species identification. A decision on the physical appearance of an animal is currently a matter of professional judgment, and even qualified professionals can disagree. But in the future, a decision on the species identity of a novel food animal could be supplemented by more objective methods such as those of numerical taxonomy using various morphological or biochemical data.

Morphology is the scientific study of the size, shape, and structure of organisms that are, or once were, alive, and of the relationships between the organisms and the parts that comprise them. Morphological characteristics can be quantified, scaled, and coded for machine computation that may provide objective support for the zoological identification of animals. Such an approach has been used for a number of years for the taxonomic classification of nonfood animals and other living organisms (16).

There are also a number of biochemical tests that may prove useful for probing and confirming the zoological identity of genetically engineered animals. Such tests include protein/amino acid sequence comparisons, electrophoretic comparisons of mixtures of proteins or nucleic acids, immunological tests such as microcomplement fixation, and nucleic acid hybridization (6). Biochemical tests such as these appear to be more useful when one is attempting to determine differences rather than to demonstrate similarities. For example, the USDA has had reasonable success distinguishing between beef and kangaroo tissue using immunological tests when appropriate antisera are available. When antisera are not available, electrophoretic methods, such as thin layer isoelectric focusing, have been used with some success also.

When the results of biochemical tests indicate similarities rather than differences, they may have to be interpreted with some caution. As an example not involving food animals, human and chimpanzee proteins are more than 99% identical based on electrophoretic, immunological, and sequencing methods (12). This finding led the original authors to an estimate of the genetic distance between human and chimpanzee comparable to that of closely related or sibling species. And yet, based on anatomical and morphological differences, biologists not only consider humans and chimpanzees to be different species, but they also place them in different genera and different families. This paradox led the authors to postulate that humans and chimpanzees have undergone different rates of rearrangement in their regulatory genes, those genes that regulate the expression of other genes (12). Fortunately, genetically engineered animals of commercial importance will probably not have sufficient time to accumulate such genetic regulatory rearrangements. Nevertheless, the near identity of human and chimpanzee proteins by biochemical tests, despite their morphological differences, may indicate that biochemical data for the zoological identification of genetically engineered animals should be supplemented with other biological data such as chromosome studies.

Thus, biochemical tests alone may suggest similarities between animals when there are recognizable differences by other criteria such as morphology. Similarly, the phenotypic approach alone, based on the cattalo and beefalo experiences, may indicate similarities between animals when there are, in fact, genetic differences. The nature and magnitude of genetic changes induced in animals by humans may affect how an animal is inspected and how food products prepared from it are inspected and labeled.

One of the first distinctions to be made regarding the nature of artificially induced genetic changes in animals is the difference between somatic and germ cell changes. Somatic cell changes affect only nonreproductive cells and therefore, if they are to be present, they must be induced separately in each generation. Germ cell changes, on the other hand, are intended to affect reproductive cells and, therefore, they may be propagated into future generations. Artificially induced germ cell changes in animals may raise certain questions of a technical, moral, or philosophical nature which are addressed elsewhere in this Volume. For purposes of this discussion, it is anticipated that legal and regulatory questions will arise, at least initially, in the context of somatic cell changes that are artificially induced in animals.

There are several other distinctions to be made in the type and magnitude of genetic changes induced in food animals. These differ in the amount of genetic change as a fraction of the total genome, and it may be necessary to handle them differently from a legal and regulatory standpoint. In the first and simplest type, the genetic change is more or less a point change in the total genome of the animal. This may alter one or a few genes and may change the animal's growth rate or disease resistance, for example, without significantly altering its basic physical appearance, general health, or behavior. Such a point genetic change might be called a "microgenomic" change, and the animal might be called a "microengineered" animal. In other words, a microengineered cow would still be a cow for purposes of slaughter and inspection. But this is not to say that the eligibility of a microengineered cow for slaughter and inspection would be automatic. Such an animal may still have to be carefully evaluated as to its identity, its potential effects on health and the environment, and the safety of food products prepared from it. It is possible that in a gene transfer experiment that is not properly controlled, the artificial vector, the foreign gene, or the corresponding gene product could cause either an adverse veterinary medical condition or the presence of a biologic residue (11). Either of these conditions may result in adulteration of the host animal if it is intended for human food use. Similarly, a transplanted gene may be integrated at an abnormal chromosomal position and expressed in an inappropriate tissue. This was the case in at least one experimental transfer of a rabbit β-globin gene into mice (14). Therefore, the demonstration of controlled or appropriate expression of a transplanted gene in a host animal may be a key step in determining whether the tissues of the animal are safe for human food use (11).

An intermediate level of genetic change with which we are all familiar is the recombination of equal amounts of genetic material from 2 parents of the same species. This is the genetic basis of sexual reproduction and it is a prolific source of genetic variation, both in the natural biological world and in traditional plant and animal breeding. In seeking to address the potential risks of artificial gene transfer in plants, it has been pointed out that new strains produced by traditional plant breeding differ from the parental strains to a much greater extent than would a microengineered strain as described above (4).

A more complicated situation arises in the case of food animals whose cells or tissues contain different genetic information. This might arise, for example, in the case of an animal with 2 or more cell lines derived from a single zygote resulting in a mosaic (1). Alternatively, it might arise from cell populations derived from different zygotes resulting in a chimera (2). A chimera, it is recalled, is an organism containing tissues of different genetic origin; a griffin, for example, is a mythical chimera with the hindquarters of a lion and the head and wings of an eagle. In 1984, laboratories in the United Kingdom and West Germany reported the occurrence of real-life chimeras between sheep and goat (5,15). These animals were produced by embryo fusion, and the adult animals had the horns of a goat and the fleece of a sheep. Animals displaying such marked genetic changes might be referred to as "macroengineered" to distinguish them from the microengineered animals described above. If a macroengineered animal, such as a sheep/goat chimera, were ever presented for slaughter and inspection at an official establishment, there might be some serious head scratching. This would indeed be a novel food animal for which inspection criteria and procedures do not exist and might have to be worked out. One important question is whether a sheep/goat chimera would be covered by the FMIA. One could argue that the chimera was produced from 2 animals, sheep and goat, that are both inspected under the FMIA, and therefore the chimera should be inspected under the FMIA. On the other hand, one could argue that a sheep/goat chimera is neither a sheep nor a goat and, therefore, cannot be inspected under the FMIA. This then might lead to the question of whether the FMIA or PPIA should be amended to require mandatory inspection of such animals and their products. The answer to this question would presumably be worked out in the social and legislative arenas. One final inspection note is that if a chimera were sound, wholesome, and fit for human food, it might be inspectable under the voluntary inspection program at the owner's expense.

LABELING PRODUCTS OF GENETICALLY ENGINEERED ANIMALS

A final provocative question concerns how products prepared from novel food animals should be labeled. For example, does the label of a product prepared from a food animal that was genetically engineered need to disclose that fact? On one hand, the consuming public may be so apprehensive about genetic engineering that it would want the labels of all food products produced by that technology to disclose its use so that individual purchasing decisions could be made. On the other hand, genetic engineering might be considered merely an alternate method of manufacture of food products (13) that would not require special labeling as long as it did not breach the boundaries of species or families listed in the food statutes.

A final resolution of the labeling for food products prepared from genetically engineered organisms may be some time in coming. In the interim, we might look to the labeling of the cattle/buffalo hybrid called beefalo for guidance. Under current meat inspection policy, meat from beefalo must be labeled "beef," presumably because of the close phenotypic resemblance of beefalo to beef cattle (26). However, the USDA has permitted food processors to capitalize on the pedigree of these hybrid animals by using a qualifying phrase such as "beef from beefalo" on food product labels.

The kind of labeling problem that might be encountered as a result of genetic engineering is exemplified in a whimsical sort of way by the sheep/goat chimera referred to earlier. If we form a hybrid name for the sheep/goat chimera as was done previously for beefalo, we get the name "shoat," which is a young hog. This would clearly be an example of misbranding under the food statutes and we would have to go back to the drawing board for a better name.

In addition to the technical challenges of genetic engineering, there may be significant regulatory, inspectional, and labeling challenges as well. I am confident that if government, agriculture, and industry work together on the mutual resolution of these challenges, we can surmount them successfully.

SUMMARY

The commercialization of genetically engineered food animals will pose a number of legal and regulatory questions. These may be grouped into questions of process and questions of products. The process of animal genetic engineering with artificially constructed vectors will probably be regulated in much the same manner as other veterinary procedures. There may be some discussion, however, as to whether animal drug or animal biologic regulations are more applicable. The products of animal genetic engineering, i.e., transgenic food animals and food products made from them, also raise important questions about product safety and identity. These include whether and how genetically engineered food animals will be subject to federal inspection for wholesomeness, whether artificial vectors, foreign genes, or gene products will adulterate recipient animal tissues, and how food products made from such animals will be labeled. Prior federal experience with the inspection of interspecific hybrids of cattle and buffalo provides a useful basis for further policy developments in the inspection and labeling of genetically engineered food animals. In particular, the inspection of cattle/buffalo hybrids has established a phenotypic (based on appearance) criterion for deciding how novel food animals should be inspected. As the genetic engineering of food animals on a production basis draws nearer, it may be necessary to supplement the phenotypic criterion with genetic (based on pedigree) criteria to assure that the essential characteristics of animals slaughtered under current food statutes are maintained.

REFERENCES

1. _Dorland's Illustrated Medical Dictionary_, 26th ed. (1981) W.B. Saunders Co., Philadelphia, p. 838.
2. _Dorland's Illustrated Medical Dictionary_, 26th ed. (1981) W.B. Saunders Co., Philadelphia, p. 254.
3. Ensminger, M.E. (1983) _Animal Science_, 8th ed., Interstate Printers and Publishers, Danville, Illinois, p. 213.
4. Fedoroff, N., quoted in L. Tangley (1985) New biology enters a new era. _BioScience_ 35:270-275.
5. Fehilly, C.B., S.M. Willadsen, and E.M. Tucker (1984) Interspecific chimaerism between sheep and goat. _Nature_ 307:634-636.
6. Ferguson, A. (1980) Higher-category systematics. In _Biochemical Systematics and Evolution_, John Wiley and Sons, New York, pp. 131-150.
7. Food and Drug Administration (1982) Section 510.3 Definitions and interpretations. _Code of Federal Regulations, Food and Drugs_ 21 CFR 500.25: 74-75.
8. Food and Drug Administration (1982) Memorandum of understanding with the United States Department of Agriculture, Animal and Plant Health Inspection Service. _Federal Register_ 47:26458-26459.
9. Jeffrey, C. (1977) _Biological Nomenclature_, 2nd ed., Crane, Russak and Co., New York, p. 44.
10. Jones, D.D. (1983) Genetic engineering in domestic food animals: Legal and regulatory considerations. _Food Drug Cosm. Law J._ 38:273-287.
11. Jones, D.D. (1985) Commercialization of gene transfer in food organisms: A science-based regulatory model. _Food Drug Cosm. Law J._ (in press).
12. King, M.C., and A.C. Wilson (1975) Evolution at two levels in humans and chimpanzees. _Science_ 188:107-116.
13. Korwek, E.L. (1982) FDA regulation of biotechnology as a new method of manufacture. _Food Drug Cosm. Law J._ 37:289-309.
14. Lacy, E., S. Roberts, E.P. Evans, M.D. Burtenshaw, and F.D. Costantini (1983) A foreign β-globin gene in transgenic mice: Integration at abnormal chromosomal positions and expression in inappropriate tissues. _Cell_ 34:343-358.
15. Meinecke-Tillman, S., and B. Meinecke (1984) Experimental chimaeras-- Removal of reproductive barrier between sheep and goat. _Nature_ 307: 637-638.
16. Sneath, P.H.A., and R.R. Sokal (1973) _Numerical Taxonomy_, W.H. Freeman and Co., San Francisco.
17. U.S. Congress (1982) Viruses, serums, toxins, antitoxins, and analogous products. _U.S. Code 1982 Ed._ 8:720-722.
18. U.S. Congress (1982) Distribution and marketing of agricultural products. _U.S. Code 1982 Ed._ 2:608-615.
19. U.S. Congress (1982) Meat inspection. _U.S. Code 1982 Ed._ 8:849-873.
20. U.S. Congress (1982) Section 360b. New animal drugs. _U.S. Code 1982 Ed._ 8:786-792.
21. U.S. Congress (1982) Poultry and poultry products inspection. _U.S. Code 1982 Ed._ 8:831-849.
22. U.S. Congress (1982) Insecticides and environmental pesticide control. _U.S. Code 1982 Ed._ 2:103-140.
23. U.S. Department of Agriculture (1973) 3.1 Game animals. In _Meat and Poultry Inspection Manual_, U.S. Department of Agriculture, Washington, D.C., p. 2.
24. U.S. Department of Agriculture (1973) 3.2 Catalo or cattalo. In _Meat_

and Poultry Inspection Manual, U.S. Department of Agriculture, Washington, D.C., p. 2.

25. U.S. Department of Agriculture (1973) 3.3 Beefalo. In Meat and Poultry Inspection Manual, U.S. Department of Agriculture, Washington, D.C., p. 2.

26. U.S. Department of Agriculture (1973) 17.13(n) Beefalo. In Meat and Poultry Inspection Manual, U.S. Department of Agriculture, Washington, D.C., p. 118.

27. U.S. Department of Agriculture (1982) Section 101.2 Administrative terminology. Code of Federal Regulations, Animals and Animal Products 9 CFR 1.1:332-333.

28. U.S. Department of Agriculture (1984) Food Safety and Inspection Service, Meat and Poultry Inspection, Department of Agriculture. Code of Federal Regulations, Animals and Animal Products 9 CFR 201.1:105-446.

29. U.S. Department of Agriculture (1984) Section 309.17 Livestock used for research. Code of Federal Regulations, Animals and Animal Products 9 CFR 201.1:137.

30. U.S. Department of Agriculture (1984) Subchapter A--Mandatory meat inspection. Code of Federal Regulations, Animals and Animal Products 9 CFR 201.1:107-291.

31. U.S. Department of Agriculture (1984) Subchapter C--Mandatory poultry products inspection. Code of Federal Regulations, Animals and Animal Products 9 CFR 201.1:343-444.

32. U.S. Department of Agriculture (1984) Section 381.75 Poultry used for research. Code of Federal Regulations, Animals and Animal Products 9 CFR 201.1:381.

33. U.S. Department of Agriculture (1984) Subchapter B--Voluntary inspection and certification. Code of Federal Regulations, Animals and Animal Products 9 CFR 201.1:292-342.

34. U.S. Department of Agriculture (1985) Voluntary inspection of buffalo. Federal Register 50:10778-10784.

"THE FRANKENSTEIN THING": THE MORAL IMPACT OF GENETIC ENGINEERING

OF AGRICULTURAL ANIMALS ON SOCIETY AND FUTURE SCIENCE

B.E. Rollin

Departments of Philosophy
and Physiology and Biophysics
Colorado State University
Fort Collins, Colorado 80523

Shortly after I had accepted the invitation to address this conference, I remarked to a friend of mine (a nonscientist) that I was going to
address a conference on genetic engineering of animals. "Ah," he said,
"the Frankenstein thing!" I didn't pay much mind to his remark until perhaps a week later, when, while perusing the new acquisitions in our library, I encountered an extraordinary, newly published, 500-page volume en-
titled The Frankenstein Catalog: Being a Comprehensive History of Novels,
Translations, Adaptations, Stories, Critical Works, Popular Articles, Series, Fumetti, Verse, Stage Plays, Films, Cartoons, Puppetry, Radio and
Television Programs, Comics, Satire and Humor, Spoken and Musical Record-
ings, Tapes and Sheet Music Featuring Frankenstein's Monster and/or De-
scended from Mary Shelley's Novel (2). The entire book is precisely a de-
scriptive catalogue, a list and very brief description of the works mentioned in the title. Amazing though it is that anyone would publish such a
book, its content is even more incredible, for it in fact lists 2,666 such
works (including 145 editions of Shelley's novel), the vast majority of
which date from the mid-twentieth century. All of this obviously indicates
that in the Frankenstein story is an archetypal myth or category which
somehow speaks to or for twentieth-century concerns, and which could perhaps be used to shed light on the social and moral issues raised by genetic
engineering of animals. My intuition was confirmed while visiting Australia and discussing with an Australian agricultural researcher the, to him,
surprising public hostility and protest that his research into teratology
in animals had provoked. "I can't understand it," he told me. "There was
absolutely no pain or suffering endured by any of the animals." "All I can
think of," he said, "is that it must have been the Frankenstein thing."
And in its cover story on the 40th anniversary of the Hiroshima bombing,
Time Magazine again invoked the Frankenstein theme as a major voice in
post-World War II popular culture, indicating that it was society's way of
expressing its fear and horror of a science and technology that had un-
leashed the atomic bomb (10).

Given this pervasive reaction, it seems valuable to explore the social
and moral concerns about research into genetic engineering of agricultural
animals, using the Frankenstein myth as a framework for our discussion.
As I shall try to show, the social concerns and the genuine moral concerns are not always identical, and are, in fact, sometimes confounded and
not clearly separated in the public mind and, indeed, in the minds of many

scientists. Furthermore, some of the deepest and most genuine moral concerns encapsulated in the Frankenstein story are undoubtedly least discussed and explored either by the scientific community or the public.

Before pursuing this inquiry, it is worth pausing to stress that, in general, not just in the case of genetic engineering, both the scientific community and the general public often miss the mark in their attention to the ethical issues growing out of scientific activity. My good friend, the late Dr. Bernard Schoenberg, Associate Dean of the Columbia University College of Physicians and Surgeons, used to remark that while the public and the medical community alike were spilling a great deal of ink on issues like disconnecting the respirator from Karen Ann Quinlan, almost no one was discussing the far more fundamental moral issue of fee for service in medicine! By the same token, when the Baby Fae affair occurred, generating much debate, neither scientists nor the public seemed to realize that there was little moral difference between this case and any case of killing an animal for possible human benefit or for research. For that matter, it was hard to see the moral difference between harvesting hearts from baboons and harvesting heart valves from pigs--something which has been standard practice for some time and yet something which no one has raised as an ethical question. The practical rather than moral difference, of course, was in the sensational nature of the story--transplanting hearts from animals plucks at primordial emotions. As I told the press, this issue was not from a conceptual point of view worth discussing in isolation from the general question of whether science has the right, in far less dramatic cases, to expend animal lives, sometimes accompanied by far more suffering, anxiety, and fear on the part of the animal than that experienced by the anesthetized baboon.

Again, and in the same vein, I recently gave the keynote speech at an Australian conference on the moral issues in animal experimentation. In the course of my talk, I pointed out that merely citing a list of human benefits engendered by research on animals does not in itself logically serve to morally justify that invasive use of animals, anymore than a listing of benefits which emerged from medical research on political prisoners, concentration camp inmates, slaves, criminals, and the like would justify doing such research without obtaining noncoerced, informed consent. Despite this obvious point, many researchers, in their talks, continued to base their defense of animal use solely on benefits to man, as indeed the United States medical research community has tended to do.

Unfortunately, the general public is usually too ignorant about science to be able to sort out the genuine moral issues emerging from scientific activity, and in practice tends to rely on the media to do the job for it. The media, in turn, is of course less interested in conceptual or factual accuracy than in selling papers, as one reporter candidly told me during the Baby Fae case. So that, as we shall see in the case of genetic engineering, what gets presented to the public as major moral issues are often not moral issues at all. At the same time, scientists are themselves often unable to discriminate the ethical issues implicit in or arising out of their own activity, and essentially wait to have the issues defined for them by the public, or by the same people who define them for the public, so that the issues do not get adequately dealt with from the scientific side either. The failure of scientists to discriminate moral issues in science in turn arises out of what I have called "the ideology of science": in essence, the set of philosophical principles, positions, assumptions, presuppositions, and values that scientists tend to acquire unconsciously along with their scientific knowledge in the course of their training. This pervasive ideology is rooted in the logical positivism and behaviorism of the 1920s, and suggests that science deals only with what is observable and verifiable--with "facts." Since statements about values, including moral values, are not verifiable, they are alleged not to fall within the scientist's purview, at least in his or her capacity as scientist. This is often codified as the slogan that science is "value-free," and is accompanied by the claim that values perhaps enter into the use to which science is put by society, never into science itself. Value judgments, including ethical ones, are often viewed by scientists as emotive responses and matters of individual preference or taste, and hence not rationally adjudicable; after all, de gustibus non disputandum est. Thus, philosophically, many scientists see nothing wrong with ignoring moral issues, or even with being emotional on moral issues, since their unspoken philosophical training leads them to believe that moral issues are nothing but emotional issues.

In actual fact, as I have taken pains to demonstrate elsewhere (6), science is not value-free and includes ethical values. Indeed, all science is permeated with valuational presuppositions. Surprisingly, perhaps, the very notion of what will count as a fact, as a legitimate object of investigation, or as data relevant to a given question, rests squarely upon valuational presuppositions. Consider, for example, the Scientific Revolution, during which the common-sense, sense-experience-based physics and cosmology of Aristotle were replaced by the rationalistic, mathematical, geometrical physics of Galileo and Newton. The discovery of new data or new facts is not what forced the rejection of Aristotelianism--on the contrary, empirical observations all buttressed Aristotle's idea of a world of qualitative differences! What led to the rejection of Aristotelianism was essentially a change in value--a discrediting of information provided by the senses, as Descartes does so well in his Meditations, and a correlative valuing of the rational and mathematically expressible over the empirical, of Plato's philosophy over Aristotle's. This was so nicely expressed in Galileo's claim that, in essence, an omniscient deity would have to be a mathematician, and create a mathematical unity underlying apparent diversity.

Few of you would go along with one of my acquaintances, an accomplished medical researcher and Rhodes Scholar, who heatedly informed me that the question of the use of animals in science is simply a scientific, not a moral, question, and that, indeed, science has nothing to do with ethics. In an attempt to show him that he had not thought out the logic of his position, I pointed out that if science is indeed constrained only by scientific concerns, why didn't we use children for research, since they are better models for humans than are animals. His reply, amazingly enough, was "because they won't let us." And none of you who have watched the obviously morally based changes in scientific opinion on whether race differences and intelligence are legitimate objects of study, on whether homosexuality is a disease or an alternative lifestyle, and on whether alcoholism or wife-beating is sickness or badness can truly deny that science is rooted in moral valuational assumptions (3).

In any event, my main concern thus far has been with showing that our understanding of moral issues does not usually keep pace with the scientific progress that generates these issues. And if I should stress to you any urgent message at all, let it be that scientists ignore or shunt off these issues until they assume crisis proportions at their own peril. In the final analysis, public money pays for science, and ever-increasingly demands accountability. A failure on the part of any area of science to clearly define the moral issues growing out of its activity, and to deal with them, puts its very existence in peril, as the case of animal research around the world dramatically illustrates. Furthermore, in a moral variation on Gresham's Law, bad moral thinking can drive good moral thinking out of circulation. Thus a failure on the part of scientists to articulate the genuine moral issues in genetic engineering or any area leaves open the very real possibility of false and irrelevant, but sensationalistic, issues occupying the public mind and being used as a basis for social policy. And, as we shall now see as we return to "the Frankenstein thing" as a basis for discussing genetic engineering of animals, the same sort of thing can happen here.

A very nice illustration of my moral Gresham's Law may be found in the fact that probably the most socially pervasive component of the Frankenstein metaphor as it applies to genetic engineering of animals is also the least interesting morally. This component may be characterized in terms of the classic line from old Frankenstein genre movies that "there are certain things man was not meant to know" (or to do, or to explore). In other words, there is certain scientific knowledge or activity, or applications of scientific knowledge, irrespective of its consequences, that in and of itself is taboo. In the case of genetic engineering of animals, this would most likely be attributed by those who hold such a view to the creation of chimeras or crossing of species lines; to major modifications within a species which are phenotypically apparent (such as genetically manipulating for leglessness in farm animals); or even, as press and public reaction to the Fox-Rifkin lawsuit against the U.S. Department of Agriculture (USDA) indicates, to introduction of genetic material derived from humans into animals, or, presumably, to the introduction of animal-derived genetic material into humans. (As I suggested earlier, a similar strain of thought arose during the Baby Fae case; numerous people seemed to have perceived unspecified ethical difficulties in a human having an animal part.)

The pattern of thinking represented in this sort of version of "the Frankenstein thing," though widespread, does not represent a genuine moral issue, and does not raise moral questions requiring social adjudication. It appears to me to have a variety of nonmoral sources which are typically confused with moral concerns.

One such source is most certainly theological: the Judaeo-Christian notion that God created living things "each according to its own kind," with the clear implication, expressed both in nineteenth-century and contemporary opposition to Darwin, that species are fixed, clearly separated from one another, and immutable, and furthermore, ought to be. A nontheological, historically influential, philosophical vector buttressing this view in Western thought is Platonized Aristotelianism, which again postulates fixed natural kinds, again immutable and clearly demarcated from one another. Indeed, Aristotle defends this view on the grounds that its contrary would made knowledge impossible. (An opposite tendency also found in Aristotle, which suggests an infinite continuum and gradation in species, has been all but ignored.) But, of course, such theological and philosophical prejudices are not in themselves legitimate bases for moral questioning of genetic engineering, though they help explain certain people's knee-jerk bias against it. And, of course, to a religious person, anything that violates any of his or her religious tenets must be seen as morally problematic.

But reservations against "meddling with species" stem from sources beyond theology and Aristotle. They stem also from a common but scientifically unsophisticated and rather muddled understanding by a virtually scientifically illiterate public of species as being, as it were, the building blocks or atoms of the biological world, out of which the biological world is built and upon which it rests. To tinker with species is, in this view, to tinker with the stability of nature, to (in some unarticulated way) shake the entire Great Chain of Being, as Coleridge's Ancient Mariner did when he killed the Albatross. The fact that species are, in current biological theory, dynamic rather than static, stop-action views of a continuing evolutionary process is ignored by such critics. These critics also ignore the fact that the notion of (genetic) species is highly complex and problematic, and that it has been rejected by some biologists such as Rensch in favor of notions like subspecies, races, Rassenkreis, or Formenkreis as not being the fundamental taxonomic unit (1). [On the other hand, the fact that most biologists do treat species as the fundamental taxonomic unit and as being "more real" than other such units, as Michael Ruse puts it, lends support to such critics (9).] If subspecies is the fundamental unit, incidentally, then we have been genetically engineering biological reality with no fuss for thousands of years. For that matter, if one takes seriously the currently standard definition of a species as a naturally interbreeding population, then one could argue that certain subspecies we have genetically engineered by breeding, such as the Great Dane and Chihuahua, in fact constitute separate species.

Incidentally, as I have argued elsewhere (5), much of the debate about the reality or nonreality of species rests upon a deep and ancient philosophical mistake, the attempt to classify all phenomena as being either nomos or physis, nature or convention. In actual fact, it appears that species represent something of both; what species we find in the world depends on the scientific-theoretical lenses with which we examine the world. Given current theories of evolution and molecular genetics, such procedures as DNA matching, serological evidence from protein matching, etc., give us an objective method of species classification. But, at the same time, we must recognize that these objective tests are based on accepted biological theories, and that given an alternative biological theory, say one oriented far more to whole-organism function or ecological place than to the molecular basis of life, we would probably generate a completely different taxonomy, complete with a totally different set of objective tests (5).

Another factor which appears to me to foster the belief in inviolability and sacredness of species is the environmental movement. It is a psychologically small, albeit conceptually untenable and logically vast step, to go from concern that species not be allowed to become extinct to the idea that we ought not change them. Or perhaps, a bit more reasonably, the movement of thought is rather from the idea that species ought not be allowed to vanish as a result of what humans do, to the idea that they ought also not change at our hands. Built into the environmental movement is, in short, a "nature knows best, hands off nature" mindset, but more as an attitude than as a reasoned position.

In my view, as I have argued elsewhere (4), species are not the sorts of things which are legitimate objects of moral concern. It makes little sense to me to assert that it would be permissible to shoot 10 Siberian tigers as long as there were plenty of Siberian tigers, or to suggest, as one of my environmental ethicist colleagues has written, that if a species of endangered moss is in the migratory path of a species of plentiful elk, it is not only permissible but obligatory to save the moss by shooting the elk (8). In my view, as I shall discuss later in detail, only sentient individuals are legitimate objects of moral concern; species only count morally insofar as they represent a group of individuals, and the last 10 Siberian tigers are no different morally than any other 10 Siberian tigers (4). There is certainly a great loss in species becoming extinct, but it is fundamentally, perhaps, an aesthetic one, analogous perhaps to our repulsion at trampling a flower. Ethics is relevant only insofar as one is morally obligated not to destroy aesthetic objects, or to deprive future generations of having them in their umwelt.

In any case, I think we can conclude from all of the above that the first aspect of "the Frankenstein thing," namely that "there are certain things we simply ought not do, and species modification by genetic engineering is one of them," does not represent a defensible moral claim, even if it may be so perceived by large numbers of people. To respond to this pervasive idea, however, the research community needs to do a great deal of public education, necessarily preceded by self-education in ethical issues.

Any rational attempt to extract a genuine moral issue from the first aspect of "the Frankenstein thing" we have discussed must be based in a second aspect of "the Frankenstein thing" to which we now turn. Crucial to most versions of the Frankenstein myth is the danger to humans that grows out of unbridled scientific curiosity. Thus, the dictum that "there are certain things that are just wrong to do" becomes replaced in this aspect of the myth by the dictum that "there are certain things that are wrong to do because they must or will inevitably lead to great harm to human beings." The archetypal image of this is Dr. Frankenstein's monster on a rampage--terrorizing, hurting, killing, and harming the innocent. Despite the scientist's noble intentions (Dr. Frankenstein's purpose, in the novel, was to help humanity), his activity was morally wrong not (or not merely) because of hybris, but because of his unjustifiable failure to foresee the dangerous consequences of his actions, or even to consider the possibility of such consequences and take steps and precautions to limit them. And to this objection, of course, twentieth-century science and technology is quite vulnerable. We have tended to believe that if we can do something, we should, and forge ahead as quickly as possible, damn the torpedoes. And we have also tended to believe, as part of the ideology of science discussed earlier, that scientists are not morally responsible for the pernicious uses to which their explorations are put; the responsibility for these consequences allegedly belongs to politicians, governments, military agencies, or corporations. There are, of course, notable exceptions to this claim, as Asilomar nobly illustrates, but in the main, scientists are vulnerable to this criticism, as any of us who have served on university biosafety or surveillance committees knows all too well. The recent discovery of killer bees on the loose in California represents another example of unjustifiable negligence on the part of scientists, who of course imported and bred these insects apparently without proper regard for the dangers involved.

What, if any, are the potential dangers inherent in genetic manipulation of animals in agriculture? This is certainly a legitimate issue which should be addressed by all of those working in the area. Even a cursory examination of the area suggests a number of possibilities that should be raised, explored, and assessed in terms of likely risk, and for which mechanisms of minimizing the risk should be devised before embarking upon genetic engineering of animals utilizing new principles of biotechnology.

I would suggest, in any country contemplating such work, the establishment of formal mechanisms to ensure that the social questions associated with potential risks growing out of genetic engineering of animals be fully evaluated and made known to the public, much in the way recombinant DNA work has been dealt with in the United States. I have recommended to the USDA the establishment of something analogous to the National Institutes of Health's (NIH's) Recombinant DNA Advisory Committee (RAC) to assess potential risks and other ethical and social issues associated with genetic engineering of agricultural animals. This ought to proceed in a number of stages. First, a fairly large committee consisting of scientists, attorneys, public policy people, ethics people, and members of the public should delineate the issues and suggest broad guidelines for assessing and minimizing risk. If possible, levels of risk should be identified, and broad characterizations of types of research and applications thereof delineated. Subsequently, local committees analogous to human research committees, animal research committees, and biosafety committees, with significant public membership, should be appointed at institutions engaged in research or application of genetic engineering of animals. As much accurate publicity as possible should accompany all aspects of this process, both to dispel irrational components of "the Frankenstein thing," and to show responsiveness to legitimate concerns. Such committees should also engage an entirely different set of ethical questions which we will outline shortly in discussing the last component of the Frankenstein myth.

The sorts of hazards, risks, and potential dangers associated with genetic engineering of agricultural animals appear to be the following (doubtless most of you could supplement my list significantly). At this stage, I believe that it is vital to err on the side of caution, and look at and consider every possible danger, however apparently unlikely. It is usually far easier to prevent than to amend, especially in an area like agriculture, in which vast amounts of money or food are at stake when a technological tool or procedure becomes integral to an operation and is later found questionable or unsuccessful. The use of antibiotics in feeds provides a clear example, as do overly intensified and overly capitalized systems in pork production and crop decimation growing out of unanticipated disease and genetic uniformity.

The first set of potential dangers emerging from the new forms of genetic engineering of agricultural animals obviously stems from the rapidity with which such activity can introduce wholesale change in organisms. Traditional genetic engineering, of course, was done by selective breeding over long periods of time, during which time one had ample opportunity to observe the untoward effects of one's narrow selection for isolated characteristics. But with the techniques we are discussing here, we are doing our selection "in the fast lane." This leads to two sorts of potential danger.

First of all, there may be untoward consequences affecting the organism which one is rapidly changing. The characteristic one is genetically engineering may have implications that are unsuspected. Thus, for example, when wheat was genetically engineered for resistance to blast, that characteristic was looked at in isolation, and the genetic basis for this resistance encoded into the organism. The back-up gene for general resistance was, however, ignored. As a result, the new organism was very susceptible to all sorts of viruses which, in one generation, mutated sufficiently to devastate the crop.

Second, the isolated characteristic being engineered into the organism may have unsuspected harmful consequences to humans who consume the resultant animal. Thus one can imagine genetically engineering, for example, faster growth in beef cattle in such a way as to increase certain levels of hormones, which when increased in concentration, turn out to be carcinogens for human beings over a 30-year period, or teratogens, in a manner similar to diethylstilbestrol. The deep issue here is that one can of course genetically engineer traits in animals without a full understanding of the mechanisms involved in phenotypic expression of the traits, with resulting disaster. This in turn suggests that it would be prudential to be cautious in one's engineering until one has at least a reasonable sense of the physiological mechanisms affected.

A second set of risks growing out of genetic engineering of the sort we have been discussing replicates and amplifies problems already inherent in selection by breeding, namely the narrowing of a gene pool, the tendency towards genetic uniformity, the emergence of harmful recessives, the loss of hybrid vigor, and, of course, the greater susceptibility of organisms to devastation by pathogens, as has been shown in genetic engineering in crops. (On the other hand, genetic engineering can have the opposite effect in making available to the gene pool greater variety than ever before, as in the case of artificial insemination making new genetic material available to beef breeders.)

A third set of risks arises out of the fact that in certain cases when one changes animals, one can thereby change the pathogens to which they are host. This can occur in 2 conceivable ways. First, if one were genetically engineering for resistance to a given pathogen in an animal, one could thereby unwittingly be selecting for new variants among the natural mutations of that microbe to which the modified animal would not be resistant. These new organisms could then be infectious to these animals, other animals, or humans. Second, even if one were changing the animal in nonimmunological ways, one could be changing the pathogens to which it is host by changing the microenvironment where they live. This in turn could result in these pathogens becoming dangerous to humans or to other animals. Thus, in changing agricultural animals by accelerated genetic means, one runs the risk of affecting the pathogenicity of the microorganisms that inhabit the organism in unknown and unpredictable ways. And the more precipitous the change, the more inestimable the effects on the pathogens are likely to be.

A fourth set of risks is environmental and ecological and is associated with the possibility of radically altering an animal and then having it get loose in an environment which was not anticipated. While this certainly seems like a minimal danger when dealing with intensively maintained cattle or chickens, it could surely pose a real problem with extensively managed swine, or with rabbits, or even with extensively managed cattle. Bitter experience teaches us that such dangers cannot be estimated, even with species whose characteristics are well-known (witness what happened with rabbits and cats in Australia and with the mongoose in Hawaii); a forteriori, an ignorance of what would happen with newly engineered creatures is even more certain.

We have talked briefly of the potential risks of genetically engineering agricultural animals on the animals themselves, on the general human population, on other animals, and on the environment. A fifth set of risks is relevant to a special subgroup of the human population, namely those who will actually be doing the experimentation on and genetic manipulations of the animals. Common sense tells us, and there is ample evidence to support this claim, that people working directly with dangerous materials are at greater risk than the general population. The last smallpox death in England came from a laboratory, and standard precautions are taken universally to protect people working with dangerous substances. But the need for extra vigilance in dealing with new situations is well illustrated by the deaths caused 20 years ago by Marburg virus, which did not affect laboratory workers who had been handling live monkeys, but which killed those people who had been collecting cells from dead animals for cell culture. In the case of genetic engineering, people handling the vectors used to introduce the genes could conceivably be at risk.

For any significant risks which we have discussed, or for others I may have omitted and which might pose real dangers, the imperative for their management can be generated without recourse to ethical considerations; rational self-interest and prudence would dictate that one not be cavalier about them. Thus even if a person has absolutely no concern for anyone but himself and his loved ones, he would wish to see anything that might do massive harm controlled, since he and his might just as easily as anyone else fall victim to its effects. Thus, in my view, following up an insight of Kant's, it is difficult to separate moral from prudential reckoning in such areas. Only when we consider the third and final aspect of the "Frankenstein thing" do we in fact encounter something that requires purely moral deliberation and decision, because morality and self-interest are very unlikely to coincide in these cases. In other words, one is unlikely to do the right thing for prudential reasons and, in actual fact, moral behavior in this area is likely to exact costs in self-interest. It is to these questions we now turn.

The final aspect of the Frankenstein myth is more difficult than the others to find in many of the popular renditions of the myth, but was in fact a central theme in Mary Shelley's novel. This dimension concerns the plight of the creature engendered by abuse of science. In the novel, the creature is innocent, yet isolated; shunned, mocked, abused, and persecuted in a plight not of its own making. Seeking love and companionship, it finds only hatred and rejection. One can find traces of this concern for the monster in the classic Frankenstein movie, and it is in fact a central theme in the recent remake of King Kong. Translated into the arena of genetic engineering of agricultural animals, this aspect of the myth, in essence, raises the question of the moral status of animals, of the rights of these animals, certainly the most difficult of the moral questions we have looked at in our discussion. And it is so difficult for a complex of reasons worth briefly detailing.

In the first place, when it comes to trying to get a purchase on our obligations to other creatures, we get little help from common sense, our intuitions, ordinary practice, the law, or even traditional moral philosophy. Common sense and ordinary practice say little about our obligations to animals, other than enjoining us to avoid cruelty, hardly a great help since most animal suffering and death is not the result of cruelty. (The great emphasis on cruelty to animals and love for animals is the major failing in the traditional animal welfare movement. Most scientists and agriculturists are not cruel, yet they invasively use countless numbers of animals. On the other hand, loving something is neither necessary nor sufficient for treating it morally. I certainly don't love most of the human surgeons I know; I don't even like them; yet, I am bound to treat them morally. By the same token, many people who love their pets mistreat them in countless ways, from providing improper diets to denying them exercise.) Our intuitions on animals are incoherent; often the same people who condemn branding of cattle will dock the ears and tails of their dogs. The law is of no help--in the eyes of the law animals are property, either private property or community property. The Animal Welfare Act, reflecting irrational social prejudice, does not consider rats, mice, or domestic farm animals to be animals; for purposes of the Act, a dead dog used in research is an animal, a live mouse is not. And traditional moral philosophy is of no help either, since for most of its history it was virtually mute on the subject of our obligations to other creatures. More has been written on this subject, in fact, in the past 10 years than in the previous 3,000.

All of this is further complicated by a major component of the same ideology of science that we discussed earlier, and that I have explored in detail elsewhere (7). From about 1920 until the mid-1970s, behaviorism was a major component of scientific ideology, and it was dogma to assert that we could not scientifically know that animals were conscious or even that they felt pain. Indeed, this is still dogma in many quarters--a USDA inspector recently told me that a medical researcher had informed him that dogs lack a sufficiently highly developed cerebral cortex to experience pain, and I have heard variations on this theme over and over. A leading veterinary pain expert told me that the majority of veterinarians still view anesthesia as a way of restraining the animal. (This was confirmed for me when I was lecturing at a leading veterinary school early on in my involvement with this issue and naively remarked that at least veterinarians could not doubt that animals felt pain, or else why would they study anesthesia and analgesia. Up jumped the associate dean, livid with rage.

"Anesthesia and analgesia have nothing to do with pain," he shouted, "They are methods of chemical restraint.") Analgesia is virtually never used on laboratory animals, and very rarely used in clinical veterinary practice. Ironically, rodents are the most infrequent recipients of analgesia, yet most pain and analgesic research is done on rodents, so the dose response curves are well-known.

Obviously, much scientific research, agricultural practice, and, indeed, ordinary activity rest on exploiting animals, so that it is far easier and more comfortable not to think about animals in moral categories. Nonetheless, common sense has never denied that what we do to animals matters to them, that they have needs and interests, physical and psychological, and that they can suffer, physically and psychologically, when those needs and interests are thwarted and infringed upon. For the past decade, society has just begun to realize the implications of its own assumptions about animals, and has begun to be aware that we do have moral obligations to them. Hence the rise of the animal rights movement, a massive international stirring which cannot be ignored, which questions much of our traditional treatment of animals, and which has been called "the Vietnam of the 80's."

At all events, a growing number of people in the scientific community are beginning to think seriously about the moral status of animals. In the past 8 years, I have lectured to over 30 veterinary schools all over the world on these issues, as well as to biomedical scientists of all sorts, attorneys, agriculturalists, psychologists, government officials, farmers, ranchers, and scores of other groups. I have testified before Congress and state legislatures, and served as a consultant to various agencies of three national governments. In the course of this decade, I have tried to develop an ethic to guide us on the uses of animals, one that I believe follows logically from moral assumptions we all share by virtue of living in democratic societies. In other words, rather than generate my own ethic and attempt to force it on others, following Socrates I have attempted to extract from others what their own moral assumptions entail about animals, though they may not and often do not realize it. Such an ethic is necessary, not as a blueprint for instant social change in all areas, but as Aristotle put it, as a target to aim at, and as a yardstick to measure our current conduct. Without such an ethic, as my colleague Dr. Harry Gorman, surgeon and researcher, has beautifully put it to me, we tend to confuse what we are doing with what we ought to be doing.

Given the constraints of time, I can only present the briefest sketch of this ideal for animals. For those of you who wish to pursue the topic more deeply, I would refer you to my book, Animal Rights and Human Morality (4). Stated boldly, I ask people to consider whether they can present any rationally defensible grounds for excluding animals from the moral arena, or from the scope of moral concern and deliberation. Surely animals are more like children than like wheelbarrows, in that they can be hurt, and that what we do to them matters to them. None of the standard, historically pervasive differences that have been cited to exclude animals from the moral arena will bear rational scrutiny. The claims that man has a soul and animals do not, that man is evolutionarily superior to animals, that man is superior in force to animals, that man is rational and animals are not, do not suffice to exclude animals from the moral arena and from falling under the purview of our socially pervasive moral concepts. In other words, given the logic of our moral ideas, there is no way to preclude extending them to animals. And this is not difficult to do.

In democratic societies, we accept the notion that individual humans are the basic objects of moral concern, not the state, the Reich, the Volk, or some other abstract entity. We attempt to cash out this insight, in part, by generally making many of our social decisions in terms of what would benefit the majority, the preponderance of individuals, i.e., in utilitarian terms of greatest benefit to the greatest number. In such calculations, each individual is counted as one, and thus no one's interests are ignored. But such decision-making presents the risk of riding roughshod over the minority in any given case. So democratic societies have developed the notion of individual rights, protective fences built around the individual which guard him or her in certain ways from encroachment by the interests of the majority. These rights are based upon plausible hypotheses about human nature, i.e., about the interests or needs of human beings that are central to people, and whose infringement or thwarting _matters most_ to people (or, we feel, _ought_ to matter). So, for example, we protect freedom of speech, even when virtually no one wishes to hear the speaker's ideas. Similarly, we protect the right of assembly, the right to choose one's own companions and one's own beliefs, and also the individual's right not to be tortured even if it is in the general interest to torture, as in the case of a criminal who has stolen and hidden vast amounts of public money. And all of these rights are not simply abstract moral notions, but are built into the legal system.

The extension of this logic to animals is clear. Animals too have natures, i.e., fundamental interests central to their existences, whose thwarting or infringement matters to them. This set of needs and interests, physical and psychological, genetically encoded and environmentally expressed, which make up the animal's nature, I call the animal's _telos_, following Aristotle--it is the pigness of the pig, the dogness of the dog. Such a notion is not mystical; it follows, in fact, from modern biology. Thus, it ill serves the issue at hand when scientists sneer at this notion, as one person at the NIH did recently, by suggesting that an animal's only nature is to serve us and die. According to the logic of our position, animals' basic interests as determined by their _telos_ ought to be morally and legally protected as well; this is the cash value of talking about rights. This, then, is what I take to be the logical extension of our socially sanctioned, moral notions when applied to animals, and when one cannot cite a morally relevant difference between people and animals which would forestall such application. Obviously animals do not have the same rights as humans, even ideally, since they do not have the same natures. So it will not do to ridicule the position by saying that I am urging that turtles have the right to vote or dogs have freedom of bark.

I have devoted much of my recent activity to attempting to actualize this ethic as far as is practically possible into veterinary medicine and research uses of animals, where its relevance is evident. But what does it tell us about genetic engineering of animals? Let me first of all clear up a misconception which has arisen about my notion of _telos_. It has been asserted by some opponents of genetic engineering that on my view _telos_ is inviolable, and it is immoral to change it. I have never said that. What I argue is that _given_ an animal's _telos_, certain interests which are part of that _telos_ ought to be inviolable. Thus given a burrowing animal, it is wrong to cage it so that it can't burrow. But I have never asserted that there is anything wrong with changing the _telos_ of a burrowing animal so that burrowing no longer matters to it.

The proper application of these ideas to genetic engineering of farm animals is made quite interesting by the fact that so much of our current intensive agricultural use of animals involves forcing animals into environmental contexts for which their natures are not suited. As a result, we must perpetually depend on highly artificial devices like debeaking in chickens and extensive uses of drugs and chemicals, and contend with "production diseases." While extensive agriculture has its own problems, at least the problems are, as it were, natural to the animal, rather than being created by the humanly devised management system. Ideally, from the point of view of the animals' welfare, I would like to see society back off from ever-increasing intensification. (This would, I think, have certain social and economic benefits as well.) But in all likelihood, increasing intensification is here to stay. So the main moral challenge to those involved in genetic engineering of agricultural animals is to avoid modifying the animal for the sake of efficiency and productivity at the expense of the animal's happiness or satisfaction of its nature. Economic pressures, of course, in the main, militate against my recommendation. This is why I asserted earlier that this is truly a _moral_ challenge. Also militating against this is the fact that hitherto the animal's welfare (except insofar as it affects economic productivity of an entire operation) has not entered into intensive agricultural decisionmaking or into research serving it. (This was freely admitted to me by a group of high-ranking USDA officials.)

Nonetheless, given the increasing public concern about the welfare of all animals, including agricultural animals, as well as the strong moral arguments in favor of concern for animals, it is imperative that this moral vector enter into agriculture. And certainly genetic engineering is an excellent place for this vector to be felt. The basic principles that should guide thinking in this area are not hard to see. Obviously, as a minimal principle, the animals should suffer no more as a result of genetic intervention than they would have without it. Ideally, they should suffer less and be happier. Thus, in my view, it would be grossly immoral (as has actually been suggested) to use genetic engineering to change chickens into wingless, legless, and featherless creatures who could be hooked to food pumps and not waste energy. Similarly, as has also been suggested, it would be wrong to manipulate the genome of pigs to produce leglessness, with the animals after all still having all the psychological urges to move. On the other hand, if genetic engineering is used to genuinely suit the animal to its stipulated environment, and therefore eliminate the friction between _telos_ and environment which clearly results in suffering, boredom, pain, stress, and disease, and this conduces to the animal's happiness, it does not appear morally problematic. Thus, if one were to genetically alter chickens' physical and psychological needs so that all evidence (such as results of preference testing, physiological signs of stress, behavioral signs of stress, individual animal productivity, and health) indicated that the animals were happy, this would be morally acceptable according to the theory I have been expounding, though many people, myself included, would certainly not be quite comfortable with it, probably on aesthetic grounds.

Obviously, therefore, these considerations of the animal's welfare, independently of the effect on humans, should be weighed and considered before a piece of genetic engineering is undertaken. And such consideration should be part of the formal charge of the committee we discussed earlier. Thus, if someone were to suggest using genetic engineering to create larger beef cattle, the researcher should be required to show that there is good reason to believe that the animal's joints could withstand the extra stress, and that no new suffering would be engendered by such genetic manipulation. In this way, we can at least begin to assure that the animals' interests are weighed along with ours. In the case of totally virgin territory, as in the creation of chimeras, an even stronger burden of proof should be put on the proposer to demonstrate that this manipulation would not lead to suffering.

In sum, in my view, the genetic engineering of animals in and of itself is morally neutral, very much like the traditional breeding of animals or, indeed, like any tool. If it is used judiciously to benefit humans and animals, with foreseeable risks controlled, and the welfare of the animals kept clearly in mind as a goal and a governor, it is certainly morally non-problematic and can provide great benefits. On the other hand, if it is used simply because it is there, in a manner guided at most only by considerations of economic expediency and "efficiency," or by quest for "knowledge for its own sake," with no moral thinking tempering its development, it could well instantiate the worst rational fears encapsulated in "the Frankenstein thing." To those of you upon whom the primary responsibility for this choice rests, let me conclude by reminding you that though Frankenstein was in fact the name of the scientist, virtually everyone thinks it is the name of the monster.

ACKNOWLEDGEMENTS

I wish to thank Linda Rollin, M. Lynne Kesel, David Neil, Murray Nabors, Robert Ellis, George Seidel, and Dan Lyons for dialogue and criticisms.

REFERENCES

1. Baker, J.R. (1984) Race, Oxford University Press, London, pp. 65ff (see the discussion and references thereto).
2. Glut, D.F. (1984) The Frankenstein Catalog, McFarland, Jefferson, North Carolina, p. 525.
3. Rollin, B.E. (1979) On the nature of illness. Man and Medicine 4(3): 157ff.
4. Rollin, B.E. (1981) Animal Rights and Human Morality, Prometheus, Buffalo, 185 pp.
5. Rollin, B.E. (1981) Nature, convention, and genre theory. Poetics 10: 127-143.
6. Rollin, B.E. (1985) The moral status of research animals in psychology. Am. Psychologist 40(8):920-926. See also B.E. Rollin (1983) The Teaching of Responsibility, Universities Federation for Animal Welfare, Potters Bar, Hertfordshire, England, 30 pp.
7. Rollin, B.E. (in press) Animal consciousness and scientific change. In New Ideas in Psychology, R. Kitchener, P. Moessinger, and J. Broughton, eds. Elsevier, Amsterdam. See also B.E. Rollin (in press) Animal pain. In Advances in Animal Welfare Science. 1985/1986, M. Fox and L. Mickley, eds. Martinus Nijhoff, The Hague.
8. Rolston, H. (1984) Duties to endangered species. A version of this paper is forthcoming in BioScience, November or December, 1985.
9. Ruse, M. (1973) The Philosophy of Biology, Hutchinson, London, 127 pp.
10. Time, July 29, 1985, pp. 54-59.

CHARACTERISTICS OF FUTURE AGRICULTURAL ANIMALS

George E. Seidel, Jr.

Animal Reproduction Laboratory
Colorado State University
Fort Collins, Colorado 80523

INTRODUCTION

Writing a short chapter on a topic such as "Characteristics of Future Agricultural Animals" by necessity entails considerable arbitrariness. One is confronted immediately with the need to define or at least delimit the words agriculture, animals, and future. Furthermore, one must decide whether to limit considerations to a particular country or group of countries. One also must make tacit assumptions about economic growth, international cooperation, lack of major destructive wars, sociological phenomena such as animal rights movements, and the nature of our own species including curiosity, selfishness, intelligence, etc. Worse yet, this is not an appropriate task for one whose training and subsequent employment have been in mammalian reproductive physiology. Probably my best credential is having grown up on a farm.

I have decided first to concentrate on the year 2025. This coincides with my life expectancy, which means that I have a 50:50 chance of living to see what has occurred by about that time. Furthermore, longer times seemed impossible to aim at with any accuracy, and shorter times seemed less interesting.

Second, I will limit the discussion to birds and mammals for several reasons. Currently they constitute the bulk of the animal agriculture enterprise, and this probably will be true in 2025. Most other domesticated food animals are fish and shellfish. Clearly, aquaculture is providing an increasing proportion of our food supply, and this trend almost certainly will continue, exploiting such theoretical advantages of poikilotherms as not needing to use energy to maintain body temperature as well as other factors (3).

I also have decided to limit my consideration to North America. However, it must be emphasized that over 90% of the world's population lives outside of North America, and these people cannot be ignored, even in the context of agricultural animals in North America, due to trade, flow of information, etc.

A final problem is defining animal agriculture. I have chosen to use a broad definition which includes the horse industry, as well as cattle, sheep and goats, swine, and poultry. Other speciality species such as rabbits, bison, mink, and musk oxen will be considered only superficially.

HISTORY OF NORTH AMERICAN FARM ANIMALS

With exception of turkey, all major species of farm animals were introduced into North America via Europe. Most were originally domesticated in Asia and Africa. Domestication of animals must be ranked along with use of fire, invention of the wheel, and development of writing as major human achievements. Although some species undoubtedly were more amenable to domestication than others, there probably was a huge element of serendipity among the reasons that some species were domesticated and others were not. Undoubtedly, there was considerable unconscious selection in the process of domestication, e.g., for docility or overt estrous behavior (1), but there was little scientific basis for selection until the last century when breeding societies were founded in England. Selective breeding is nothing more than choosing the parents of the next generation. This involves identification and propagation of superior parents; techniques such as artificial insemination, superovulation, and embryo transfer are particularly useful for this purpose (2), although enormous progress also is possible by arranging for selected males to mate naturally at a high frequency.

Progress in breeding lines for selected purposes has been astonishing. Spectacular early examples include ponies and draft horses. More recent examples include specialized lines of poultry, hogs, and cattle. A complication in evaluating genetic progress is the changing goals of animal breeding. Decades ago, when lard was in demand, exceptionally fat hogs were produced for market, which are very unlike today's meat-type hogs. To a lesser extent, similar changes have occurred with beef cattle. While fatness is dictated to a great extent by feeding policies, there is also a huge genetic component. Clearly, it is possible to select for the number of adipocytes in the body.

Cattle represent an interesting case of differing selection objectives in time and space. The major uses of cattle are for meat, draft power, milk, and hides. Each of these four has been the predominant use in one culture or another. Interestingly, use of cattle for milk production was minor until several centuries ago. The main use until then in Europe was for draft power, with meat and hides as by-products. In North America, oxen greatly outnumbered horses as a source of draft power prior to the Civil War, and the North American dairy industry only got started in the last century. Milk was clearly in use earlier, but it was a relatively minor product of cattle in this country.

There are other uses of cattle, too, primarily in other parts of the world: dung for fuel, blood for food, various religious purposes, as a buffer for use of excess feed supplies, as a means of currency (e.g., dowries in certain African tribes), recently as a tax shelter in North America, and as a source of prestige. In fact, there are remarkable similarities between certain African tribes and North American businessmen and professionals with regard to owning cattle for prestige. (For a more detailed treatment, see Ref. 11.)

Only in the last few centuries have cattle been selected for milk production to any extent. In the last few decades, the mathematical, statistical, and computational techniques for such selection have become remarkably sophisticated and specialized. Currently, the genetic trend in milk production per cow is on the order of 40 kg/yr. Note that this is a cumulative function. Similar advances have occurred with swine and, even more remarkable advances, with poultry. Selection of beef cattle and sheep has been slower. Essentially all of this genetic progress has been the result of empirical selection of parents of the next generation based on the phenotype of either the animals themselves or their relatives, primarily progeny. Now, however, it is possible to move genes one at a time, even across species, whereas in the past one always was dealing with haploid sets of genes in selection programs. Thus, to some extent, it will be possible to design the parents of future generations of livestock.

NEEDS FOR ANIMAL PRODUCTS IN THE YEAR 2025

Technically, it would be possible to dispense with agricultural animals completely in North America by 2025. We could fill all of our food needs using plants, fossil fuels, minerals, air, water, and tissue culture techniques. There would be considerable social, ecological, and economic costs to such a course of action, but it could be done without compromising the health of the population. This will not occur for a variety of reasons, especially reasons of economy. On the other hand, the ratio of plant, animal, and synthetic organic products is bound to change, probably to less animal and more synthetic. These changes are likely to be dramatic in some instances, as occurred, for example, with the switch from wool to synthetic fibers. This had a lot to do with the drop in numbers of sheep in the United States to one-fourth of their peak numbers (13).

An especially interesting question is whether substantial amounts of milk and meat products will be made by cells in tissue culture. The main components of meat are actin and myosin proteins, and water. Perhaps the proteins can be made in tissue culture, seasoned with fats and flavors, and sold without bones, tough connective tissue, blood vessels, etc. Cheese, the main ingredients of which are caseins and water, is already being synthesized industrially. Sodium caseinate is considered an industrial chemical, and artificial cheeses are made from it, particularly for use in pizza and the like. Currently, the sodium caseinate is extracted from milk, but it is a small step to obtaining it from transformed mammary epithelial cells growing in vitro. The amino acids required for casein synthesis by such cells can easily be obtained from plant byproducts. These enterprises will be industrial, not agricultural.

However, animal products will be displaced only partially. In fact, the population will still be growing in 2025 in North America, and the increased numbers of consumers probably will make up for loss of markets to nonanimal products. The net effect probably will be little change in the total consumption of animal products by 2025. Many other factors also will affect animal product consumption. For example, the American dairy industry now is one of the nation's largest advertisers, and sales of dairy products have increased demonstrably as a result.

Probably the largest determinant of animal product usage will be cultural inertia. All cultures have remarkable fetishes about their food. These tastes take generations to change. The culture and subcultures of North America are no exception. Ham and eggs, hot dogs, and ice cream will not soon disappear from menus. That is not to say that patterns of consumption will not change, but simply that there is much inertia. Another cultural determinant of consumption patterns is "snob appeal." One can impress a client or potential spouse with a steak dinner; soup and salad bar usually will not do. So-called "health foods" and "organically grown produce" are another factor to consider.

One area of animal agriculture that has grown enormously in recent years and will continue to grow is the recreational horse industry. In our undergraduate animal science program at Colorado State University, more students are interested in horses than all other species combined, and jobs are available when they graduate.

Ownership of beef cattle often has a recreational component. A huge proportion of the brood cows that produce the calves that are fattened for slaughter is owned by people whose major source of income is from other sources. These people include professionals, people who live in the country but work in the city, farmers whose main livelihood comes from sources such as crops or poultry, businessmen who may or may not use cattle as a tax shelter, etc. The net effect of this is that much of the industry is not affected by the classical laws of economics, and it is one of the reasons that the beef cow/calf business has been able to survive while being extremely unprofitable for the past 15 years. Even those who thought that they were making a living with cow/calf operations really were living on land appreciation until a few years ago. When land started to depreciate substantially, bankruptcies were declared by these enterprises on a scale not seen since the Great Depression, and a lot of those still hanging on are doomed. The point is that rules of profitability are severely distorted in the cow/calf segment of the beef cattle industry. Another way of thinking of it is that many property owners have cattle to keep the weeds down, and the beef production is a byproduct. Depending on precise definitions, as many as 25% of beef cows are owned by people in this situation. This trend probably will increase.

As can be seen, animal agriculture, particularly with beef cattle and horses, is a complex sociological phenomenon. This is one of the myriad of reasons that North American animal agriculture will still be a thriving enterprise in 2025.

CHARACTERISTICS OF FARM ANIMALS IN THE YEAR 2025

Despite the remarkable cultural complexity of animal ownership, except for horses, the driving force determining the kind of animals on farms in 2025 will be profitability. This means producing animal products that meet consumer demands as efficiently as possible. In response to animal welfare concerns, there will be more attention paid to the way animals are kept; to some extent animals will be bred with personalities to fit their environments. Fortunately, in most cases the well-being of animals is closely related to efficient production. Table 1 includes some of the characteristics that may be found in farm animals in 2025. Some of these will be discussed in more detail below.

Disease Resistance

Many knowledgeable people advocate breeding animals for disease resistance. Chickens can be selected for resistance to Marek's disease, and cattle for resistance to a variety of tropical diseases and pests. However, for many diseases such as tuberculosis in cattle, the approach is to keep the environment free of the causative organism by maintaining closed herds and by rigorously destroying all infected animals. Specific pathogen-free swine herds are another approach. It is not always possible to keep the environment free of certain microorganisms, but good husbandry practices increasingly rely on quarantine of incoming animals until disease status is ascertained. For many diseases, vaccination is effective and inexpensive. The new biotechnologies will improve the efficacy of vaccines even more (6).

Clearly, there are some diseases for which breeding for genetic resistance will be more effective than other methods of control. One possibility in North America is foot rot in sheep, which has been shown to respond

302

Tab. 1. Ideas for improving efficiency of animal agriculture.

- Modify animals genetically so that they no longer are seasonal breeders.
- Add double muscling and growth genes to the Y-chromosome.
- Develop in vitro oogenesis and spermatogenesis. This results in unlimited gametes and eliminates the need to keep animals for breeding purposes except for gestation.
- Increase appetite to get more production relative to maintenance costs.
- Develop methods or genes for hibernation to overwinter animals.
- Modify animals genetically for earlier puberty and markedly shorter gestation.
- Control timing and rate of ovulation precisely to produce twins in cattle, sheep, and goats, large litters in swine, and daily ovulation in chickens.
- Karyotype and test embryos for enzyme deficiencies before embryo transfer. This may result in markedly less embryonic death and obviates the need to sex semen.
- Rejuvenate hens by simulating molting.
- Add genes for generalized disease resistance.
- Decrease turnover of gut epithelium to markedly decrease maintenance costs.
- Develop high protein milk, low fat meat, and low cholesterol eggs.
- Dispense with difficult births by having young born at much smaller sizes.
- Exploit mitochondrial and other cytoplasmic inheritance.
- Improve biotechnical techniques such as cryopreservation, pregnancy tests, artificial insemination, in vitro fertilization, embryo transfer, gene transfer, etc.

to selection. Mastitis in dairy cattle is another possibility. However, both of these diseases can be alleviated to a great extent by proper management.

What about manufacturing disease-resistant animals by gene injection? Unfortunately, little is known about the molecular basis of resistance to disease in farm animals. Undoubtedly, as such information accumulates, there will be several opportunities for modifying the genome for disease resistance. I expect several spectacular successes in this area by 2025. On the other hand, this will not be the method of choice for dealing with most diseases.

Size

A common misconception is that a major objective in animal engineering is to increase the size of animals. In most instances the opposite is true, simply because smaller animals cost less to maintain. Consequently, chickens and pigs have become dramatically smaller in recent years. Beef cattle have become larger, but that is primarily due to introduction of larger breeds from Europe for crossbreeding. Because of the huge maintenance costs, most people agree that it is an economic disaster to have breeding cows get too large. Indications are that dairy cattle within breeds are getting smaller. These smaller animals do not win cattle shows, but they are efficient. We can expect that most farm animals will be slightly smaller in 2025 than in 1985.

Age

The average age of farm animals has been decreasing for years and this trend will continue. One reason that crossing cattle with bison has been uneconomical is their slow growth rate and late maturity. It is true that such animals have slightly better survival characteristics in adverse circumstances, but interest rates on capital that does not turn over with reasonable rapidity destroy potential profit margins. For meat purposes, one wants rapidly growing animals that reach market weight quickly. For milk and eggs, rapidly maturing animals that come into production quickly are the objective. Most layers are made into soup by the time they are 1½ years old because younger animals are more efficient. With dairy cattle, the mean culling age currently is sometime in the third lactation. Dairy cows seem to succumb to one problem or another at an earlier age than seems desirable, especially considering the huge investment required to bring a cow into production at about 2 years of age. Nevertheless, the average age of dairy cows is only about 5 years, and it seems likely that the national herd will become younger still.

How much younger will animals be? Huge gains can be made by decreasing the age at puberty if growth rates are increased concomitantly. For dairy cattle, layers, and breeding animals of other strains and species, one also desires longevity so that costs of raising the animals can be spread over a long, productive life. Probably earlier puberty and longevity are both attainable in the same animal. A major driving force to younger animals is genetic improvement. With a good breeding program, each generation is better than the previous one, and older generations must be replaced to take advantage of this improvement. The net result of these forces will be younger animals, except for horses.

Numbers of Animals

Recently, the number of farm animals has changed dramatically, but the pattern has varied with the species. In the United States there are about 11,000,000 dairy cattle, in contrast to the 25,000,000 peak in 1945. Today's dairy cow produces 2.5 times more milk than one did during World War II, and the national herd produces considerably more milk on about two-thirds of the nutrients because there are 14,000,000 fewer cows to maintain.

In the past, beef cattle numbers have followed a 10-year cyclical pattern. Numbers peaked in 1975 but have declined by 22 million in less than a decade (8). It is unlikely that the national beef herd will ever be as large as it was in 1975. Sheep numbers have declined to 25% of their peak numbers when wool was in demand. There are many fewer hogs than there were 5 years ago, and numbers will continue to decline in the long run. Goat numbers are small and unlikely to change much.

Numbers of laying chickens have also declined over the years. With the exception of horses, it is only the number of poultry for meat that has increased, primarily because lower-cost chicken and turkey have replaced lamb, beef, and pork in the diet. Except for horses and poultry for meat, the trend for the numbers of all farm animal species to decline will continue. The main reason is that the animals are becoming more efficient, and it simply will take fewer of them to produce the same amount of food. The movement to poultry, fish, and nonanimal products for food is a second reason for declining numbers of cattle, hogs, and sheep.

In the absence of major wars or social upheaval, as suggested earlier, the total amount of animal products consumed in North America in 2025 probably will be similar to the amount consumed in 1985. The number of horses

304

may double by then, but numbers of dairy and beef cattle and swine probably will drop to 60% of the current national herd size. Numbers of sheep and goats probably will not decrease much, but many of them will be in "hobby" flocks and herds. Poultry numbers will drop to 70% of 1985 levels; numbers of layers will decline more than meat birds. It is unlikely that new species will replace current species to any extent on this continent with the possible exception of increased fish. We will continue to have deer and buffalo, geese, ducks, quail, rabbits, and numerous other speciality species; however, they are unlikely to become a major source of food by 2025. On the other hand, there will be profitable niches for some of these.

Other Considerations

I expect that I have conveyed the impression that animal agriculture will not change dramatically over the next four decades. In fact, I have barely dealt with economics, production systems, patterns of ownership, etc. These may change markedly. I probably have left the impression that agricultural animals themselves would not change much. In fact, the animals will change considerably for two reasons: (a) the new methods of genetic engineering, and (b) the relentless cumulative effects of selection with each generation. We have no experience with combining these two strategies for improving animals, but it is likely that they will be synergistic.

Probably cows will still look like cows, chickens like chickens, etc., but their metabolic patterns will change considerably. For example, the current strategy with dairy cows is to accumulate on the order of 100 kg of fat during a dry period prior to calving, and then have the cow in negative energy balance for some weeks because nutrients in the milk produced plus maintenance requirements exceed feed intake capability. Probably a sounder strategy would be to start milking a cow at 1½ years of age and continue for 5 years without the annual cycle of reproduction including dry periods, negative energy balance, etc. We need to learn considerably more about initiation and maintenance of lactation to make this a functional reality. With sexed semen, embryo transfer procedures, and other management changes, there would be no need for each cow to have a calf each year. Even now the value of the average dairy calf is considerably less than 10% of the value of the milk produced during the resulting lactation.

Numerous other changes in biologic characteristics may evolve, for example, circumventing seasonal breeding, changing gestation lengths, improving fertility, etc. Probably other useful characteristics will be bred into animals too, e.g., bulls that produce only X-bearing sperm possibly with identical twin brothers that produce only Y-bearing sperm. Analogously, chickens may be available that produce only Z- or W-chromosome-bearing eggs.

HOW TO GET FROM 1985 ANIMALS TO 2025 MODEL ANIMALS

While today's farm animals will be the ancestors of those in 2025 in most respects, some of the genes will come from other sources. Probably some genes will be custom-designed by computer and introduced into the species by injection of genes or chromosomes into embryos. Some genes will be moved from one species to another. In some cases, the objective will be to remove or inactivate a gene. Other ideas are presented in Tab. 1.

Although such techniques will be used, animals are much more complicated than computers or spaceships, and adding new genes will, in most instances, create more problems than it solves. In some cases, however, genes from sources other than the species of interest will make important contributions to productivity.

Let's consider how introduction of a new gene might be accomplished by examining some case histories. Genetic traits such as the polled (no horns) condition can be moved around without much consequence to the basic physiology of cattle. Horns are a nuisance. Good husbandry practices require their removal. This really becomes a chore if not done at a very early age. Genes for the polled condition have arisen in various populations of cattle. In most cases, the trait is inherited in a simple dominant manner. All cattle in a number of breeds are naturally polled. Although considerable efforts have been made to promote polled cattle, most cattle in North America continue to be born with horns. It would not be particularly difficult to wipe out the genes for horns in cattle. Why has this not occurred for such an obviously useful trait?

One reason is that the economic value of this trait is probably only a few dollars per head at most, perhaps around $100,000,000 per year in North America. However, the gene pool of polled animals is only on the order of 25% of the breeding population. The cost in productivity of selecting future parents only from this pool would be enormous, particularly in a breed such as Holstein dairy cows. Thus, the problem is not that the polled gene is not desirable, but that the genes that come with it are not good enough. Thus, introduced genes should be placed on appropriate genetic backgrounds, especially in species with a long generation interval such as cattle. This is an opportune point to indicate that the average generation interval in cattle is about 6 years (average of males and females), and thus there will be an average of only seven generations of cattle between now and 2025 in this species unless there are marked changes in breeding practices (see, e.g., Ref. 14).

Perhaps a more relevant example of introducing new genes is illustrated by fecundity in sheep. Under some circumstances, it is an absolute disaster for sheep to have more than one offspring per breeding season. For example, on the western range twins frequently will die. On the other hand, under more intensive management, ewes that produce only one lamb per year are unprofitable; twins can be very profitable for producing lamb. Fortunately, there are several breeds of sheep that routinely have multiple births, e.g., the Finnish Landrace. Furthermore, a single gene for multiple births, the Booroola gene, has been identified in Australian Merino sheep, which usually have single births (see Bindon and Piper, this Volume). Again, it is a huge effort to get this useful genetic trait into the sheep population. The time-frame will be more than a decade. This is in part because the average generation interval in sheep is 4 to 5 years, but more importantly because the carcass characteristics of the Finnish Landrace and Booroola Merino do not measure up to those of breeds traditionally used to produce lamb. It has been difficult to breed for increased fecundity and maintain the quality of the animals. This is not to suggest that such genes are of no value; they are of tremendous value. Gene injection procedures offer a method of circumventing these problems.

It may be instructive to speculate what might happen if extra copies of a gene that increases growth rates were introduced into meat-producing species such as cattle, hogs, sheep, or poultry. Unfortunately, such speculation requires a myriad of assumptions that almost certainly will not be met exactly. However, assume that an active promoter is attached to the extra gene(s), which is regulated rather nonspecifically. Further, assume that growth is more rapid. Unfortunately, this probably would lead to economic disaster with sheep and cattle, because more rapid growth almost always brings with it the undesirable correlated trait of larger mature size. Probably with diligent selection procedures over a period of years, this tendency to larger mature size could be dealt with effectively, essentially by various modifying genes. Large mature size would be disastrous with sheep and cattle because of the huge maintenance costs for the breeding females. This would be less of a problem with hogs or poultry since feed costs for maintenance of breeding females are a much lower percentage of total feed costs than with sheep or cattle. Moreover, to take advantage of rapid genetic progress it will probably become economically advantageous to slaughter sows after the first litters, rather than maintain them for breeding purposes.

Suppose that the introduced growth gene(s) has a promoter that can be turned on specifically by an orally fed molecule, e.g., a derivative of melatonin. In this case, breeding animals might never have increased rates of growth; only those to be fattened would be fed the melatonin derivative. Of course, side effects, such as reduced fertility, might occur even in animals that are not induced to express a gene. This would have to be sorted out on a case-by-case basis. Feeding substances such as heavy metals or a derivative of melatonin also would have to be tested for safety. In beef cattle, since males grow more efficiently than females, the breeding animals could be kept small if the introduced growth gene were on the Y chromosome.

One point that seems to be clear to those who have thought about it is that the methodologies for changing from current to future farm animals will be a mixture of traditional and new technologies. At the present time, there is no more powerful technology for making genetic progress than selecting the best animals in the population and making them parents. Once superior animals are identified, the most effective tool for propagating their genes almost always is artificial insemination, in the case of females, via their sons. Techniques like embryo transfer, and splitting, sexing, and freezing embryos are important but ineffective relative to artificial insemination. Combining some of these technologies with artificial insemination can be extremely effective (10,14).

Techniques, such as gene injection, definitely will be important in the future, but the relatively few animals produced in this way will be propagated by the more conventional methods already in use. Embryo transfer coupled with artificial insemination can be very useful to build up a nucleus of desirable homozygous animals that can be propagated conventionally. Just making animals homozygous for a desirable trait is a huge effort in a species like cattle, which usually do not produce young until 2 years of age. One has the other interesting problem of whether the homozygote is a normal, fertile animal, especially since some inbreeding will be required to fix the new gene in the homozygous state.

Man has numerous tools for improving or changing animals to suit his needs. Frequent questions that arise include: "What new technologies are on the horizon?" and "How soon will the majority of cattle be born from embryo transfer?" and "When will sexing of sperm become available?". Of course, no one has the answers to these questions. I have speculated on them before (9,12). Several times per decade it seems that truly novel approaches come along that quickly become incorporated into the repertoire of biological scientists. Examples include monoclonal antibodies, restriction endonucleases, and gel electrophoresis. At a more macro level, tools such as embryo transfer, nuclear transplantation, and cryopreservation of embryos are examples. All of these techniques have a continuum of uses, from basic research to applied agriculture. We can be certain that additional techniques will evolve, such as, for example, transplantation of individual chromosomes or development of artificial chromosomes for mammals (5). However, there is an awful lot left to learn. For example, we do not even know the molecular basis of hybrid vigor. Also, it is embarrassing to relate that fewer than 5% of beef cattle, sheep, swine, and horses become pregnant by artificial insemination in North America. That this simple, inexpensive, effective technology is not used more widely attests to the complexity of applying technology (4).

It seems to me that one area in which many agricultural scientists and administrators have things backwards is in the application of the new technologies. The main short-term value of many technologies lies in obtaining information, which then can be put to use, rather than for direct use of technology to increase production. The production of transgenic animals, for example, provides unique models about effects of genetic changes on physiological processes. Much of the information gained from this very useful technology probably will be applied without using transgenic animals at all.

PHILOSOPHICAL CONSIDERATIONS

In designing animals for the future, the most common objective is to provide inexpensive food, or more food for the hungry masses. This is an admirable goal, and our efforts certainly will contribute to this objective as they have in the past. However, in many, perhaps most, situations, political, sociological, and especially economic considerations have much more to do with food production than new technology. We need to remember that technology is only a part of the solution to mankind's problems, and sometimes only a small part. Furthermore, reasons for development of a given technology are exceedingly complex. Gomory (4) has written an excellent paper on this subject.

A common misconception is that technology is beneficial for farmers. This essentially is never true in the long run, because farmers are forced to adopt new technologies or go out of business because their more efficient neighbors have already adopted them (7). Consumers benefit from such increased efficiency, but it puts those farmers who do not adapt out of business.

A final philosophical point concerns certain aspects of the utility of agricultural animals. At times it is painful to some to see resources used in breeding food animals that seem to have more of a recreational than a production objective. One could make the same argument for most aspects of the horse industry. However, production of meat from mammals is a very biologically inefficient process in any case, and under most circumstances in North America the nutrients used in this enterprise would go much further if eaten directly by man. There are exceptions such as ruminants grazing rangeland and hogs eating garbage, but these represent a minority of nutrients used.

Should we feel guilty for eating a steak or riding a horse? I think not. Currently there are huge surpluses of grains in North America, and getting rid of animal agriculture would have a decided negative effect on plant agriculture for years to come. Getting rid of animal agriculture would have little impact on world hunger; in most cases there would be no impact. In the long run, less grain would be produced without animals, not the same amount that somehow would find its way to starving people. I do not wish to imply that we should be wasteful or unphilanthropic. On the other hand, dispensing with agricultural animals is not a solution to anything.

In the future, new species will be created, possibly extinct ones re-created, and probably new hybrids will come into use. From the animal agriculture standpoint, the gene pools in use likely will become narrower within any geographical area. However, reservoirs of genes will be available from frozen semen and frozen embryos. Will such prognostications mean anything a century or a millennium from now? Will current agricultural animals be kept primarily in zoos or, for all practical purposes, as pets? Will various species be kept primarily as models to satisfy scientific curiosity? It is clear that answers to these questions will depend heavily on social, political, and demographic factors. From the scientific standpoint, it will be possible to make absolutely dramatic changes. Breeding zebras with feathers, even in a striped pattern, will be easy. Breeding animals with extra or fewer limbs will be possible. Chicken wings could be dispensed with. Limits to such changes include laws of nature, laws of society, and allocation of resources. It is quite clear that even a millennium of scientific advances, however, will not answer all questions. In fact, the number of questions may increase. Hopefully, we will no longer be asking questions to relieve human misery, although the problem of keeping the body and mind going to older and older ages seems an infinite one. Except for this, the best of all possible worlds is that work with animals will be done primarily because it is interesting, not because it is morally necessary to relieve human misery. Interestingly, most science of previous centuries also was done primarily to satisfy curiosity.

REFERENCES

1. Baker, A.E.M., and G.E. Seidel, Jr. (1984) Why do cows mount other cows? Appl. Anim. Behav. Sci. 13:237-241.
2. Brackett, B.G., G.E. Seidel, Jr., and S.M. Seidel, eds. (1981) New Technologies in Animal Breeding, Academic Press, New York.
3. Clark, Jr., W.H., and A.B. McGure (1981) Fish and aquatic species. In New Technologies in Animal Breeding, B.G. Brackett, G.E. Seidel, Jr., and S.M. Seidel, eds. Academic Press, New York, pp. 91-106.
4. Gomory, R.E. (1983) Technology development. Science 220:576-580.
5. Murray, A.W., and J.W. Szostak (1983) Construction of artificial chromosomes in yeast. Nature 305:189-193.
6. National Research Council Committee on Biosciences Research in Agriculture (1985) New Directions for Biosciences Research in Agriculture, National Academy Press, Washington, D.C.
7. Niswender, G.D. (1985) How is research in animal agriculture funded? In Technology in Animal Agriculture: A Seminar for Investment Strategists, G.E. Seidel, Jr., ed. Colorado State University, Fort Collins, Colorado, pp. 304-313.
8. Riley, J. (1985) Beef cattle industry. In Technology in Animal Agriculture: A Seminar for Investment Strategists, G.E. Seidel, Jr., ed. Colorado State University, Fort Collins, Colorado, pp. 48-74.
9. Seidel, Jr., G.E. (1980) Management of reproduction in cattle in the 1990's. (Visiting Scholar Lectures.) Univ. Arkansas Agric. Exper. Station Special Report, 77, Carl B. and Florence E. King Visiting Scholar Lectures, pp. 51-59.
10. Seidel, Jr., G.E. (1984) Applications of embryo transfer and related technologies to cattle. J. Dairy Sci. 67:2786-2796.
11. Seidel, Jr., G.E. (1985) Embryo transfer in cattle--A special case of high technology application. In Technology in Animal Agriculture: A Seminar for Investment Strategists, G.E. Seidel, Jr., ed. Colorado State University, Fort Collins, Colorado, pp. 159-178.
12. Seidel, Jr., G.E., and S.M. Seidel (1981) The embryo transfer industry. In New Technologies in Animal Breeding, B.G. Brackett, G.E.

Seidel, Jr., and S.M. Seidel, eds. Academic Press, New York, pp. 41-80.

13. Silva, W.J., and G.D. Niswender (1985) Overview of sheep industry. In Technology in Animal Agriculture: A Seminar for Investment Strategists, G.E. Seidel, Jr., ed. Colorado State University, Fort Collins, Colorado, pp. 84-101.

14. Van Raden, P.M., and A.E. Freeman (1985) Potential genetic gains from producing bulls with only sires as parents. J. Dairy Sci. 68:1425-1431.

WHERE DOES GENETIC ENGINEERING LEAD?

Charles C. Muscoplat

Molecular Genetics, Inc.
10320 Bren Road East
Minnetonka, Minnesota 55343

INTRODUCTION

Genetic engineering technology is the major scientific revolution of the century. Rapid developments are being made, especially in veterinary medicine. Genetically engineered vaccines and monoclonal antibodies are already in the marketplace for veterinary use. Many other animal health care products are being tested.

In agriculture, genetic engineering is focusing on manipulation of organisms to produce animal vaccines, hormones, amino acids, chemicals, and drugs. These technologies will have the greatest impact on improved livestock production by: (a) reducing animal losses through prevention of infectious diseases, using effective genetically engineered vaccines and antitoxins; (b) increasing production of meat and milk through use of growth promotants; and (c) improving the nutritional value of animal feed.

THE TECHNOLOGY

Two major developments in genetic engineering, specifically recombinant DNA and monoclonal antibody technology, have been used to address problems in animal health care and production.

Recombinant DNA technology, the essence of genetic engineering, is not a single discipline; it represents a fusion of ideas and techniques from biochemistry, molecular biology, genetics, organic chemistry, immunology, and medicine. This scientific breakthrough involves restructuring and editing genetic information and constructing microorganisms with new genetic information.

The technology allows us to isolate genes from any source (viruses, bacteria, fungi, plants, or animals) and amplify these genes to unlimited quantities. It also allows us to manipulate genes by mutating or rearranging their components to develop hybrid or novel gene products.

THE NEED

In the United States, products from the dairy and beef cattle industries account for approximately one-third of the total farm income derived from livestock.

Each year, the dairy industry increases production, providing more milk from fewer cattle. Costs of production have increased dramatically during the past decade. Therefore, the most serious challenge facing the dairy industry is not how to improve productivity, but, rather, how to improve efficiency of production.

In well-managed herds using existing technologies of feeding and breeding, the most serious limitation to efficient production is the presence of disease. Control of reproductive, digestive, and respiratory diseases is essential to realize maximum productivity and profitability in the dairy and beef industries.

LIMITATIONS

The American Veterinary Medicine Association estimates the economic loss of livestock due to infectious diseases to be several billion dollars each year. Of the 45 million calves born annually in the United States, approximately 7% die of infectious diseases in the first 6 months of life. Approximately 15% of the 100 million pigs born annually die from infectious diseases within the first few months of life. These losses occur despite the availability of hundreds of vaccines, drugs, hormones, vitamins, feed additives, and antibodies.

It is widely accepted that many devastating outbreaks of disease occur because antibiotics are ineffective in controlling or limiting the severity of infectious disease, and because effective vaccines are not available to producers.

APPLICATIONS

Vaccines and novel antibodies were the first genetically engineered products to reach the marketplace. Vaccines or antibodies for calf scours, parvoviruses, wart disease, Rift Valley fever, foot-and-mouth disease, transmissible gastroenteritis virus, adenoviruses, infectious bovine rhinotracheitis virus, and bovine viral diarrhea virus are already under development. Genetically engineered subunit vaccines will exhibit features of potency and efficacy, plus safety, ease of manufacturing, and economy.

Other major genetically engineered products will be natural growth hormones for livestock and poultry. Many scientists think these hormones hold greater potential for agriculture than vaccines do. Other scientists foresee hormones replacing existing steroids or other growth promotants.

The feed industry will gain enormously from developing genetic technology. Using recombinant DNA, researchers will develop microorganisms for production of less expensive and more nutritious feed ingredients. Genetic engineering will eventually help increase crop yields; produce more nutritious crops; and produce less expensive vitamins, amino acids, and single-cell proteins.

Finally, some organisms produce antibiotics in such low concentrations that it is not practical to recover them. Through recombinant technology, these organisms can be made to produce antibiotics in commercial quantities.

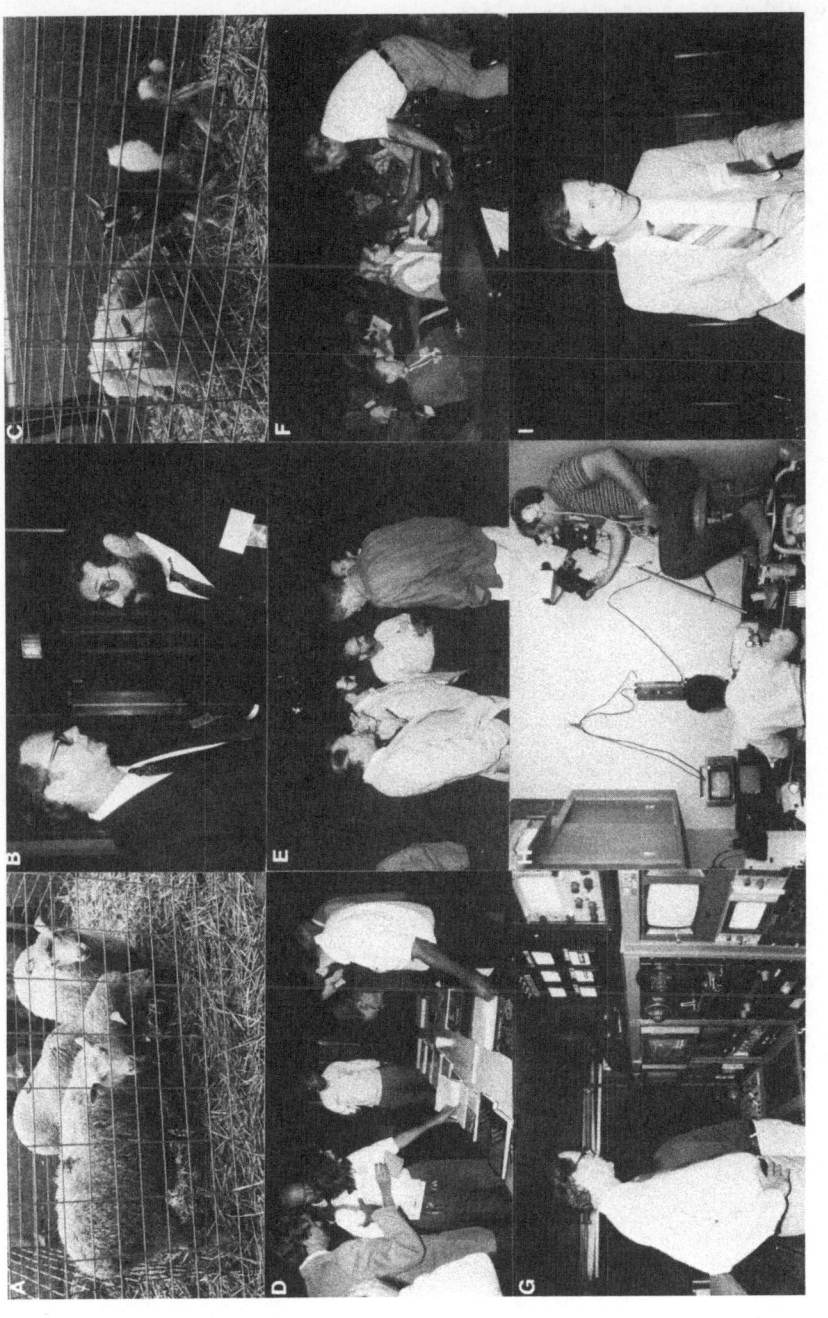

LEGEND: (A) Display of sheep chimeras; (B) J. Warren Evans (l) and Bernard Rollin (r) discussing moral issues of genetic engineering of animals; (C) Goat and sheep cross (r) or "geep"; (D) Conference participants examining publications display from Plenum Press and others; (E) (l to r) Bill Hansel, Caird Rexroad, Douglas Bolt, Gary Anderson, Keith Bitteridge; (F) Robert Foote talking with students; (G) J. Warren Evans monitoring sound and video production in laboratory; (H) Video crew in laboratory filming embryo manipulation techniques; (I) Gary Anderson answering questions following live demonstrations. (Photos B,D–I by Cathy Closson, U.C., Davis; Photos A,C by C.M. Wilson.)

Similarly, some antibiotics are produced naturally, but, in environments so hostile that the antibiotic is rapidly destroyed. It is possible to use recombinant DNA technology to produce these antibiotics from transformed organisms in more conducive environments. Amplification of productive capability through recombinant technology could be used to increase concentrations of existing antibiotics in culture media, thus, decreasing their cost and widening their availability.

POTENTIAL

The industrialization of recombinant DNA technology will lead to many useful products and processes. The underlying science of molecular biology and molecular genetics is dynamic; it is reasonable to assume that new opportunities will be created as the depth of scientific understanding increases. Genetic engineering technology is not a panacea, but it carries the realizable potential of solving some of the most difficult problems facing agriculture today.

ROSTER OF SPEAKERS,* SCIENTIFIC ORGANIZING COMMITTEE,**

LOCAL ORGANIZING COMMITTEE,*** AND PARTICIPANTS

ABBOUD, SAADO, University of Minnesota, St. Paul, MN
ADAMS, BETTY M., University of California, Davis, CA
AGUEL, CARLOS, Louisiana State University, Baton Rouge, LA
ALCIVAR, ACACIA, Iowa State University, Ames, IA
ANDERSEN, BERNT BECH, National Institute of Animal Science,
    Orum-Sondbryling, DENMARK
ANDERSON, CLARK, University of Minnesota, St. Paul, MN
ANDERSON, GARY B.,*/*** University of California, Davis, CA
ANDERSON, W. FRENCH,*/** National Institutes of Health, Bethesda, MD
ANDERSSON, LEIF, Swedish University of Agricultural Sciences,
    Uppsala, SWEDEN
ANDREW, DEBRA, University of California, La Jolla, CA

BACHRACH, HOWARD L.,* Plum Island Animal Disease Center, U.S. Department
    of Agriculture, Greenport, NY
BAILE, CLIFTON A., Monsanto Company, Chesterfield, MO
BALLACHEY, BRENDA E., Oregon State University, Corvallis, OR
BARTON, JOHN, International Technology Management, Los Altos, CA
BAUMGARTNER, ANTHONY, Purdue University, W. Lafayette, IN
BEAL, BILL, Virginia Polytechnic Institute, Blacksburg, VA
BENDER, ROBERT, Meiogenics Research Corporation, Ottawa, Ontario, CANADA
BETTERIDGE, KEITH J., University of Montreal, St-Hyacinthe, Quebec, CANADA
BIDWELL, CHRIS, University of California, Davis, CA
BINDON, B.M.,* CSIRO, Armidale, NSW, AUSTRALIA
BLAKEWOOD, GRIFF, Louisiana State University, Baton Rouge, LA
BOLAND, MAURICE, University College, Newcastle, Dublin, IRELAND
BOLT, DOUGLAS J.,** U.S. Department of Agriculture, Beltsville, MD
BOUJENANE, ISMAIL, University of California, Davis, CA
BOWLING, ANN T., University of California, Davis, CA
BOYD, JAMES B., University of California, Davis, CA
BRADFORD, ERIC,*** University of California, Davis, CA
BRAY, T.M., University of Guelph, Guelph, Ontario, CANADA
BREM, GOTTFRIED, LMU University of Munich, Munich, FEDERAL REPUBLIC
    OF GERMANY
BROWN, KEITH I., Ohio State University, Wooster, OH
BUCHANAN, DAVID S., Oklahoma State University, Stillwater, OK
BUNCH, THOMAS D., Utah State University, Logan, UT
BUSCH, ROBERT E., California State University, Chico, CA
BUTLER, JAMES E.,* University of California, Davis, CA

CAHOON, BARBARA E., Alameda Company Public Health Lab, Oakland, CA
CALVERT, CHRIS, University of California, Davis, CA
CALVI, BETH R., U.S. Department of Agriculture, Washington, D.C.
CAMPTON, DONALD E., University of California, Davis, CA
CARDIFF, ROBERT,*** University of California, Davis, CA
CARLSON, DON M., University of California, Davis, CA

CATTELL, MARGUERITA B., Colorado State University, Fort Collins, CO
CHAVEZ, LAWRENCE A., Cetus Corporation, San Ramon, CA
CHEN, C.L., University of Florida, Gainesville, FL
CHEN, HOWARD Y., Merck, Sharp & Dohme Research Laboratories, Westfield, NJ
CHITKO, CAROL G., Oregon State University, Corvallis, OR
CHRISMAN, C. LARRY, Purdue University, W. Lafayette, IN
CHUNG, BON-CHU, University of California, San Francisco, CA
CHURCH, R.B., University of Calgary, Calgary, Alberta, CANADA
COBERLY, SUZANNE K., University of California, Davis, CA
CREMER, KENNETH J., U.S. Department of Agriculture, Washington, D.C.
CROY, ANNE, University of Guelph, Guelph, Ontario, CANADA

DANFORTH, DOUGLAS R., Eastern Virginia Medical School, Norfolk, VA
DAVIES, CHRIS, Cornell University, Ithaca, NY
DE MAYO, FRANCESCO J., Baylor College of Medicine, Houston, TX
DENARDO, SALLY,*** University of California, Davis, CA
DE NISE, SUE, University of Arizona, Tucson, AZ
DICKERSON, PATTY SUE, Michigan State University, East Lansing, MI
DICKEY, JOSEPH F., Clemson University, Clemson, SC
DOI, ROY,*** University of California, Davis, CA
DONAHUE, SUSAN E.,* University of California, Davis, CA
DUNNING, DAVID, University of California, Davis, CA

EBERT, KARL M., Tufts University, North Grafton, MA
ELDRIDGE, FRANKLIN, University of Nebraska, Lincoln, NE
ENAB, AHMED ABDEL WAHAB, Ohio State University, Columbus, OH
EVANS, J. WARREN,*/** Texas A&M University, College Station, TX
EWING, SOLON A., Iowa State University, Ames, IA

FABRICANT, JILL D., Biosyne Corporation, Houston, TX
FANGUY, ROY C., Texas A&M University, College Station, TX
FLAGLOR, BETTY JO, University of California, Orland, CA
FOOTE, ROBERT H.,** Cornell University, Ithaca, NY
FOOTE, W. DARRELL, University of Nevada, Reno, NV
FORREST, DAVID W., Texas A&M University, College Station, TX
FORSBERG, NEIL E., Oregon State University, Corvallis, OR
FOSTER, DOUGLAS N., Ohio State University, Wooster, OH
FUQUAY, JOHN W., Mississippi State University, Mississippi State, MS

GALEHOUSE, DONNA, Ohio State University, Wooster, OH
GARDINER, CATHERINE S., Oregon State University, Corvallis, OR
GARDNER, MURRAY B.,* University of California, Davis, CA
GARNER, DUANE L., University of Nevada, Reno, NV
GILBERT, JAMES H., Cetus Corporation, Emeryville, CA
GILES, R.E., Advanced Genetics Research Institute, Oakland, CA
GODKE, ROBERT A., Louisiana State University, Baton Rouge, LA
GOLDMAN, HOPE, Purdue University, W. Lafayette, IN
GOLDSPINK, GEOFFREY, Tufts School of Veterinary Medicine, Boston, MA
GRAY, PETER N., Life Sciences International Minerals and Chemical,
  Terre Haute, IN
GROET, SUZANNE, Integrated Genetics, Farmingham, MA
GUISE, KEVIN, University of Minnesota, St. Paul, MN
GWAZDAUSKAS, F.C., Virginia Polytechnic Institute, Blacksburg, VA

HAMER, DEAN H.,* National Cancer Institute, Bethesda, MD
HAMMER, ROBERT E.,* University of Pennsylvania, Philadelphia, PA
HANSEL, WILLIAM,** Cornell University, Ithaca, NY
HARVEY, JEANNE P., University of California, Emeryville, CA
HAUGE, JENS G., Norwegian College of Veterinary Medicine, Oslo, NORWAY
HAWK, HAROLD,** U.S. Department of Agriculture, Beltsville, MD
HENCKE, JAN, University of California, Davis, CA

HESS, CHARLES E.,* University of California, Davis, CA
HODGETTS, ROSS, University of Alberta, Edmonton, Alberta, CANADA
HOLLAENDER, ALEXANDER,** Council for Research Planning in Biological
  Sciences, Washington, D.C.
HSU, CHAO-KUANG, Merck & Company, Inc., Rahway, NJ
HUMES, PAUL E., Louisiana State University, Baton Rouge, LA

JENSEN, JOSEPH, Advanced Genetics Research Institute, Oakland, CA
JOHNSON, BRYAN, North Carolina State University, Raleigh, NC
JONES, DANIEL D.,* U.S. Department of Agriculture, Washington, D.C.
JONES, GWETHALYN, New England BioLabs, Inc., Salem, MA

KASSAM, AMIR, University of California, Davis, CA
KAY, DAVID, University of California, Davis, CA
KILLIAN, GARY J., Pennsylvania State University, University Park, PA
KITCHEN, HYRAM, University of Tennessee, Knoxville, TN
KOPCHICK, JOHN J.,* Merck Institute for Therapeutic Research, Rahway, NJ
KOSUGE, TSUNE,** University of California, Davis, CA
KRAEMER, DUANE C., Texas A&M University, College Station, TX
KRIVI, GWEN G., Monsanto Company, Chesterfield, MO
KROHN, INGRID, Butterworth Publishers, Stoneham, MA
KUNY, GREGORY, Council for Research Planning in Biological Sciences,
  Washington, D.C.

LEIBO, S.P.,* Rio Vista International, Inc., San Antonio, TX
LEUNG, FREDERICK,* Merck, Sharp & Dohme Research Laboratories, Rahway, NJ
LINDNER, GARY M., Pacific Embryonics, Lodi, CA
LIU, KATIE Y.S., University of California, Davis, CA
LOSKUTOFF, NAIDA M., Texas A&M University, College Station, TX
LOTHROP, C.D., University of Tennessee, Knoxville, TN
LU, YAO-CHI, Office of Technology Assessment, U.S. Congress,
  Washington, D.C.
LYUP, MARLA CARTES, University of Vermont, Burlington, VT

MAIJALA, KALLIE, Agricultural Research Center, Institute of Animal
  Breeding, Jokioinen, FINLAND
MARTIN, JIM, University of California, Davis, CA
MC WHIR, JAMES, University of Calgary, Calgary, Alberta, CANADA
MEDRANO, JUAN, University of California, Davis, CA
MENINO, FRED, Oregon State University, Corvallis, OR
MERSMANN, HARRY J., Hruska U.S. Meat Animal Research Center, U.S.
  Department of Agriculture, Clay Center, NE
MICHEL, RAYMOND, University of California, Riverside, CA
MIKUCKIS, GENE, University of California, Davis, CA
MILLER, ANN M., Texas A&M University, Bryan, TX
MILLER, ROBERT H., U.S. Department of Agriculture, Beltsville, MD
MINHAS, BRIJINDER S., Biosyne Corporation, Houston, TX
MOBERG, GARY P., University of California, Davis, CA
MOORE, CHRIS C.D., University of California, Davis, CA
MOORE, KATHIE, University of California, Davis, CA
MOSES, PHYLLIS B., National Research Council, Washington, D.C.
MUGGLI, NOELLE E., Oregon State University, Corvallis, OR
MUNGER, RANDY, College of Great Falls, Manteca, CA
MURPHY, CLIFTON N., University of Missouri, Columbia, MO
MURTAUGH, MICHAEL P., University of Minnesota, St. Paul, MN
MUSCOPLAT, CHARLES C.,* Molecular Genetics, Inc., Minnetonka, MN

NANCARROW, COLIN, CSIRO, Blacktown, NSW, AUSTRALIA
NANGALAMA, ANDREW W., University of California, Davis, CA
NEILSON, JOHN T., University of Florida, Gainesville, FL
NELSON, ALISA C., Simi Valley, CA

NEMEC, LORI A., Texas A&M University, Bryan, TX
NOTTER, DAVID, Virginia Polytechnic Institute, Blacksburg, VA
NYBORG, JENNIFER K., University of California, Riverside, CA

O'BRIEN, STEPHEN J.,* National Cancer Institute, Frederick, MD
O'CONNOR, ANITA, University of Florida, Gainesville, FL
OHLSON, DANNY L., University of Idaho, Moscow, Idaho
OSBORN, RUSSELL G., North Dakota State University, Fargo, ND
OSBURN, BENNIE I.,* University of California, Davis, CA
OVERSTROM, ERIC W., Tufts University, Boston, MA

PARK, CHUNG S., Stanford Medical Center, Stanford, CA
PENEDO, CECILIA TORRES, University of California, Davis, CA
PICARD, LOUIS, University of Montreal, St-Hyacinthe, Quebec, CANADA
PIEDRAHITA, JORGE A., University of California, Davis, CA
PISENTI, JACQUELINE M., University of California, Davis, CA
POLLARD, JOHN WILLIAM, Iowa State University, Des Moines, IA
POLZIN, VICTORIA, University of California, Davis, CA
PRICE, JENNIFER, Salk Institute Biotechnology, San Diego, CA
PURUSHOTTAM, MAINALI, Alcorn State University, Lorman, MS

RADKE, KATHRYN,*** University of California, Davis, CA
RAJAB, MOHAMMAD H., Texas A&M University, College Station, TX
REED, MICHAEL L., Texas A&M University, College Station, TX
RENDEL, JAN E.R., Swedish University of Agricultural Sciences,
  Uppsala, SWEDEN
REXROAD, JR., CAIRD E.,* U.S. Department of Agriculture, Beltsville, MD
RICE, LARRY, Oklahoma State University, Stillwater, OK
RICHARDSON, THOMAS,* University of California, Davis, CA
RICKS, CATHERINE A., American Cyanamid, Princeton, NJ
ROEDER, RICHARD A., University of Idaho, Moscow, ID
ROLLIN, BERNARD E.,* Colorado State University, Fort Collins, CO
RORIE, RICK, Louisiana State University, Baton Rouge, LA
RUDDLE, F.H.,* Yale University, New Haven, CT
RUDER, CARLA A., University of Idaho, Moscow, ID
RUFFING, NANCY, University of California, Davis, CA

SABOUR, MOHAMMAD P., Animal Research Center, Agriculture Canada, Ottawa,
  Ontario, CANADA
SAINZ, ROBERTO D., University of California, Davis, CA
SALMON, KEITH, University of Alberta, Edmonton, Alberta, CANADA
SALTER, DONALD WAYNE, U.S. Department of Agriculture, East Lansing, MI
SANBUISSHO, ATSUSHI, Ohio State University, Columbus, OH
SARKAR, SIDDHARTHA, Center for Reproductive Biology, Memphis, TN
SATAYAPUNT, CHAMNEAN, Kasetsart University, Bangkok, THAILAND
SCHIEWE, MITCHEL C., National Zoological Park, Smithsonian Institute,
  Washington, D.C.
SCIBIENSKI, ROBERT,** University of California, Davis, CA
SEIDEL, JR., GEORGE E.,* Colorado State University, Fort Collins, CO
SEITZ, ANNA, University of Massachusetts, Durham, NH
SENYEI, ANDREW, University of California, Irvine, CA
SHIFRINE, MOSHE, University of California, Davis, CA
SHOFFNER, R.N., University of Minnesota, St. Paul, MN
SINGHAJAN, SAMPHAN, Ratchaburi A.I. Station, Ratchaburi Province, THAILAND
SKJAERLUND, DAVID, Michigan State University, East Lansing, MI
SKJERVOLD, HARALD, Agricultural University of Norway, Aas-NLH, NORWAY
SLANGER, WILLIAM, North Dakota State University, Fargo, ND
SMITH, ALBERT L., Colorado State University, Fort Collins, CO
SMITH, STEPHEN B., Texas A&M University, College Station, TX
SPEAROW, JIMMY L., University of Wisconsin, Madison, WI
SPRUILL, DAVID G., University of Georgia, Athens, GA

SRIKUMARAN, SUBRAMANIAM, University of Nebraska, Lincoln, NE
STANDAL, NILS, Agricultural University of Norway, Aas-NLH, NORWAY
STEPONKUS, PETER,* Cornell University, Ithaca, NY
STONE, J.B., University of Guelph, Guelph, Ontario, CANADA
STOTISH, RONALD L., American Cyanamid, Princeton, NJ
STUART, CATHERINE V., Miner Institute, University of Vermont, Chazy, NY
STULL, CAROLYN L., University of California, Davis, CA
SUJARIT, VANDA, Kasetsart University, Bangkok, THAILAND
SWANSON, LLOYD V., Oregon State University, Corvallis, OR
SYED, MOHASINA, Norwegian College of Veterinary Medicine, Oslo, NORWAY

TAM, ALBERT,* Genentech, Inc., South San Francisco, CA
TEAGUE, HOWARD S.,** U.S. Department of Agriculture, Washington, D.C.
TERRILL, CLAIR E., U.S. Department of Agriculture, Washington, D.C.
THOMPSON, NEAL P., University of Florida, Gainesville, FL
TOUCHBERRY, ROBERT, University of California, Davis, CA

VAN KUYK, ROBERT W., University of California, Davis, CA

WAGNER, THOMAS E.,* Ohio University, Athens, OH
WAGNER, WILLIAM C., University of Illinois, Urbana, IL
WALDBIESER, GEOFFREY C., Purdue University, W. Lafayette, IN
WALL, ROBERT J.,** U.S. Department of Agriculture, Beltsville, MD
WALTON, JOHN S., University of Guelph, Guelph, Ontario, CANADA
WATSON, GARY, University of California, Davis, CA
WELSH, THOMAS H., Texas A&M University, College Station, TX
WHEELER, MATTHEW B., Colorado State University, Fort Collins, CO
WHITE, KEN, University of California, Davis, CA
WHITE, THOMAS J., Cetus Corporation, Emeryville, CA
WHITTINGHAM, DAVID E.,* Medical Research Council, Surrey, UNITED KINGDOM
WIESEHAHN, GARY P., Advanced Genetics Research Institute, Oakland, CA
WIGGLESWORTH, KAREN, Huff 'n Puff Embryo Transfer, Vincentown, NJ
WILLADSEN, S.M.,* Granada Genetics, Marquez, TX
WILLIAMS, TIMOTHY J., CSIRO Tropical Cattle Research Center, Rockhampton,
   Queensland, AUSTRALIA
WILSON, CLAIRE M., Council for Research Planning in Biological Sciences,
   Washington, D.C.
WOLFORD, JOHN H., Virginia Polytechnic Institute, Blacksburg, VA
WOMACK, JAMES E., Texas A&M University, College Station, TX
WOOD, WILLIAM I., Genentech, Inc., South San Francisco, CA

YANG, XIANGZHONG, Cornell University, Ithaca, NY
YOUNGS, CURTIS R., Louisiana State University, Baton Rouge, LA
YUN, JEUNG S., Ohio University, Athens, OH

ZELINSKI, MARY, University of California, Stockton, CA